Conservation and Production of Woodlands
Modern Iberian Peninsula

Brill's Series in the History of the Environment

Series Editors

Matteo di Tullio (*University of Pavia*)
Simona Boscani Leoni (*University of Lausanne*)

VOLUME 9

The titles published in this series are listed at *brill.com/bshe*

Conservation and Production of Woodlands and Royal Forests in the Early Modern Iberian Peninsula

Edited by

Koldo Trapaga-Monchet
Félix Labrador Arroyo

BRILL

LEIDEN | BOSTON

The realisation of this book was supported by the Gerda Henkel Stiftüng (through the project "Protection, production and environmental change: the roots of Modern Environmentalism in the Iberian Peninsula (XVI–XVIIIth centuries)"; project reference AZ 60/V/19), by the Comunidad de Madrid (through the project "Madrid, Real y Cortesano (MARCO)"; project reference PHS2024/PH-290-HUM), and by the CINTER research group of the Universidad Rey Juan Carlos.

The Open Access publication of this book was funded by the Universidad Rey Juan Carlos.

Cover illustration: Pedro Núñez del Valle, *Vista del real sitio de la Casa de Campo* (*c.* 1637). Painting. Courtesy of Museo de Historia de Madrid. CER.es Colecciones en Red - Búsqueda general.

Library of Congress Control Number: 2025951210

Typeface for the Latin, Greek, and Cyrillic scripts: "Brill". See and download: brill.com/brill-typeface.

ISSN 1876-6595
ISBN 978-90-04-71234-8 (hardback)
ISBN 978-90-04-74569-8 (e-book)
DOI 10.1163/9789004745698

This book is printed on acid-free paper and produced in a sustainable manner.

Contents

Figures, Charts and Tables

Figures

Charts

Tables

Notes on Contributors

Ángel Ignacio Aguilar Cuesta
is a lecturer at the International University of Valencia. His main line of research has focused on the analysis and study of geohistorical sources from the Modern Age, particularly the Ensenada Cadastre.

Concepción Camarero Bullón
is Full Professor of Human Geography at the Universidad Autónoma de Madrid. She has devoted her long career to the study of geohistorical sources, in particular the Ensenada Cadastre and the General Planimetry of Madrid (18th century) and the Spanish Cadastre of 19th century.

Ignacio Ezquerra Revilla
is Assistant Professor of Early Modern History at Universidad de Cantabria. His main line of research focuses on the extended royal domestic government and its administrative quality in the Early Modern Age, both in Portugal and in Spain. Author of more than fifty papers at international conferences, and more than a hundred articles and contributions to collective works, as well as three books, including *El Consejo Real de Castilla en el espacio cortesano* (*Siglos XVI–XVIII*), Madrid: Polifemo, 2017.

Francisco Fernández Izquierdo
is Scientific Officer at the Spanish National Research Council (CSIC) in the Early Modern History Department, Institute of History, and collaborating professor in the Autonomus University of Madrid. His research topics are nobility, water, natural resources and economics in early modern Spain, and IT applied to historical work.

José Eloy Hortal Muñoz
is Full Professor of Early Modern History at the University Rey Juan Carlos in Madrid. His main research interests are the political history of the Netherlands at the XVIth and XVIIth centuries, the Courts of Brussels and Madrid at both centuries, the royal guards of the Spanish Habsburgs and, lastly, the Royal Geographies. Among his numerous works, we can highlight the last book he has co-edited with Merlijn Hurx, *Building the Presence of the Prince. The Institutions Responsible for the Construction and Management of the Buildings of European Courts* (*14th–17th centuries*), Turnhout, Brepols, 2025.

Alfredo José Martínez González
is an associate professor of the History of Law and Institutions at the University of Seville where he obtained his Ph.D. (2013) that was awarded with the prize of best Ph.D. dissertation. He has addressed various research topics within the fields of Law in the Medieval and Early Modern Period. Currently, his two main research lines are juridical and institutional elements of the woodlands devoted to shipbuilding from the 16th to the 19th century, and the history of the Spanish administration in Cuba and Puerto Rico.

Francisco J. Moreno Díaz del Campo
is Associate Professor in the UCLM Faculty of Arts in Ciudad Real. His main field of research is social history, principally in relation with the analysis of the morisco minority in Castile. Nevertheless, issues related to the agrarian history have been one of their object of study, especially the ones connected to water utilizations in the Early Modern period.

Cristina Joanaz de Melo
is a contracted senior researcher at NOVA University of Lisbon (IHC-Lab: in2past) and Invited Professor at the Autónoma University of Lisbon, lecturing on 'Environmental History'. She is a founding member of the Portuguese Network of Environmental History (REPORTHA 2015) and was awarded her Ph.D. on Hydrological and Forestry Policies in Portugal (and southern Europe) in the 1830s–1880s by the European University Institute (Florence–2010). She is currently working and publishing on natural resources, hunting and forests since the 1990s, her current interests lie in restoring and renewing natural resources and compensating for the human agency ongoing between the 1400s and 1800s across the Iberian Peninsula and naval empires.

Félix Labrador Arroyo
(Ph.D. 2007, Universidad Autónoma de Madrid) is Full Professor of Early Modern History at Rey Juan Carlos University. His research focuses on the royal sites in Spain, the administration and materiality in the early modern age. He has published more than 150 articles and contributions to collective works, as well as nine books, such as *La casa real en Portugal (1580–1621)* (2009), *La Casa de Borgoña: la casa del rey de España* (2014), *La extensión de la corte: los Sitios reales* (2017), *La configuración de la imagen de la Monarquía Católica. El ceremonial de la capilla real de Manuel Ribeiro* (2020) and *"Las legumbres del Rey". Mesa y alimentación en la Corte (siglos XVI–XIX)* (2020).

Ana Luna San Eugenio
is a researcher in the Department of Geography at the Universidad Autónoma de Madrid. Her research interests are focused on the study of non-technical cartography and the integration of geohistorical sources in various geospatial tools.

Raúl Romero-Calcerrada
is a full professor at Rey Juan Carlos University. He holds a Ph.D. in Geography from the University of Alcalá. His academic career has been shaped by interdisciplinary collaboration, particularly with Forestry and Agricultural Engineers, and by a strong focus on the interactions between humans and nature, as well as pressing environmental issues. His work emphasises the application of innovative techniques to better understand and address these challenges from a landscape and sustainability perspective. His research has resulted in the publication of approximately twenty scientific articles in national and international journals, as well as contributions to books and book chapters.

Koldo Trapaga-Monchet
(Ph.D. 2015, Universidad Autónoma de Madrid) is Professor of Early Modern History at Rey Juan Carlos University. His research mainly centres on the environmental history of early modern Portugal, with a special focus on forest management and shipbuilding. He published and coordinated monographs, articles, and book chapters. Among the publications related to this publication are "Revisiting the Narrative of Deforestation in the Central and Southern Mainland Early Modern Portugal as a 'Ruined Landscape': The Case of Shipbuilding in Lisbon" (*History*, 2025), and the coordination of the book (together with Álvaro Aragón-Ruano and Cristina Joanaz de Melo) *Roots of Sustainability in the Iberian Empires: Shipbuilding and Forestry, 14th–19th centuries* (Routledge, 2023).

Introduction

Koldo Trapaga-Monchet and Félix Labrador Arroyo

'We shall use words like 'live-oakery' and 'pinery' to mean an area of trees, without committing ourselves to whether it was forest, savanna or maquis. It is essential to read the original documents and not to trust others' translations and interpretations'.[1]

This statement comes from the book by Alfred T. Grove and Oliver Rackham, who specifically highlight the importance of being careful about the meanings given to some words, because they are not always obvious. The Spanish words *monte* and *bosque* and the Portuguese *mata* are clear examples of this. In our minds (the writers of this introduction) the word 'monte' has been equated to 'woodland',[2] which nowadays refers to *bosque*/forest, understood as a landscape covered with trees. One of the original purposes of this book was to try to put together a history of the concepts of 'royal forest', 'forest' and 'woodland' that were used by Early Modern societies and monarchies on the Iberian Peninsula.

The task seemed difficult enough to start off with, but it turned out to be downright impossible to compile a history of the concepts listed in this introduction, despite the fact that some of them are paramount to all the contributions. The decision not to pursue this avenue stemmed partially from a lack of knowledge of the land uses and landscapes of the inhabitants of the Early Modern Iberian Peninsula, especially the areas with a Mediterranean climate. Although it may be argued that historical records provide insights, they are not sufficient. The Spanish and Portuguese Crowns have provided misleading impressions about the landscapes, land uses, and, very likely, the daily life of the local societies. These impressions derived from the tunnel vision of the Iberian Crowns,[3] which caused their officers and administrative bodies to

1 Alfred T. Grove and Oliver Rackham, *The Nature of Mediterranean Europe: An Ecological History* (New Haven: Yale University Press, 2001), 18.

2 Among other publications, see Cristina Joanaz de Melo, "Menos coutadas melhores pinhais: impériodo, inundações, fisiocracia, guerra e especialização das matas reais em Portugal (1777–1824)," *Tiempos Modernos: Revista electrónica de Historia Moderna*," 9, no. 39 (2019) 465–6; Cristina Joanaz de Melo, "Floresta em movimento: usar, regenerar, cuidar (séculos XIV–XIX)", in *Como Fénix Renascida – Matas, bosques e arvoredos (séculos XVI–XX). Representações, gestão, fruição*, ed. Cristina Joanaz de Melo (Lisbon: Colibrí, 2020), 92–3.

3 Here we use the meaning given to the term by James C. Scott, *Seeing Like a State, How Certain Schemes to Improve the Human Condition Have Failed* (New Haven and London: Yale University Press, 1998), 9–21.

grasp only a (tiny) part of the reality of the humans and landscapes of their time, disregarding the things that did not fall directly within their interests.

However, there is a bright side. The quality and variety of the contributions to this book have surpassed the editor's original aims. The Iberian Peninsula is diverse and complex in its ecosystems, climate, and societies. These complexities are multiplied when coupled with the different human societies, ecosystems and climates. Therefore, historical analyses necessarily need to approach human–nature interactions from multiple perspectives, not solely from multiple disciplines. On many occasions, this demand for a multidisciplinary, interdisciplinary, or transdisciplinary approach has led to bias or failed to take into account the slow pace of historical research based on primary sources, especially, but not only, archival resources. It is very likely that scientific contributions based primarily on archival sources are currently not very popular, but they can hold their own when challenging established narratives.

There is a very large body of literature that has approached the human-nature interrelations of the Early Modern Iberian Peninsula from the perspective of an ongoing destruction of the woodlands.[4] Although a detailed state of the art on this matter goes beyond the purposes and possibilities of this introduction, it is worth noting the contribution of Grove and Rackham to unravelling some of the strands of the theory of 'ruined landscapes' or a 'lost Eden', according to which the human societies of the Mediterranean basin basically destroyed nature, especially woodlands, whose devastation, so they claimed, triggered soil erosion, floods and desertification.[5]

It is very likely that part of this narrative stemmed from the well-known *Physical Geography* (1861) of John Frederick W. Herschel, who stated that one of the reasons for the aridity of inland Spain was the lack of vegetation, especially trees. Without providing any evidence, he plainly stated that 'the hatred of a Spaniard towards a tree is proverbial.'[6]

In 1864, George P. Marsh not only reproduced the statement about Spaniards' pathological hatred of trees, but went one step further when he asserted that Spain had been the only European country that had not adopted policies for

4 An updated state of the art is available in Koldo Trapaga-Monchet, Álvaro Aragón-Ruano and Cristina Joanaz de Melo, "Introduction: The Game of Demiurge in the Garden of Chronos: Woods play hide-and-seek in the long run through sustainable management", in *Roots of Sustainability in the Iberian Empires: Shipbuilding and Forestry, 14th–19th centuries*, ed. Koldo Trapaga-Monchet, Álvaro Aragón-Ruano and Cristina Joanaz de Melo (New York: Routledge, 2023), 1–25.

5 Grove and Rackham, *The Nature*, 8–12.

6 John F. W. Herschel, *Physical Geography* (Edinburgh: Adam and Charles Black, 1861), 244.

protecting trees.[7] He displayed a surprising honesty when he admitted 'I do not know at what period the two Castiles were bared of their woods, but the Spaniard's proverbial 'hatred of a tree' is of long standing.'[8] Civilized countries were thus equated to those with large stands of trees (forests), while uncivilized nations were those – like Spain – that were held to fight against trees.[9] This idea dated back at least to the Enlightenment, in a period when the wood-pasture biome began to be neglected.[10] Two important elements need to be taken into account: 1) Marsh relied on the writings of Antonio Ponz, who according to Rackham and Blondel hated the forests and blamed the *Mesta* (guilds of shepherds) for deforesting Extremadura, while favouring agriculture in areas not suitable for this purpose.[11] 2) He extended the situation of bare trees of the two Castiles (plateaus) to all of Spain, a territory with a highly diverse climate, ecosystems, and societies.

Carl von Berg, the director of the Tharandt Forestry School (near Dresden, Germany), also mentioned the idea of the hatred of trees among Spaniards, who were believed to be more interested in pasture and agriculture than in woodlands.[12] Curiously enough, it was during the 1860s and 1870s that so-called Modern Silviculture was introduced to Portugal by Bernardino Barros Gomes, who had studied at the Tharandt School. Under the principle of 'repairing and improving', Martín de los Heros, the intendent of the Spanish royal household, was aware of the need to give a scientific orientation to the management of the royal sites and decided to fund two scholarships at the Tharandt School in late 1842.[13] He was by no means the only Spanish or Portuguese forester trained in this school, as German forestry was the principal source of knowledge for Spanish and Portuguese silviculturists.[14] The application of the idea of forest

7 George P. Marsh, *Man and Nature, or Physical Geography* (London: 1864), 187–8, 279.

8 Marsh, *Man and Nature*, 187–8, 279.

9 This perception was also embedded by nineteenth century foresters. See, for instance, José Bonifacio de Andrada e Silva, *Memoria sobre a necessidade de utilidades e do plantio de novos bosques em Portugal* (Lisbon: Typografia da Academia Real das Sciencias, 1815), 9–11.

10 Oliver Rackham, *Woodlands* (London: Harper Collins, 2006), 134–6.

11 Grove and Rackham, *The Nature*, 14–15.

12 This is mentioned by Joachim Radkau, *Nature and Power: A Global History of the Environment* (Cambridge: Cambridge University Press, 2008), 189–90.

13 Martín de los Heros de las Bárcenas, *Memoria que acerca del Estado de la Real Casa y Patrimonio, y su administración en los últimos cinco meses de 1841, presenta al Excmo. Señor Tutor de S. M. D. Agustín Argüelles, el intendente general en comisión de la misma* (Madrid: Aguado, Impresor de Cámara de S.M., 1842), p. 21.

14 Ignacio García Pereda, "Creando el boque matemático en la década de 1860. Barro Gomes en la *Mata Nacional da Machada* (Barreiro, Portugal): testigos cartográficos", in *Estudos*

envisaged by German technical forestry to more complex landscapes resulting from multiple convergent land uses of multiple actors, not only the burgeoning nineteenth-century centralized liberal nation-states, further strengthened the narrative of forest destruction by showing bias against the areas that did not fit in with the concept of monocultural woodland in which the local inhabitants and other land uses were prohibited.[15] This implementation was so antinatural that even António Arala Pinto, head of the Marinha Grande forest district from 1927 to 1957, complained about the prohibitions: local inhabitants were not allowed to gather dry wood or bring goats into the pinewoods of Leiria.[16]

In 1911, Bernhard E. Fernow stated that Spain was a wealthy country during the Middle Ages but had turned into a desert by the early twentieth century 'through neglect, indolence, ignorance, false pride, lack of communal spirit, despotism of church, and misrule by a corrupt bureaucracy.'[17] More surprisingly, he concluded that Spain lacked woodlands even though the official Spanish statistics stated otherwise.[18] His misunderstanding of the Mediterranean ecosystems becomes obvious when he addressed the case of Portugal, ten percent of which was covered with forests, half of which could not be regarded as woodland because they 'are wooded with olive, fig, almond plantations, or open woodlands [probably the savanna-like *dehesas* or *montados*] and brushwood.'[19] He continued to express these biases towards the *montados* and *dehesas* or wood-pastures in the following pages, concluding that the pinewood of Leiria was the only well managed forest in Portugal.[20] In this connection, Grove and Rackham have correctly stated that savannas are categorized as 'degraded forest' without evidence that they are not, in fact, grassland that has become invaded by trees.[21] Traces of this narrative are even found in more recent publications. To mention but a few authors, in 1981 J. V. Thirgood commenced his book on the destruction of the Mediterranean forest by quoting George P. Marsh. His research continues the narrative popularized

da Paisagem, ed. Pedro Fidalgo (Lisbon: IHC, 2017), 223–40, esp. 226–29. For the study of the Portuguese case in-depth see, Ignacio García Pereda, *Experts Florestais: Os primeiros silvicultores em Portugal* (Ph.D dissertation: Universidade de Évora, 2018).

15 Scott, *Seeing like a State*, 14–20, and Paul Warde, "The Invention of Sustainability", *Modern Intellectual History*, 8/1 (2011), pp. 153–70, on pp. 162–3.

16 António Pinto, *O Pinhal do Rei – Subsídios* (Alcobaça: da oficina de J. de Oliveira Júnior, 1938), vol. 2, 314–18.

17 Bernhard E. Fernow, *History of forestry in Europe, the United States and other countries* (Toronto: Toronto University Press, 1911), 349–50.

18 Fernow, *History of forestry*, 352–53, 57–59.

19 Fernow, *History of forestry*, 361.

20 Fernow, *History of forestry*, 361–63.

21 Grove and Rackham, *The Nature*, 16.

by Marsh on the accumulative destruction of the Mediterranean forests by human societies.[22] Even David Attenborough has reproduced and spread to the general public Marsh's idea of the Mediterranean basin as the First Eden which humans have relentlessly destroyed.[23]

The implications of such negative statements extended beyond the woodlands to the management of natural resources, as well as to political management. The management of natural resources and the landscapes societies produced were viewed as a reflection of their social and political organization. Largely to justify their own socio-political system, nineteenth-century authors tended to disregard former political organizations, including the Early Modern socio-political organization based on a multiplicity of powers and corporations that clashed with the liberal conception, according to which all the powers were concentrated in the state. These historians thus praised the liberal state project and dismissed Early Modern plurality and corporativism.[24] The Court was the main political organization of the Early Modern period, but differed greatly from the State as we know it today.

Multiplicity and heterogeneity were present in several elements throughout the Early Modern period. Ignacio Ezquerra Revilla's essay evidences how the articulation of the court jurisdiction at the royal sites was a 'complex patchwork' whose negative analysis has resulted from 'a biased view of the concept [court] that is more focused on aspects of magnificence such as the pomp and cultural traces associated with palace life, the solemnity and ceremonial duties related to the royal person, or the role played by the nobility'.[25] This arose from a negative vision created by 19th-century writers who set out to justify the new political regime based on the nation-state and citizenship. They regarded Early modern courts and royal sites as simply the royal palaces where the kings, royal family and the courts were based, overlooking the multifaceted realities of the royal sites.[26] As José Eloy Hortal Muñoz notes from a spatial

22 J. V. Thirgood, *Man and the Mediterranean forest: A history of resource depletion* (New York: Academic Press, 1981).

23 David Attenborough, *The First Eden: Mediterranean World and Man* (Boston: Little, Brown & Co., 1987). It is worth watching the BBC four-episode series, David Attenborough, *The First Eden: Mediterranean World and Man* (BBC: 1987).

24 António Manuel Hespanha, *As vésperas do Leviathan. Instituições e poder político. Portugal* (*séc. XVII*) (Lisbon: Pedro Ferreira, 1987), 2 vols, especially vol. 1, pp. 18–25.

25 See the contribution of Ignacio Ezquerra Revilla in this essay.

26 Félix Labrador Arroyo and Magdalena Merlos Romero, "Introducción", in *Pragmatismo e ilusión: El agua y la gestión del espacio y territorio en Aranjuez y otros sitios cortesanos* (*siglos XVI–XIX*), directed by Félix Labrador Arroyo and Magdalena Merlos Romero (Madrid: Sílex, 2023), 15–19.

perspective, the royal sites were much more than just royal palaces, as they encompass -among other things- gardens, woodlands and royal forests. Therefore, he coined the term 'royal geographies' to reflect the multifaceted reality of the royal belongings.[27]

The simplification of the political and administrative realities also extended to landscapes. Landscapes that were not given over to agriculture, forestry, or cattle exclusively but resulted from various land uses in the same land unit were often disregarded. It is likely that this contempt arose from the need (or conviction) to justify the superiority of private property devoted mainly to one marketable economic activity backed by a strong central liberal state. Bastias Saavedra recently stated that land ownership and land use in the Early Modern Iberian Empires needs to go beyond the discussion between commons and private property.[28] The interrelations between 'cultivation, enclosure and improvement of the land could generate property rights' dating back at least to John Locke.[29]

The contribution by Alfredo José Martínez González demonstrates that from the early 19th century, the Spanish liberals aimed to disband the Ancien Regime 'in accordance with the demands of a revolutionary bourgeoisie who were socially *two sided*: on the one hand, they advocated freeing workers so that their strength and activity could be harnessed; on the other hand, they encouraged the expropriation of land rights in favour of private property'.[30] The bourgeoisie relied on private property that, in turn, would have brought about prosperity and development (economic growth of the country). Therefore, it was necessary to eliminate all the obstacles to freely enjoying private property, including, but not only, communal rights.[31] Enclosures and private properties were regarded as catalysts of prosperity and they were legally implemented across different land tenures, ecosystems, and land uses.[32] Congost correctly pointed out that Early Modern societies had regarded the 'commons' both as commons and individuals.[33]

---

27 See the contribution of José Eloy Hortal Muñoz in this volume.

28 Manuel Bastias Saavedra, "Beyond Private and Common. Ownership Regimes in the Iberian World (1500–1800)", in *Ownership Regimes in the Iberian World (1500–1850). The Normative Role of Kinship and Communited*, ed. by Manuel Bastias Saavedra (Leiden-Boston: Brill, 2024), 1–35.

29 Bastias Saavedra, "Beyond Private", 2–3.

30 See the contribution of Alfredo José Martínez González in this essay.

31 Rosa Congost, *Tierras, leyes, Historia. Estudios sobre «la gran obra de la propiedad»* (Barcelona: Crítica, 2007), 31–34.

32 Congost, *Tierras, leyes*, 219–226.

33 Congost, *Tierras, leyes*, 225–226.

A similar process occurred with woodlands. James C. Scott maintained that state-backed 19th-century German forestry aimed 'to transform the real, diverse, and chaotic old-growth forest into a new, more uniform forest that closely resembled the administrative grid of its techniques. To this end, underbrush was cleared, the number of species was reduced (often to monoculture), and plantings were carried out simultaneously and in straight rows on large tracts.'[34] The woodlands were thus regarded as the sum of equally sized trees that were exclusively managed – under geometrical and mathematical methods – to produce revenue, which clashed with the multiple uses of the trees.[35] Nature was simplified and landscapes were designed by increasing agricultural yields by using artificial fertilizers, and natural grazing in grasslands or woodlands was replaced by keeping cattle in sheds.[36] Cristina Joanaz de Melo's case study is a good example of this landscape design implemented from the government.[37]

Similarly, Warde noted that 19th-century German Forestry had its roots in 17th-century cameralism.[38] Concerning woodlands, cameralism aimed to achieve the same wood production every year without diminishing the woodlands by means of scientific calculations and standardization. This idea caught on not only among foresters but also among 'general cameralist works on the fiscal state [...] notably in the work of Johan Heinrich Gottlied von Justi in the 1750s and 1760s'. The purpose of the management was straightforwardly to work out how much 'wood can be annually felled sustainably, economically and without ruin to the woodlands'.[39] Thus, the cameralists and foresters preferred tree species whose growth could be easily predicted, which led to the creation of the 'normal tree', the *Normalbaum*; and 'the forest became the aggregate of the individual tree'.[40]

This landscape design simplified the complex natural, social and political reality of Early Modern Culture. It is likely that from the perspective of Cameralist Science and German scientific forestry, agriculture, forestry, and cattle were not compatible economic activities in the same land unit, which had led to their spatial separation. Early Modern socio-ecological realities were far more complex, and so were land uses, as Bastias Saavedra has recently pointed

---

34 Scott, *Seeing like a State*, 15.

35 Scott, *Seeing like a State*, 12–22.

36 Erich Bauer, *Los montes de España en la Historia* (Madrid: Servicio de Publicaciones Agrarias, 1980), pp. 29–30.

37 See the essay of Cristina Joanaz de Melo.

38 This paragraph is based on Warde, "The invention", 161–164.

39 Warde, "The invention", 163.

40 Warde, "The invention", 162.

out: 'Until the late 18th century, land tenure in Europe was organized through different forms of reciprocal obligations between kings and subjects and lord and tenants, as well as being tied to cities and towns, kinship and marriage and various forms of communal usage or ownership'.[41]

In the past few decades some researchers in the fields of natural sciences (mainly ecology and biology) and social sciences (geography) have claimed that humans have not only had a positive impact on the sustainability and resilience of the Mediterranean ecosystems, but have sometimes even contributed to enhancing biodiversity.[42] Early modern land-use practices resembled what we refer to nowadays as 'an agroforestry system'. Agroforestry system is the new name given to an age-old land-use practice where trees are combined with agricultural and/or livestock activities.[43] Agroforestry systems have three main attributes: productivity, adoptability, and sustainability.[44] A recent study has assessed the environmental and economic services provided by European landscapes with and without agroforestry systems. The findings of the article are clear: the agroforestry systems of the Mediterranean region are not only economically more profitable than the agricultural systems, but also provide more environmental services. Conversely, the economic profitability and environmental advantages of agroforestry systems are less obvious in Atlantic and Continental Europe.[45]

In this connection, *dehesas* and *montados* are some of the most characteristic agroforestry systems of the Mediterranean region. *Montados* and *dehesas* or savanna-like landscapes have been studied by scholars, who have found cork oaks (*Quercus suber*) and holm oaks (*Quercus ilex*) to be the best adapted and most representative tree species of the *montados* or *dehesas*. These agro-silvopastoral systems or agroforestry systems are manmade landscapes, and they are good examples of sustainability, resilience, and biodiversity conservation. The conservation or sustainability of *montados* or *dehesas* was attained through

41 Bastias Saavedra, "Beyond Private and Common", 3–4.

42 Jacques Blondel, "The 'Design' of Mediterranean Landscapes: A Millennial Story of Humans and Ecological Systems during the Historic Period", *Human Ecology* 34 (2006): 713–29. Grove and Rackham, *The Nature*, 190–215.

43 P. K. R. Nair, *An Introduction to Agroforestry* (Dordrecht: Kluwer Academic Publishers, 1993); Antonio Rigueiro-Rodríguez, Jim McAdam and María Rosa Mosquera-Losada (eds.), *Agroforestry in Europe. Current Status and Future Prospect* (Dordrecht: Springer, 2009).

44 Nair, *An Introduction*, 16–7.

45 Sonja Kay *et al.*, "Agroforestry is paying off – Economic evaluation of ecosystem services in European landscapes with and without agroforestry systems", *Ecosystem services* 36 (2019), 1–10.

a balanced equilibrium of a broad range of complementary products such as agriculture, grazing, and forest products like charcoal or cork production.[46]

It is very likely that the key point for sustainable land use and resource management of the Mediterranean region is to find the 'intermediate disturbance' of the human impact on the environment, of which the agro-silvo-pastoral systems of the *montados* and *dehesas* are a paramount example.[47] For us, the historians who have spent several years working with primary sources, it makes sense to think that Early Modern Iberian societies put the same land unit to multiple uses (agriculture, wood products, grazing or pasture land), but our discourse does not employ the terminology used by the so-called hard or accurate science. Ofelia Rey Castelao's essay on the woodlands of Galicia[48] and Álvaro Aragón Ruano's studies on forestry policies in Early Modern Gipuzkoa that brought about a change in the forest landscape due to different management techniques are but a few.[49]

These two studies are based on the Atlantic climate region of the Iberian Peninsula, but there are countless examples of how the multiple land uses that characterize agroforestry systems were normal practice in the Early Modern Iberian Peninsula across periods and types of land tenure.[50] The fact is that these authors' studies do not employ the terminology used in the fields

46 Teresa Pinto-Correia, Nuno Ribeiro and Paulo Sousa, "Introducing the montado, the cork and holm oak agroforestry system of Southern Portugal", *Agroforestry systems* 82 (2011), 99–101; Augusta Costa and Helena Pereira, "Montados e sobrerais: uma espécie, duas perspectivas", in *Os montados: muito para além das árvores*, ed. Joaquim Silva (Lisbon: Fundação Luso-American para o Desenvolvimento, 2007), 17–35.

47 Jacques Blondel and James Aronson, *Biology and Wildlife of the Mediterranean Region* (New York: Oxford University Press, 1999), 199–201, 216–17.

48 Ofelia Rey Castelao, *Montes y política forestal en la Galicia del antiguo régimen* (Santiago de Compostela: Universidad de Santiago de Compostela, 1995).

49 Álvaro Aragón-Ruano, *El bosque guipuzcoano en la Edad Moderna: aprovechamiento, ordenamiento legal y conflictividad* (Donostia: Sociedad de Ciencias Aranzadi, 2001); Álvaro Aragón-Ruano, "Los robles trasmochos guiados o ipinabarros: una apuesta sostenible de futuro para una técnica forestal olvidada", *Cuadernos de la Sociedad Española de Ciencias Forestales* 30 (2009), 137–142.

50 The amount of literature is particularly huge, and we therefore list solely some essays that encompasses multiple case-studies: Francisco Fernández Izquierdo and Francisco Javier Moreno Díaz del Campo (coords.), *Montes, pastos y caza a la vera del Guadiana en las Tablas de Daimiel: la Real Dehesa de Zacatena en la edad moderna* (Granada: Comares, 2022); Félix Labrador Arroyo and Koldo Trapaga-Monchet: "Recursos hídricos y forestales en la Europa mediterránea (siglos XV–XIX)", special issue of the journal *Manuscrits. Revista d'Història Moderna* 42 (2020); Cristina Joanaz de Melo (coord.), *Como a Fénix Renascida. Matas, bosques e arvoredos (séculos XVI–XX): representações, gestão, fruição* (Lisbon: Edições Colibri, 2020).

of ecology or agroforestry systems, even though they address a very similar research topic. This basic assumption is true for this book, which we decided to set within the academic discussion or framework of 'ruined landscapes' throughout the entire Early Modern Iberian Peninsula across different types of land tenures (that is, municipalities, royal forest, royal site or *sitio real* in Spanish) that resulted from multiple convergence and conflicts arose by multiple agents.[51]

The essay by Francisco Fernández Izquierdo and Francisco J. Moreno Díaz del Campo discusses a significant example of sixteenth-century *dehesas*, savanna-type landscapes combined with pasture and agriculture that were neither royal sites nor private property: they enjoyed a 'special status halfway between the Royal Woodlands and the Woodlands of the Realm. They were directly managed by royal officials and monitored by administrators or foresters, who were in charge of enforcing regulations specifically designed to protect, maintain, and increase the natural resources that directly or indirectly benefited the Crown. The particular features of these wooded areas and the demarcation of open spaces where trees of the oak family – *Quercus robur, ilex,* and *suber*, which yielded acorns, firewood and timber – grew side by side with grasses and dryland crops were a key element of that policy'.[52] Similarly, the contribution of Félix Labrador Arroyo demonstrated how at the end of the 18th century and the beginning of the 19th century the royal site of La Florida had dispersed trees, which were integrated into a land unit with agriculture and cattle. Whereas agriculture was of high importance, the administrators consciously integrated agriculture and cattle to improve the economic viability of this space, which was deeply impacted by the war.

Consequently, while the Iberian Peninsula includes different climates and land tenures, both the Crown and the local inhabitants purposely integrated (either from a bottom-up or a top-down approach) different land uses into the same land unit. This reality was perfectly reflected in the Ensenada Cadastre, a survey drawn up in mid-eighteenth-century Spain, as the contribution of Ángel Ignacio Aguilar Cuesta, Ana Luna San Eugenio and Concepción Camarero Bullón evidence. Rather than mere woodlands, the responses to the questionnaire noted that the wooded areas were key for the peasant economy across different land -tenures and climates of Spain. The absence of a question that directly addressed woodlands did not mean that woodlands were overlooked. Rather, it is a reflection of how wooded areas were integrated with other

51 Bastias Saavedra, 'Beyond private', 1–28.

52 See the contribution of Francisco Fernández Izquierdo and Francisco Javier Moreno Díaz del Campo on this essay.

socio-economic activities: 'In view of these antecedents and consequences, does the absence from the cadastral questionnaire of a direct question focused solely and exclusively on wooded areas mean that the woodlands were left out of the major survey promoted by Ensenada or that they were of no interest to the minister? Not at all – quite the opposite, in fact: they are there, but it is necessary to know where to look for them and for information about their floral composition, uses, area and value, bearing in mind that the Ensenada Cadastre is not a single record but a set of documents with different territorial levels, different purposes and differing degrees of data aggregation'.[53]

Therefore, one of the main contributions of this book is to continue overturning the idea of the Iberian royal sites as places of economic squandering, corruption, and mismanagement.[54] From the 15th to the 19th centuries the royal sites and palaces played a major role in the social, political and cultural organization of the European dynastic monarchies.[55] These spaces were not only residences for the monarchs and their court but also centres of power and administration, places of representation with symbols that conveyed an image of monarchs' magnificence and authority through the use of architecture, painting, sculpture, gardens, and fountains, among other features. They attested to a domination of nature and an ability to create beauty and order and made it posible to tell the story of the dynasty and of the policy aimed at playing a significant role in building the royal image and shaping a rhetoric of power.[56]

These spaces were likewise heavily ritualized and hierarchized by a barrage of rules and etiquette that regulated life there and the image they were to convey through ceremonies and royal imagery where every single detail was designed to underline the monarchs' authority and prestige.[57] They unquestionably reflected the evolution of the monarchy's political organization, and

53 See the contribution of Ángel Ignacio Aguilar Cuesta, Ana Luna San Eugenio, Concepción Camarero Bullón in this essay.

54 Labrador Arroyo and Merlos Romero, "Introducción", 15–23.

55 José Eloy Hortal Muñoz, "Los Sitios Reales como elementos clave de las monarquías europeas de la edad Moderna: una aproximación", *Studia Histórica. Historia Moderna* 42/2 (2020), 200–201. José Eloy Hortal Muñoz and Merlijn Hurx (eds.): *Building the Presence of the Prince: The Institutions Responsible for the Construction and Management of the Buildings of European Courts* (*14th–17th centuries*) (Turnhout: Brepols, 2024).

56 Malcolm Smuts & George Gorse, "Introduction", in Marcelo Fantoni, George Gorse & Malcom Smuts (eds.), *The Politics of Space: European Courts ca. 1500–1750* (Rome: Bulzoni, 2009), 16.

57 See for instance, Janette Dillon, *The Language of Space in Court Performance, 1400–1625* (Cambridge: Cambridge University Press, 2015); Marcello Fantoni, *Il potere dello spazio. Principe e cittá nell'Italia dei secoli XV–XVII* (Rome, Bulzoni, 2002).

also structured and articulated a great court space which polarized territories and transformed and restructured the landscape.[58] They were also places for resting and engaging in leisure activities, namely hunting and courtly festivities held in both the gardens and outdoor areas and the palace interiors: chiefly banquets, balls (some masked) and performances of plays.[59]

María del Pilar García Rodríguez and Beatriz Álvarez García have noted that the royal sites in the provinces of Madrid and Segovia have played a major role in the conservation of the environment.[60] These spaces were not only areas managed for leisure activities but also had an economic output such as agriculture, grazing, wood commodities.[61] The great majority of the Spanish royal sites were economically exploited from the outset: mainly agriculture and forestry to cover part of their upkeep. For instance, timber for the fleet, artillery gun carriages, and royal works in the Alhambra complex was extracted from the Soto de Roma in Granada, while the Juanelo Turriano mill on the river Tagus was maintained out of the revenues of the country estate (Cigarral) in Toledo, and the fruit and vegetables grown in the gardens and orchards of the Casa de Campo were used by the palace kitchens and surpluses were sold in the Madrid markets.[62]

However, in the case of the Spanish Monarchy, the royal site of Aranjuez is the best example of economic rationality. This point is highlighted in the introduction not only because of its uniqueness, but also because this volume does not include a specific contribution revolving around this royal site. Throughout the 16th and 17th centuries, the royal site of Aranjuez was not only conceived as an unit of agricultural production, but also generated income that was used to fund other royal sites.[63] Greater prominence was given to this economic aspect during the reign of Ferdinand VI (1746–1759) by his minister, the

58 Fernando Checa Cremades, "Felipe II y la ordenación del territorio en torno a la corte", *Archivo Español del Arte* 58 (1985), 392–398.

59 The multifaceted reality of the royal sites has been pointed out in several essays, such as Concepción Camarero Bullón and Félix Labrador Arroyo (dirs.), *La extensión de la Corte: los Sitios Reales* (Madrid: Ediciones UAM, 2017).

60 María del Pilar García Rodríguez and Beatriz Álvarez García, "Preservación del medio natural en los Reales Sitios del entorno de Madrid", *Investigaciones Científicas* 76 (2021), 221–242.

61 There is a large literature on this matter, and it goes beyond the scope of this introduction.

62 José Miguel Morán Turina, "Los sitios reales entre los Austrias y los Borbones", *Madrid: Revista de arte, geografía e historia* 5 (2002), 211–212. For the royal site of Soto de Roma see Félix Labrador Arroyo and Koldo Trapaga-Monchet, "La configuración del espacio y la explotación forestal de un enclave singular: el Real Sitio del Soto de Roma durante la dinastía Hasburgo", *Studia Histórica, Histora Moderna* 39, nº 2 (2017), 293–327.

63 Morán Turina, "Los sitios reales", 201.

Marquis of Ensenada, who set out to ascertain the people, assets, revenues and liabilities of the Crown of Castile in his project to draw up a cadastre, begun in 1749. As part of this process he also attempted to enquire into the situation of the royal sites with a view to modernising them and making them profitable on learning of the economic revenues they contributed to the royal treasury.[64] However, it was during the reign of Ferdinand's brother, Charles III (1759–1788), that the greatest importance was attached to the economic value of the royal patrimony, owing no doubt to the physiocratic ideas embraced, among others, by his enlightened ministers Floridablanca (1728–1808), Grimaldi (1710–1789) and Campomanes (1723–1802), according to which farming and nature were sources of wealth and economic organization was based on natural principles. Charles III clearly incorporated the natural environment of the royal patrimony by conducting projects that experimented with production. As Virginia Tovar has noted, during his reign plans were carried out to 'urbanize and prepare the territory to seek a way of boosting agricultural, livestock, industrial and other yields'.[65] The sovereign thus undertook new actions directed at the royal sites, including starting up pioneering crop and livestock farming activities as examples of the economic reform he wished to implement in the kingdom, which were compatible with leisure, representation and court festivities.

Notable among the new crop farming initiatives carried out at Aranjuez were the Cortijo de San Isidro at Aranjuez, with grapevines and olive bushes; the Jardín de los Deleites, where fruit and vegetables and vines were grown; and the Campo Flamenco, on the left bank of the River Tagus. Projects in other areas included the Huerta Valenciana where vines were dry farmed, as well as the Infantas and Mira el Rey and the extensión of the Serrano and Castillejo estates.[66] Similarly, the Casa de Campo in Madrid was given a greater productive emphasis with dry-farmed crops, vegetable gardens, vines, fruit tres and olive groves, and pursuant to a royal decree of 3 April 1761 the Albufera in Valencia, one of the most productive and representative estates of the

64 Concepción Camarero Bullón and Laura García Juan, "Geografía histórica de los espacios reales: Alóndiga, Aceca y Barciles, despoblados del rey en la vega del Tajo", *Estudios Geográficos* 284 (2018), 209–235.

65 Virginia Tovar Martín, "Consideraciones al valor de los 'rústico' en los Sitios Reales (reinado de Carlos III)", *Fragmentos* 12–14 (1988), 220.

66 José Luis Sancho Gaspar, "El urbanismo del Real Sitio de Aranjuez", in *Riada. Estudios sobre Aranjuez* (Aranjuez: Doce Calles, 1991), Nº 3, p. 14; Tovar Martín, "Consideraciones al valor", 224–225.

Valencian royal patrimony, became Crown property again after being sold by Philip V (1700–1746).[67]

As far as livestock farming is concerned, notable among other cases are the construction in 1762–63 of the dairy (*casa de vacas*) established chiefly to supply butter and milk to the court and to sell surplus production on the market, based on the model of the Reale Vaccheria di Alifreda founded at Caserta (Campania, Italy); the introduction of buffalos in the Casa de Villamayor in Aranjuez to produce mozzarella; and the encouragement of horse breeding by building the Sotomayor stables in 1761 and the subsequent division of the royal stud farm into three sections. The first of these sections, the Andalusian, was geared to breeding saddle horses; the second, the Neapolitan or draft horse farm, concentrated on Friesian horses; and the third bred mules and stud jackasses.[68]

Greater attention was likewise paid to conservation and forestry at the royal sites. A notable development is the ordinance of 15 October 1761 for the administration of the royal pinewoods of Valsaín, Pirón and Riofrío,[69] which highlighted the transformation of a royal space: San Ildefonso, which was essentially a luxurious palatial residence surrounded by a large hunting ground until 1760 and thereafter also became a major source of forest products. The contributions of Koldo Trapaga-Monchet and Félix Labrador Arroyo in this essay provide two examples of the materiality of the royal sites in Portugal (the pinewood of Leiria from 1580 to 1640), and of the royal site of La Florida in the late 18th century and early 19th century.

However, although this book focuses especially on the royal forests of the Iberian Peninsula, we have deliberately included essays that: 1) show how the idea of conservation was embedded in State forestry policies, as well as reflected in the landscapes that resulted from multiple land uses across different land tenures; 2) confirm the idea that jurisdictional and institutional mechanisms played a major role in the shaping of the royal sites as patchworks as complex as the landscapes of Early Modern Iberia; 3) make it possible to understand the concepts of 'royal site' and 'royal forest'. The following section accordingly describes the contents of the book in greater detail.

---

67 Carmen García Monerris, *La Corona contra la Historia. José Canga Argüelles y la reforma del Real Patrimonio valenciano* (Valencia: Publicacions de la Universitat de València, 2005), p. 59.

68 Archivo General de Palacio (AGP), Reinados, Carlos III, caja 24, exp. 1.

69 *Real Cédula, Instrucción y Ordenanzas, que su Majestad (Dios le guarde) manda observar, para la Custodia, Administración, Conservación, y Cría de los Reales Pinares, y Matas de Robledales de Valsaín, Pirón, y Riofrío, desde quince de Octubre de mil setecientos sesenta y uno, en que se incorporaron a en la Corona* (Madrid: Imprenta de Juan de San Martín, 1761).

1 Contents of the Book

Early Modern sovereigns relied on legislation and their administrative staff to enforce their objectives on the ground.[70] However, the Early Modern Period was characterized by multiple conflicting and overlapping jurisdictions, which are of great importance to understanding the strategies and objectives pursued by Iberian sovereigns with respect to the royal sites. Although this is a highly complex task, in his contribution Ignacio Ezquerra Revilla unravels the articulation of the jurisdiction of the royal sites during Philip II's reign as a complex patchwork. Spanish kings exercised over the royal sites a more restricted jurisdiction which was embedded in the *Junta de Obras y Bosques* and, due to the corporate nature of the Spanish Monarchy, often clashed with other jurisdictions. The author provides extensive archival evidence of the interdependences between the *alcaldes de casa y corte* (judges of the royal household and court) who were responsible to the Royal Council of Castille and the *jueces de bosque* (judges of the forests) with respect to the creation and consolidation of the restrictive and special jurisdiction of the royal sites of the Early Modern Spanish Monarchy in the kingdom of Castile.

José Eloy Hortal Muñoz's contribution provides a thought-provoking framework for studying the royal forests and woodlands of the Spanish Monarchy, which are addressed from the perspective of territory, because it is a key element in the making of the term 'Royal Geographies' that the author coined. This conceptual framework facilitated the complex reality of the royal sites of the Early Modern Spanish Monarchy as it embraced palaces, forests, woodlands, stables, royal archives, and even a mint. After explaining the general approach, the author examines in depth the importance of the royal forests and woodlands in enhancing and consolidating the presence and power of the sovereign over the territory, as well as providing an overview of the multiple material outputs of the different royal forests and woodlands. Nevertheless, the core of this article revolves around the different offices entrusted with taking care of the royal forests and woodlands linked to the *Junta de Obras y Bosques* (board of works and woodlands): their nature, the tasks they involved, who held them, and how the social status of the incumbents evolved from the late 1500s and throughout the 1600s. This analysis helps understand the role

70 Just to mention two case studies, Karl Appuhn, *A forests on the Sea: Environmental Expertise in Renaissance Venice* (Baltimore: John Hopkins University, 2009); Paul Warde, *Ecology, Economy and State Formation in Early Modern Germany* (Cambridge: Cambridge University Press, 2006).

of the royal forests and woodlands in the configuration of the courtly space of the Spanish Monarchy.

The essay by Alfredo José Martínez González addresses the implementation and evolution of the institutional mechanisms put in place by the Spanish Monarchy to establish a sustainable forestry system for the conservation of the woodlands of Northern Spain for imperial shipbuilding from the reign of Philip II (1556–98) to the *Cortes of Cádiz* (1808–14) when the Ancien Regime was abolished in Spain. Beginning in the early 1560s, Philip II reinforced the activity of the local and central governing bodies. During the 1560s and 1570s, a forestry management system was introduced in an attempt to harmonize the felling of trees with the planting of new ones throughout the north of the Iberian Peninsula, as well as in the principality of Catalonia. Among other measures, this resulted in the founding of the *Superintendencia de Montes y Plantíos*, the institution that for nearby two centuries held responsibility for ensuring the conservation of the existing woodlands and guaranteeing that the local institutions complied with the obligations of plantings in Atlantic Spain (from Galicia to Gipuzkoa). Whereas at first the Spanish Crown focused on the provinces of Biscay and Gipuzkoa, from the late sixteenth century onwards the institution spanned the entire Cantabrian area. The next step forward made by the *Superintendencia de Montes y Plantíos* occurred in the mid-seventeenth century when Toribio Pérez Bustamante drew up a new set of regulations that can be regarded as a precedent for what came to be known as sustainable development centuries later. Subsequently, the author surveys the forest management institutions during the Bourbon dynasty with the benchmark of the Ordinance of 1748, until the disbandment of the *Montes of Marina* (woodlands designated for naval construction) during the Cortes of Cádiz.

As has been mentioned, this book sets out to demonstrate the idea that the sustainability – as well as the resilience – of the Mediterranean ecosystems was achieved through the convergence of agricultural, livestock, and forestry land uses in the same land-unit management: what it is known as agroforestry systems. The idea of conservation to meet societies' material needs went beyond the municipalities, encompassing different land tenures throughout the Early Modern Period. The contribution of Francisco Fernández Izquierdo and Francisco Javier Moreno Díaz del Campo unravels the implementation of management methods and systems for agricultural, animal and wood resources in three selected case studies: Alcobaça, Otos and Aceca, Zacatena, which were managed by the Council for the Military Orders (*Consejo de las Órdenes Militares*) from the early sixteenth century onwards. Although the authors mainly look at forestry resources, they correctly highlight the fact that these spaces were chiefly used as sources of pastureland and agriculture.

They were *dehesas*, which constitutes a paramount example of an agroforestry system or savanna-type landscape. The convergence of multiple uses in the same areas of land was due to the need to fulfil a variety of requirements for multiple actors, which triggered what the authors regard as 'sustainability in the Ancien Régime Way'.

Koldo Trapaga-Monchet's contribution revolves around the royal pinewood of Leiria during the Habsburg period (1580–1640). Although the pinewood of Leiria is the best-known woodland in Portuguese forestry, until a few years ago its history prior to the eighteenth century had hardly been explored. Therefore, this essay begins with a description of the administrative staff and forest policies issued by the Portuguese Crown from the early sixteenth century in order to introduce the period from 1580 to 1640, which is analysed by combining existing literature and archival and cartographical sources with a document that was hitherto unknown, but has revealed new insights on the history of the pinewood. The two fires of 1613 triggered a harsh response from the Portuguese Crown, which conducted major inquiries to clarify their origins, and also had long-standing administrative and management consequences. All the members of the administrative staff were dismissed for two years, and the Crown built new firebreaks alongside the pinewood. This essay furthermore provides new insights into the following aspects of the pinewood of Leiria from 1580 to 1640: 1) the administrative staff of the pinewood; 2) the evolution of the supply and prices of timber for the royal navy; 3) the relations between the Crown officers and local inhabitants with respect to the administration and protection of the pinewood; and 4) social and ecological constraints on timber for imperial shipbuilding supplied from the pinewood of Leiria to the Lisbon dockyard.

A huge amount of literature in the fields of the humanities and social and natural sciences has blamed imperial shipbuilding for being the main agent of forest depletion in Early Modern Portugal. Whereas the majority of such studies are qualitative analyses, the contribution by Raúl Romero-Calcerrada and Koldo Trapaga-Monchet breaks away from existing scholarship by bringing a quantitative approach to this discussion. Theirs is an interdisciplinary contribution, as it combines GIS, historical and cartographical sources with the approach of Integrated Suitability for Tree Species, building three-layer interconnected research in order to evaluate the sustainability of forest resource management and human activity (shipbuilding) on these resources. The authors begin with a cartographical proposal that envisages the geographic space covered by the *monteiro-mor* ordinance of 1605. This makes it possible, secondly, to estimate the forest area and number of homegrown trees available per year to supply the *Pinus pinea*, *Pinus pinaster* and *Quercus suber* timber demanded by the Royal Navy in accordance with three silvicultural rotations for each species. This is achieved by applying the Integrated Suitability for Tree Species approach, which makes it possible to assess the potential distribution for each tree species. Thirdly, this data is crossed-referenced with archival evidence on the amount of timber demanded by the Portuguese royal navy from 1621 to 1634 for imperial shipbuilding in Lisbon. This quantitative approach shows there was a sufficient forestry capacity to address the timber needs for shipbuilding reflected in literature and archival sources without compromising the viability and future sustainability of this area, suggesting new topics of research.

The dynamics of the Early Modern Iberian Peninsula during the eighteenth century are addressed in three contributions. Firstly, the essay by Ángel Ignacio Aguilar Cuesta, Ana Luna San Eugenio and Concepción Camarero Bullón delves into the importance of the woodlands in the peasant economy of mid-eighteenth-century Spain through the analysis of the so-called *Catastro de la Ensenada* (Ensenada Cadastre). The authors remind us of the importance of bearing in mind that the Ensenada Cadastre is a large array of documents with different territorial levels, different purposes, and differing degrees of data aggregation. They have mainly worked with the *Respuestas Generales* (general answers), *Memoriales* (declarations), and *Libros de lo real* (records of real estate) of which they provide a useful description before moving onto their case studies that encompasses different geographical areas of large parts of Spain (i.e., current provinces of Almeria, Burgos, Jaén, Lugo, Madrid, Soria). This first-hand information clearly demonstrates the importance of woodlands in the socio-economic structure of the local inhabitants, mainly as part of larger land-use structures and different landscapes which included agriculture, wood-pasture and uncultivated lands. This essay furthermore provides new data on several elements such as forest management, land-use patterns, tree biodiversity, planting policies, and the interrelations between the local municipalities and the State on the supply of materials for imperial shipbuilding. All these sets of data demonstrate that Spain was far from being desertified due to a lack of woodlands.

The so-called Lisbon earthquake of November 1, 1755 largely destroyed the city of Lisbon, severely damaging the Portuguese fleet. Although the earthquake and its aftermath has been studied extensively, Cristina Joanaz de Melo's contribution sheds light on new aspects. Firstly, this book chapter evidences how the most urgent wood-related demands were met by domestic supplies, as the colonies were too far away to be reached rapidly. After outlining the methodological and conceptual approaches, the essay moves onto its core idea: that the Lisbon earthquake triggered a chain of new political programmes led by the Marquis of Pombal that landscaped Portugal. The economic and political recovery of mainland Portugal, especially Lisbon, entailed landscaping specific geographic areas owing in part to Portugal's difficulty in communicating with the colonies during the months that followed the earthquake. The ministers perceived the effects of the earthquake as an opportunity to promote land-use changes intended to do away with the cork oak savannas in the Douro valley in order to expand Porto wine production, as it was one of the few Portuguese commodities that easily competed in the international markets. However, this does not constitute evidence of a desire to transform cork oak savannas into agricultural lands; rather, it is part of a larger landscaping programme to intentionally concentrate areas planted with shipbuilding timber-producing trees in lowlands and sandbanks near the sea or on slopes and perimeters located close to rivers in order to provide fast, secure, and cheap transportation and storage.

The history of the royal site (*real sitio*) of La Florida is unravelled by Félix Labrador Arroyo's contribution. Although this royal site is of secondary importance if compared with the well-known royal sites of Aranjuez or El Escorial, this case study is particularly relevant to the main hypothesis outlined in this introduction. This essay begins by surveying the establishment and expansion of La Florida from the late 1780s up to early years of the nineteenth century through the acquisition of diverse lands that belonged to private individuals. The royal site of La Florida was located near the royal palace of Madrid and was created in part to ensure the territorial continuity of the royal properties of the Casa de Campo and the Monte de El Pardo. This contribution not only demonstrates that the Spanish monarch – or more likely his ministers – had good first-hand knowledge of the territory surrounding the royal palace of Madrid, but also proves that there was a rational economic management based on the combination of agricultural and livestock production with wood and forest resources. The large array of archival sources provides insights on the land uses that combined agriculture, livestock, orchards, and forest products to a lesser extent, as well as the evolution of the expenditures and income of La Florida before and during the French invasion.

Acknowledgements

This contribution has benefited from the research projects 'Protection, production and environmental change: the roots of Modern Environmentalism in the Iberian Peninsula (XVI–XVIIIth centuries)' of Gerda Henkel Stiftüng (project reference AZ 60/$_{V}$/19) and 'Madrid: real y cortesano' funded within the Call of 2024 'Ayudas a Proyectos de I+D realizados en colaboración entre grupos de investigación pertenecientes a las universidades y organismos de investigación de la Comunidad de Madrid en la modalidad de programas de actividades de I+D en Procesos Humanos y Sociales (PHS-2024/PH-HUM–290).'

Bibliography

Appuhn, Karl. *A forests on the Sea: Environmental Expertise in Renaissance Venice*. Baltimore: John Hopkins University, 2009.

Aragón-Ruano, Álvaro. "Los robles trasmochos guiados o ipinabarros: una apuesta sostenible de futuro para una técnica forestal olvidada", *Cuadernos de la Sociedad Española de Ciencias Forestales* 30 (2009): 137–142.

Aragón-Ruano, Álvaro. *El bosque guipuzcoano en la Edad Moderna: aprovechamiento, ordenamiento legal y conflictividad*. Donostia: Sociedad de Ciencias Aranzadi, 2001.

Attenborough, David. *The First Eden: Mediterranean World and Man*. Boston: Little, Brown & Co., 1987.

Bastias Saavedra, Manuel. "Beyond Private and Common. Ownership Regimes in the Iberian World (1500–1800)", In *Ownership Regimes in the Iberian Eorld (1500–1850). The Normative Role of Kinship and Communited*, ed. by Manuel Bastias Saavedra. Leiden-Boston: Brill, 2024, 1–35.

Bauer, Erich. *Los montes de España en la Historia*. Madrid: Servicio de Publicaciones Agrarias, 1980.

Blondel, Jacques. "The 'Design' of Mediterranean Landscapes: A Millennial Story of Humans and Ecological Systems during the Historic Period", *Human Ecology* 34 (2006): 713–29.

Blondel, Jacques and James Aronson. *Biology and Wildlife of the Mediterranean Region*. New York: Oxford University Press, 1999.

Camarero Bullón, Concepción and Laura García Juan. "Geografía histórica de los espacios reales: Alóndiga, Aceca y Barciles, despoblados del rey en la vega del Tajo", *Estudios Geográficos* 284 (2018), 209–235.

Camarero Bullón, Concepción and Félix Labrador Arroyo (dirs.). *La extensión de la Corte: los Sitios Reales*. Madrid: Ediciones UAM, 2017.

Checa Cremades, Fernando. "Felipe II y la ordenación del territorio en torno a la corte", *Archivo Español del Arte* 58 (1985), 392–398.

Congost, Rosa. *Tierras, leyes, Historia. Estudios sobre «la gran obra de la propiedad»*. Barcelona: Crítica, 2007.

Costa, Augusta and Helena Pereira. "Montados e sobrerais: uma espécie, duas perspectivas." In *Os montados: muito para além das árvores*, edited by Joaquim Silva. Lisbon: Fundação Luso-American para o Desenvolvimento, 2007, 17–37.

De los Heros de las Bárcenas, Martín. *Memoria que acerca del Estado de la Real Casa y Patrimonio, y su administración en los últimos cinco meses de 1841, presenta al Excmo.*

Señor Tutor de S. M. D. Agustín Argüelles, el intendente general en comisión de la misma. Madrid: Aguado, Impresor de Cámara de S.M., 1842.

Dillon, Janette. *The Language of Space in Court Performance, 1400–1625.* Cambridge: Cambridge University Press, 2015.

Fantoni, Marcello. *Il potere dello spazio. Principe e cittá nell'Italia dei secoli XV–XVII.* Rome: Bulzoni, 2002.

Fernández Izquierdo, Francisco and Francisco Javier Moreno Díaz del Campo (coords.). *Montes, pastos y caza a la vera del Guadiana en las Tablas de Daimiel: la Real Dehesa de Zacatena en la edad moderna.* Granada: Comares, 2022.

Fernow, Bernhard E. *History of forestry in Europe, the United States and other countries.* Toronto: Toronto University Press, 1911.

García Monerris, Carmen. *La Corona contra la Historia. José Canga Argüelles y la reforma del Real Patrimonio valenciano.* Valencia: Publicacions de la Universitat de València, 2005.

García Pereda, Ignacio. "Creando el boque matemático en la década de 1860. Barro Gomes en la *Mata Nacional da Machada* (Barreiro, Portugal): testigos cartográficos", in *Estudos da Paisagem*, edited by Pedro Fidalgo, 223–40. Lisbon: IHC, 2017.

García Pereda, Ignacio. *Experts Florestais: Os primeiros silvicultores em Portugal.* PhD dissertation: Universidade de Évora, 2018.

García Rodríguez, María del Pilar and Beatriz Álvarez García. "Preservación del medio natural en los Reales Sitios del entorno de Madrid", *Investigaciones Científicas* 76 (2021), 221–242.

Grove, Alfred T., and Oliver Rackham. *The Nature of Mediterranean Europe: An Ecological History.* New Haven: Yale University Press, 2001.

Herschel, John F. W. *Physical Geography.* Edinburgh: Adam and Charles Black, 1861.

Hespanha, António Manuel. *As vésperas do Leviathan. Instituições e poder político. Portugal (séc. XVII).* Lisbon: Pedro Ferreira, 1987, 2 vols.

Hortal Muñoz, José Eloy. "Los Sitios Reales como elementos clave de las monarquías europeas de la edad Moderna: una aproximación", *Studia Histórica. Historia Moderna* 42/2 (2020), 197–217. 10.14201/shhmo2020422197217.

Hortal Muñoz José Eloy and Merlijn Hurx (eds.). *Building the Presence of the Prince: The Institutions Responsible for the Construction and Management of the Buildings of European Courts (14th–17th centuries).* Turnhout: Brepols, 2024.

Kay, Sonia et al. "Agroforestry is paying off – Economic evaluation of ecosystem services in European landscapes with and without agroforestry systems," *Ecosystem services* 36 (2019): 1–10.

Labrador Arroyo, Félix and Koldo Trapaga-Monchet. "La configuración del espacio y la explotación forestal de un enclave singular: el Real Sitio del Soto de Roma durante la dinastía Hasburgo", *Studia Histórica, Histora Moderna* 39, nº 2 (2017), 293–327.

Labrador Arroyo, Félix and Koldo Trapaga-Monchet. "Recursos hídricos y forestales en la Europa mediterránea (siglos XV–XIX)," special issue of the journal *Manuscrits. Revista d'Història Moderna* 42 (2020).

Labrador Arroyo, Félix and Magdalena Merlos Romero. "Introducción." In *Pragmatismo e ilusión: El agua y la gestión del espacio y territorio en Aranjuez y otros sitios cortesanos (siglos XVI–XIX)*, directed by Félix Labrador Arroyo and Magdalena Merlos Romero, 15–23. Madrid: Sílex, 2023.

Melo, Cristina Joanaz de (coord.). *Como a Fénix Renascida. Matas, bosques e arvoredos (séculos XVI–XX): representações, gestão, fruição.* Lisbon: Edições Colibri, 2020.

Melo, Cristina Joanaz de. "Floresta em movimiento: usar, regenerar, cuidar (séculos XIV–XIX)." *Como Fénix Renascida – Matas, bosques e arvoredos (séculos XVI–XX). Representações, gestão, fruição*, coordinated by Cristina Joanaz de Melo, 79–130. Lisbon: Colibrí, 2020.

Melo, Cristina Joanaz de. "Menos coutadas melhores pinhais: imperio, inundações, fisiocracia, guerra e especialização das matas reais em Portugal (1777–1824)." *Tiempos Modernos. Revista Electrónica de Historia Moderna* 9 nº 39 (2019): 456–87.

Morán Turina, José Miguel. "Los sitios reales entre los Austrias y los Borbones", *Madrid: Revista de arte, geografía e historia* 5 (2002), 201–217.

Nair, P. K. R. *An Introduction to Agroforestry*. Dordrecht: Kluwer Academic Publishers, 1993.

Pinto, António. *O Pinhal do Rei – Subsídios.* Alcobaça: Oficina de J. de Oliveira Junior, 1938, 2 vols.

Pinto Correia, Teresa, Nuno Ribero and Paulo Sá Sousa. "Introducing the montado, the cork and holm oak agroforestry system of Southern Portugal", *Agroforestry systems* 82 (2011): 99–104.

Rackham, Oliver. *Woodlands*. London: Harper Collins, 2006.

Radkau, Joachim. *Nature and power: A global history of the environment.* Cambridge: Cambridge University Press, 2008.

Real Cédula, Instrucción y Ordenanzas, que su Majestad (Dios le guarde) manda observar, para la Custodia, Administración, Conservación, y Cría de los Reales Pinares, y Matas de Robledales de Valsaín, Pirón, y Riofrío, desde quince de Octubre de mil setecientos sesenta y uno, en que se incorporaron a en la Corona. Madrid: Imprenta de Juan de San Martín, 1761.

Rey Castelao, Ofelia. *Montes y política forestal en la Galicia del antiguo régimen.* Santiago de Compostela: Universidad de Santiago de Compostela, 1995.

Rigueiro-Rodríguez, Antonio, Jim McAdam and María Rosa Mosquera-Losada (eds.). *Agroforestry in Europe. Current Status and Future Prospect.* Dordrecht: Springer, 2009.

Sancho Gaspar, José Luis. "El urbanismo del Real Sitio de Aranjuez", in *Riada. Estudios sobre Aranjuez* (Aranjuez: Doce Calles, 1991), nº 3, pp. 9–38.

Scott, James C. *Seeing Like a State, How Certain Schemes to Improve the Human Condition Have Failed.* New Haven and London: Yale University Press, 1998.

Silva, José Bonifacio de Andrada e *Memoria sobre a necessidade de utilidades e do plantio de novos bosques em Portugal.* Lisbon: Typografia da Academia Real das Sciencias, 1815.

Smuts, Malcolm and George Gorse. "Introduction", in *The Politics of Space: European Courts ca. 1500–1750*, edited by Marcelo Fantoni, George Gorse and Malcom Smuts (Rome: Bulzoni, 2009).

Thirgood, J. V. *Man and the Mediterranean forest: A history of resource depletion.* New York: Academic Press, 1981.

Tovar Martín, Virginia. "Consideraciones al valor de los 'rústico' en los Sitios Reales (reinado de Carlos III)", *Fragmentos* 12–14 (1988), 219–231.

Trapaga-Monchet, Koldo, Álvaro Aragón-Ruano and Cristina Joanaz de Melo. "Introduction: The Game of Demiurge in the Garden of Chronos: Woods play hide-and-seek in the long rung through sustainable management", in *Roots of Sustainability in the Iberian Empires: Shipbuilding and Forestry, 14th–19th centuries*, edited by Koldo Trapaga-Monchet, Álvaro Aragón- Ruano and Cristina Joanaz de Melo, 1–30. New York: Routledge, 2023, pp. 1–30.

Warde, Paul. *Ecology, Economy and State Formation in Early Modern Germany.* Cambridge: Cambridge University Press, 2006.

Warde, Paul. "The Invention of Sustainability", *Modern Intellectual History*, 8/1 (2011), pp. 153–70.

CHAPTER 1

The Articulation of Courtly Jurisdiction at the Royal Sites: the Shaping of a Complex Patchwork

Ignacio Ezquerra Revilla

1 Introduction

The territorial jurisdiction of the Court has often been overshadowed by a biased view of the concept that is more focused on aspects of magnificence such as the pomp and cultural traces associated with palace life, the solemnity and ceremonial duties related to the royal person, or the role played by the nobility. Such a consideration has not been aided by the jurisdictional reality of the Corporate Monarchy, which has hindered the appreciation of a broader reality that provided a backdrop to political and administrative life in the kingdoms of Spain. And yet, as I hope to demonstrate, in spheres with their own doctrinal coherence, such as that formed by the Court and the kingdoms in Castile, it is legitimate to ask whether both constituted a united reality. The interpretation of the Court has been hindered by the jurisdictionalist paradigm (from an internal viewpoint) or the consideration of the whole monarchy as a corporate reality (from an external viewpoint),[1] to such an extent that it has tended to be given an interstitial meaning, articulating different actors and previous domains. But the concept of the Court was inherent in the king himself; it was the tool that enabled him to convey his decisions to the territorial sphere over which he exercised authority – a kind of semantics based on the charisma of the royal person himself, which encouraged obedience and transferred the agreements made in his most restricted sphere to the kingdoms.

When a region was conquered, all uncultivated land belonged to the king, who then legalized the transfer of part of it to the clergy and nobility for war services. Accordingly, from the Reconquest episode onwards, the proportion of territory that both powers held increased while that of the Crown decreased.[2] But this was consistent with the fact that:

1 For both concepts, see, among the extensive literature, António Manuel Hespanha, *Vísperas del Leviatán: instituciones y poder político (Portugal siglo XVII)* (Madrid: Taurus, 1989).

2 The donations given by Alfonso IX and Ferdinand III to the Order of Alcántara in exchange for its assistance in conquering Cáceres, Montánchez, Mérida, Badajoz, Elvas, La Serena, and

> Tienen los príncipes, por razón de la suprema potestad, un dominio general en todas las tierras, montes, leñas, yerbas y pastos de sus reinos, en virtud del cual todas estas cosas se entienden y presumen ser suyas e incorporadas a la Corona: de tal manera, que siempre que se ofrece duda sobre el todo o parte de ellas, entran fundando su intención contra qualquiera que no exhiba privilegio o título que acredite la legítima pertenencia.
>
> Princes have, by reason of supreme power, a general dominion over all the lands, mountains, woods, grassland and pastures of their kingdoms, by virtue of which all these things are understood and presumed to be theirs and incorporated into the Crown: in such a way that whenever doubt arises as to the whole or part of them, they intervene, grounding their intention against anyone who does not exhibit a privilege or title that proves legitimate ownership.[3]

At the time it was written, this statement aimed to reconstruct and protect royal heritage in the kingdom of Valencia.[4] However, it is broad in scope, encompassing diverse regions and periods of the Spanish Early Modern Age, such as Castile, where the aforementioned process was grounded in different laws set out in the *Partidas*, the legal text drawn up during the reign of Alfonso X the Wise to regulate the organization of the Court, governance and justice in Castile.[5] That is how the conventional concept of territory should be understood: not only as a physical or geographical actuality but also as a

---

Magacela serve as a prime example. The outcome of these contributions resulted in what Carlos de Ayala describes as the 'process of territorialization of the Alcantarine lordship' ('proceso de territorialización del señorío alcantarino'), Carlos de Ayala Martínez, "Pérez, Arias," in *Diccionario Biográfico Español*, XL (Madrid: Real Academia de la Historia), 716–17, and the sources quoted there.

3 Vicente Branchat, *Tratado de los derechos y regalías que corresponden al Real Patrimonio en el reyno de Valencia y de la jurisdicción del intendente como subrogado en lugar del antiguo Bayle General* (Valencia: Imprenta de Joseph y Tomás de Orga, 1786), vol. III, 207; Erich Bauer Manderscheid, *Los montes de España en la Historia* (Madrid: Servicio de Publicaciones Agrarias, 1980), 49.

4 Eduardo Ortega de la Torre, "Carlos III y la lucha por el Real Patrimonio en el Pais Valenciano: el informe Branchat (1784)," in *De la Ilustración al Romanticismo: IV Encuentro, Carlos III, dos siglos después. Cádiz, 7–9 de abril de* 1988, coord. Mariano Peñalver (Cádiz: Universidad de Cádiz, 1994), 111–18.

5 Title XI of the Second *Partida*, 'Quál debe el Rey ser a su tierra,' which provided the spatial basis for the task of reigning, and thus complemented the well-known dual description of the Court contained in Law XXVII, Title IX of the Second *Partida*, 'Qué cosa es Corte porqué ha assí nome e quál deue ser,' *Las Siete Partidas, del Sabio Rey don Alonso el nono, nuevamente*

political category that precisely designated a 'politically equipped space.'[6] But Covarrubias had already pointed out this meaning in his dictionary, *Tesoro de la lengua castellana*, where he defined territory as 'the space of land taken up by an estate or jurisdiction'.[7] In the case of royal jurisdiction, it referred to kingdoms in which the political equipment was undoubtedly the tendency to confuse the idea of Court with the territory itself.

Accordingly, the continuous creation of royal sites in various regions of the kingdom cannot necessarily be viewed as an expansion of the Court, but rather as a limitation on a previous, extensive continuity in which the king displayed his most patrimonial side, and in which the patrimonial essence of the dominion he exercised over his kingdoms was direct and subjected to a more restricted jurisdiction than that of the Royal Council: the *Junta de Obras y Bosques* (Board of Works and Woodlands).[8] Though as an ideal approach, it was imperfect and, paradoxically, clashed with the reality of the monarchy's governance as an extension of domestic governance, manifested in the multiplication of the *Palatium* and the values it spread, represented by the creation of royal households and sites, and translated into the dissemination of the idea of the Court. Thus, interactions between the *Junta*, on one hand, and the Council and the *alcaldes de Casa y Corte* (judges of the royal household and court, active in the courtly space), on the other, were frequent, starting with the active presence of the latter in the former.[9]

While the governance of the kingdoms resulted from the conjunction of domestic and administrative affairs, with a modulated and changing composition related to its dependence, the royal sites were the places where a patrimonial sense was most prominent. As stated above, the process from the monarch to the royal sites was not extensive but restrictive, starting with the acquisition of a more or less large portion of territory by the king. An example is the history of the Monte del Pardo, which went from being initially limited

---

glosadas por el Licenciado Gregorio López del Consejo Real de las Indias de Su Magestad (Salamanca: Andrea de Portonariis, 1555), fols. 31v–32r and 29r.

6 António Manuel Hespanha, "El espacio político," in *La Gracia del Derecho*, coord. António Manuel Hespanha (Madrid: Centro de Estudios Constitucionales, 1994), 85–121.

7 'el espacio de tierra que toma algún pago o jurisdicción', Sebastián de Covarrubias, *Tesoro de la Lengua Castellana o Española según la impresión de 1611, con las adiciones de Benito Remigio Noydens publicadas en la de 1674* (Barcelona: Alta Fulla, 1998), 959.

8 Fernando Cos-Gayón, *Historia jurídica del Patrimonio Real* (Madrid: Enrique de la Riva, 1881), 76–86.

9 Cos-Gayón, *Historia jurídica, ibidem*. On this double role of the *alcaldes*, courtly and domestic, see Ignacio Ezquerra Revilla, "La integración de la Casa en la Corte: los alcaldes de Casa y Corte," in *La Monarquía de Felipe II: la Casa del Rey*, coord. José Martínez Millán and Santiago Fernández Conti (Madrid; Fundación Mapfre-Tavera, 2005), vol. 1, 697–800.

in usage by the king to being a fully owned, 'perfectly demarcated area.'[10] Additionally, the annexation of El Campillo and Monasterio to the Royal Site of San Lorenzo resulted in their evacuation and closure.[11] However, regardless of any patrimonial aspirations harboured by past monarchs, the royal sites remained closely connected to their surroundings, particularly in the geographical, spatial and jurisdictional sense. Although their uniqueness was undeniable, they were additions to a previous reality characterized by its more or less obvious uniformity. The conventional territories delimited by the royal sites interacted unequally with their surroundings, which resulted in a clear definition of the jurisdictional specificity sought by the Crown through the *Junta de Obras y Bosques*. This fact determined the intervention of the jurisdictional agents who operated halfway between the domestic and the administrative spheres, such as the *alcaldes de Casa y Corte* and the Royal Council itself, which also had important responsibilities within a perimeter of five leagues. The direct and indirect royal patrimony interacted constantly.

The definition of the royal sites' spatial and jurisdictional boundaries played a role in the progressive development of the Court area. Acquiring, defining, and safeguarding an area with more direct patrimonial dependence on the Crown – that is, less interactive than the entire kingdom over which the monarch exerted sovereign power – was also part of the same expansion process of the governance of the royal household. In the first case, its expression can be described as immediate, and in the second as extended. Nonetheless, both instances of exerting power were responses to the same internal form, as they represented sequential stages of development rooted in the shift from smaller dimensions of kingship to larger ones. This process of expansion, however, entails similar features, modified to apply to a more complex context.

In a transversal way, this fact meant that household officials acted in both areas, regardless of their level of integration into the king's direct patrimony and the effective definition of institutional formulas to articulate it, as illustrated by the example of the *alcaldes de Casa y Corte*. That the aforementioned process followed such domestic features is not only shown by the fact that these ministers' intervention, commissional in nature, was distributed between the immediate patrimonial domestic and the extended spaces. Furthermore, in the first case they did so both in an incipient state of development of the

10 Francisco Javier Hernando Ortego, "La lucha por el Monte de El Pardo. Rey, municipio y uso del espacio en el Madrid del Antiguo Régimen," *Cuadernos de Investigación Histórica* 12 (1989): 170.

11 Gregorio Sánchez Meco, *El Escorial: de Comunidad de aldea a villa de Realengo* (Madrid: Ayuntamiento de El Escorial, 1995), 244–53.

royal sites and at a more advanced stage, when their governance depended on a more stable institutional formula, that represented by the *Junta de Obras y Bosques*.

2 Defining a Special Jurisdiction

During the initial phases of the shaping of the royal sites, sources indicate that the famous *alcalde de Casa y Corte* Rodrigo Ronquillo was involved in this branch of the royal patrimony. At the start of 1539, the judge was engaged in protecting the hunting grounds at Valsaín and other parts of the royal estate, as well as optimizing the use of the forests.[12] Three years later, he notified the emperor of the benefits of carrying out forest clearance at El Pardo and recommended putting it into practice in areas of the Madrid region, primarily for economic reasons. His words demonstrated that forest clearance was then regarded as a fitting measure to support the needs of the royal households.[13]

Likewise, when Philip II ascended to the throne, the construction of royal palaces and sites led to institutional clarification and confirmed the role of the *alcaldes* in such functions. On 10 December 1556, he issued a series of decrees to the regent Doña Juana, the president Antonio de Fonseca and Cristóbal Morillas, *alcalde de Casa y Corte*, instructing the latter to penalise people who had hunted at El Pardo and the adjacent areas. The structure of the commission entrusted to the *alcalde* is noteworthy, as it exemplified the shape of the royal household administration and the resulting impact of territorial cohesion stemming from the theological concept of "transubstantiation." For when judges were appointed to the post they were not merely assigned duties, as in modern practice; rather, these duties were considered to be performed by the royal person himself in the form of another biological entity, so to speak. Therefore, the *Cédula Real* (royal decree) establishing Morillas's commission regulated in detail his actions and particular aspects.

The king stated his intention to sign the commission himself in order to assert greater authority and maintain secrecy. This signifies a wish for his *persona* to be more faithfully embodied by his minister. Furthermore, the king did not delegate the appointment to the president of the Council. Instead, he issued it directly to Morillas, also sending notifications to Fonseca and the

12 Archivo General de Simancas (AGS), Estado, leg. 45, no. 180, letter of the Emperor Charles V, Madrid, 18 June 1539.

13 Eduardo Ruiz Ayúcar, *El Alcalde Ronquillo. Su Época. Su Falsa Leyenda Negra* (Ávila: Institución Gran Duque de Alba, 1997), 26.

regent for reinforcement. The king regarded Morillas' duties as an extension of his own criminal jurisdiction, seeking his advice on sentences to be passed before they were pronounced.

It can be inferred that the king's decision was influenced by domestic and patrimonial considerations when he suggested to President Fonseca that Morillas should be accompanied by some of the *Guardas de a pie o caballo* (guards on foot or horseback). He considered poaching at the royal sites to be a violation of the royal privilege of exploiting their resources. This fact indicates the existence of a specific structure underpinning the aforementioned style of governance, which combined both domestic and administrative aspects. For similar reasons, he urged the president to send Morillas and inform the governor (*alcaide*) of El Pardo about the locations where he should initiate his investigations. Philip II considered Torrelaguna, Colmenar Viejo and other areas of the Real de Manzanares to be appropriate.[14]

The issuance of commissions of this type also hints at other characteristic features of a transpersonalized form of governance, such as the unity of royal jurisdiction exercised at several levels by different officials (*corregidores*, *alcaldes de Casa y Corte* and *chancillerías*) or the Royal Council. This explains why the regent Doña Juana entrusted Ortiz, *alcalde* of the *chancillería* of Valladolid, with dealing with the aforementioned poaching excesses at El Pardo. Ortiz took advantage of his prior visit to the region to attend to other matters, apprehended those responsible and handed them over to the *alcaide* of El Pardo, who turned them over to the *corregidor* of Madrid. Following Ortiz's departure, the *corregidor* released the main suspect and requested further guidance on how to penalize the rest. In reply, the regent Doña Juana instructed Ortiz to return to his post,[15] where he was to oversee the mission until Morillas could take over fully.

The *alcalde de Casa y Corte* submitted his conclusions to the king on 8 February 1557. According to the report, he had failed to arrest the main culprits, Juan Vélez and the Domingos brothers, although he had sent two *alguaciles* (bailiffs) in advance. He proposed penalizing all those implicated without prior consultation. Philip II authorized Morillas to make decisions independently, without consulting the regent or the Chamber. He was, however, to delegate the confirmation of the incomplete trials to the *corregidor* of Madrid, as he would soon be moving to Seville to perform a new commission.[16] The king's

14 Archivo General de Palacio (AGP), Sección Registros (SR), libros de cédulas, II, fols. 17v–18r.

15 AGS, Estado, leg. 121, nº 89–90, Princess Joana to Philip II, 15 January 1557.

16 AGP, SR, libros de cédulas, II, fols. 29v–30r.

dissatisfaction with the lenient handling of the offenders, who were released on bail, led to the reestablishment of prior proceedings for such offences at the Royal Site of El Pardo. They were initiated by the warden and convictions were issued by the *corregidor* of Madrid.

Despite some uncertainties and setbacks, a distinct section dedicated to handling claims and concerns pertaining to *Obras y Bosques* (works and woodlands) began to emerge spontaneously, maintaining unity with the exercise of jurisdiction at the Court. In accordance with its judicial character, it inevitably called for the involvement of the *alcaldes de Casa y Corte*. Thus, on 11 August 1557, Céspedes de Oviedo was tasked with a new assignment: to evaluate the harm caused by hunting at Aranjuez, El Pardo and Valsaín.[17]

Similarly, when Philip II returned to Spain in 1559, his focus on protecting the royal sites was directed through specific assignments given to an *alcalde de Casa y Corte*. One of these commissions was issued to doctor Hernán Suárez de Toledo and secretary Pedro de Hoyo to take action against those who had been hunting in the Aranjuez area since 1552. As many of the offenders were challenging to locate, the commission was renewed in Toledo on 22 June 1560.[18] Likewise, in late February 1561, the *alcalde* Céspedes de Oviedo was entrusted with investigating poaching at El Pardo. The Martín brothers, residents of Colmenar Viejo, were charged and finally sentenced to six months' banishment. However, their term was reduced to one-third by the Royal Decree of 15 April 1561.[19] As attested by the two preceding cases, the management of *Obras y Bosques* could also show clemency, which was in line with the distributive and domestic framework of the political-administrative system.[20] This was demonstrated by the promulgation of the aforementioned decrees through the *Cámara de Castilla* (Chamber of Castile), a body that became a Council in 1588 and facilitated the king's exercise of such an important attribution reserved for him.

---

17 Historiographical attention to royal sites has increased in recent years. Apart from the work of the IULCE-UAM or the CINTER-URJC, the research group of which I am in relationship, and those cited in the text, see Magdalena Merlos Romero, *Aranjuez y Felipe II. Idea y forma de un Real Sitio* (Madrid: Comunidad de Madrid-Ayuntamiento de Aranjuez, 1998) and Carlos Manuel Valdés, *Estudio Histórico-Selvícola del Monte de Valsaín* (*Siglos XVI–XX*) (Madrid: Organismo Autónomo Parques Nacionales, 1997).

18 AGP, SR, libros de cédulas, II, fols. 219v–220v, and 473v–474r; Antonio Marichalar, "Tres figuras del XVI. Hernán Suárez de Toledo, Felipe de Borgoña y Briviesca Muñatones," *Escorial* 17 (1944): 29.

19 AGP, SR, libros de cédulas, II, fols. 98v–99r.

20 On the idea of distributive justice, see Juan de Mariana, "Del Rey y de la Institución Real," in *Obras del padre Juan de Mariana*, II (Madrid: Rivadeneyra, 1854), 560.

The involvement of the *alcaldes* in settling conflicts arising from the interaction or friction between the royal sites and their surroundings was ratified by the Royal Council through an order of 22 February 1561 stipulating that appeals against commissaries' verdicts must be directed to the *alcaldes de Casa y Corte*.[21] Following the establishment of the Court in Madrid in May 1561, the abovementioned dynamics of commissions and their judicial effects continued. In 1562, Secretary Hoyo carried out a *visita* (inspection) of the governor and officials of Aranjuez, with the assistance of doctor Hernán Suárez de Toledo.[22] Another *Cédula Real*, dated 24 July 1563, instructed *licenciado* Salazar, *alcalde de Casa y Corte*, to punish poaching in the Aranjuez district. Some clergymen were mentioned as being responsible, who theoretically were supposed to be judged by the governor of the archbishopric of Toledo.[23]

In short, the courtly nature of the royal sites meant that this aspect of their administration was mainly entrusted by commission to the *alcaldes de Casa y Corte* – though not exclusively, since other simple *letrados* continued to receive commissions of this kind until very late in Philip II's reign. On 12 November 1577, Gonzalo de Parada, a resident of Ocaña, was appointed to replace Diego de Carvajal in investigating the damage caused by royal hunting in the Aranjuez area.[24]

3 The Addition of Court Jurisdiction to Special Jurisdiction: the Definition of the *Alcalde-Juez de Bosques*

The shaping and establishment of the royal sites demonstrates the continuity and permanence of these commissions, to the point that a specialized attribution was created, exercised by a so-called *juez de bosques* (judge of the woodlands). Though, as we will see, the courtly nature of these sites meant that they always remained under the jurisdiction of the *alcaldes de Casa y Corte*, as evidenced by the fact that the latter were granted responsibility for issuing appeal judgments at first instance. The *jueces de bosques* received the additional title

21 Luis Sánchez, *Autos y acuerdos del Consejo, de que se halla memoria en los libros, desde el año de 1532, hasta el presente de 618. Mandados imprimir por el ilustríssimo señor don Fernando de Azevedo, Arçobispo de Burgos, Presidente del Consejo, y señores dél* (Madrid: Luis Sánchez, 1618).

22 Instituto Valencia de Don Juan (IVDJ), envío (e.) 99, caja 140, Juan de Ibarra to Philip II, 3 May 1580, 'Sobre la uisita del Escorial y el poco effecto de que han sido las q[ue] se han hecho.'

23 AGP, SR, libros de cédulas, II, fol. 332r.

24 AGP, SR, libros de cédulas, V, fols. 128v–129r.

of *alcalde de Casa y Corte* to enhance their jurisdiction, for the very same reasons that these ministers had been given the majority of these commissions from the time they were first established. However, although they were granted this title, this part of their jurisdiction related to the royal *Obras y Bosques* did not cease to be commission based.

3.1 *Licenciado Ortega*

As for the evolution of the figure of judge of the woodlands, we know that *licenciado* Jerónimo de Ortega undertook this commission in 1564, when he was appointed to inspect the officials of the Alcázar palace in Madrid and the residence at El Pardo, as part of the ongoing reorganization and consolidation of the royal sites. Similarly, on 20 May 1565, Ortega was appointed to investigate poaching at El Pardo, Aranjuez, and the Segovia forests, and to evaluate the ensuing damage to neighbouring estates. On 3 November of the same year, the Segovia authorities tasked him with investigating the excessive felling of pine trees in the Valsaín forest. Consequently, he initiated proceedings against residents of nearby towns, namely Cercedilla, Navacerrada, Becerril and Collado Mediano, all located in Real de Manzanares. In 1567, *licenciado* Ortega inspected the warden (*alcaide*) and officials of the Bosque de Segovia (Forest of Segovia, another name given to Valsaín), and removed some of them from their posts, which in 1580 were still being filled on a provisional basis. Soon afterwards, officials of the works (*obras*) of the Alcázar of Madrid were also subjected to *visitas*.[25]

The direct royal jurisdiction in this incipient 'territorial dominion of the royal household' received a notable boost given that in the aforementioned commission of 20 May 1565 and other similar ones jurisdiction was granted to Ortega, pre-empting that of the *corregidor* of Madrid (in relation to the forest of El Pardo) and that of Segovia (in relation to the Bosque de Segovia), and the *Cédula Real* signed in Madrid on 3 February 1569 decreed that only Ortega would hear hunting and fishing complaints relating to El Pardo.[26] There are indications that this decision stemmed from Ortega's efficient performance of his duties, as another document from the same date that mentions the need to plant pines, holm oaks, and oaks in the Bosque de Segovia for the benefit of the city of Segovia can be viewed as a consequence of his work.

With this endorsement, Ortega inspected the officials of the *obras* of the Alcázar Real in Madrid and the residence at El Pardo, starting on 31 March 1569, and appointed the *alguacil de Casa y Corte* Martín de Múgica as *alguacil*.

25 IVDJ, e. 99, caja 140, fols. 206r–207r, Juan de Ibarra to Philip II, Madrid, May 3 1580.

26 AGP, SR, libros de cédulas, III, fols. 103v–104r, Royal Decree, 3 February 1569.

But the crucial point to note is that the courtly nature of the post of *juez de bosques* required him to be given an additional appointment as *alcalde de Casa y Corte*, which Ortega received, and responsibility for hearing appeals against his sentences. This, after a period in the 1570s when this right was reserved for various members of the Cámara de Castilla (Chamber of Castile), constituted a sort of specialised board on *obras y bosques* matters. In practice, the handling of these cases by the Chamber of Castile was a more accurate representation of the domestic aspect of the *obras y bosques*. In fact, the initial work of the *Junta* was consistent with the Chamber's handling of matters involving the Alcázares and the Alhambra of Granada. As a result, the related documents contained the signature or mark of one or more members of the Chamber.[27]

This jurisdictional framework was subsequently endorsed by the Instruction for the Guarding and Preservation of El Pardo, published on 23 July 1572. This document also governed the jurisdiction of Aranjuez, which was considered an extension of El Pardo, authorizing its guards to report on the Aranjuez district. This document outlined the role of *juez de obras y bosques*, modelled on that previously held by Ortega, with jurisdiction over the Alcázar of Madrid, the palace, forest and woodlands of El Pardo, and the royal residences in the surrounding area, including Aranjuez. This minister would have exclusive authority over all matters related to royal works. This included any overindulgences in hunting within the El Pardo preserve, which extended for five leagues around, as well as hindrances from the ordinary judges of the neighbouring towns. The appeals against his sentences continued to be heard by the *alcaldes de Casa y Corte*. It was agreed that when appeals were heard this judge would sit among them, based on his seniority, with a deciding vote in cases sentenced by other judges at first instance, and a consultative vote in cases he himself had sentenced. His salary amounted to 300,000 maravedis, which was paid out of the income from the *mesa maestral* (land owned by the order of Santiago) and his court would have a notary, an *alguacil* and a *fiscal* (public prosecutor) appointed by the king.[28]

As a consequence, a more specific jurisdictional pathway was established for these affairs, as demonstrated by the Royal Council members' incapacity to bring to attention or release prisoners detained for matters related to *obras y*

27 IVDJ, e. 99, caja 140, fol. 361r–v, Juan de Ibarra to Mateo Vázquez, 27 February 1581.

28 Juan Antonio Álvarez de Quindós, *Descripción Histórica del Real Bosque y Casa de Aranjuez* (Madrid: Doce Calles, 1993), 413–14; Pedro de Cervantes-Manuel Antonio de Cervantes, *Recopilación de las Reales Ordenanzas de los Bosques del Pardo, Aranjuez, Escorial y Balsaín …*, (Madrid: Melchor Álvarez, 1687).

bosques during their *visitas* to the Court prison.[29] The jurisdictional framework in which the recently appointed judge exercised his authority meant that Ortega promptly assumed the combined roles of *alcalde de Casa y Corte* and *juez de bosques*, taking office as such on 5 September 1575 and ceasing to be paid as the latter.[30] The exercise of jurisdiction was strengthened by the introduction of a regulated penalty system designed to punish poaching.[31]

The doctrinal and transcendental questions concerning the transfer of royal attributes as a charismatic pole provide the context for a complex and potentially challenging issue. The question concerns the compatibility of simultaneously holding the formal position of *alcalde de Casa y Corte* and performing duties as *juez de bosques*, as well as continuing with special commissions to hear cases of damage caused by poaching at the royal sites or by animals beyond their boundaries. The appointment merely restricted the minister who could carry out the tasks outlined in each commission, which specified the transfer of the exercise of royal powers. All of this took place during a process of institutionalization, in which the means of processing these instruments shifted from the Royal Chamber or an informal and occasional committee of ministers to a regulated and permanent one, the *Junta de Obras y Bosques*. By the end of Philip II's reign, this board was hearing appeals against the sentences handed down at first instance by the *alcaldes*, as we will see.

However, Jerónimo de Ortega's performance of his duties based on this duality was generally affected by the political context and, particularly, by his own personal trajectory, which were interrelated. Ortega's rise to public prominence was associated with the Castilianist party, which gave priority to the kingdom of Castile in the general organization of the Spanish monarchy, but it was abruptly halted by his removal from the Court in 1577. On 1 November, a Royal Decree was issued ordering Ortega to surrender all his documents related to *Obras y Bosques*. They were to be recorded 'in an inventory and before a public notary' and given to the royal servant Antonio Montaño. The documents included those that Ortega had received from his predecessor in the post of *alcalde de Casa y Corte*, *licenciado* Céspedes de Oviedo.[32]

The cause was a sensitive and rarely excusable issue: one of his daughters had wedded an aristocrat without prior approval from the king. This action contravened the unspoken norms of familiar governance and was significant

29 *Cédula Real* of 9 July 1575, mentioned by Cos-Gayón, *Historia jurídica*, 83.

30 AGP, SR, libros de cédulas, IV, fols. 284v and 309v, *cédulas* of 29 November 1575 and 3 March 1576.

31 IVDJ, e. 99, caja 140, fol. 361r–v, Juan de Ibarra to Mateo Vázquez, 27 February 1581.

32 'por inventario y ante escribano', AGP, SR, libros de cédulas, V, fol. 126r.

enough to be recorded in the diary of imperial ambassador Khevenhuller.[33] Moreover, the episode did not portray Ortega's political loyalties in a very good light, as it brought him into contact with the papists, enemies of what was theoretically his political group: it seemed that his daughter's marriage had been engineered to satisfy the Admiral of Castile and the Dukes of Infantado, Francavila and Pastrana, members of the abovementioned political group.

Initially, the imposed sanction led to other *alcaldes de Casa y Corte* assuming responsibility for the *obras y bosques* commissions entrusted to Ortega. Among these commissions was that authorized by the king in Madrid on 24 November 1578, which was assigned to *licenciado* Juan Gómez. It involved completing the assessment of the compensation due for the incorporation of various estates into the Casa de Campo and seeing to its payment.[34] However, it is likely that Ortega was only imprisoned for a short time, while awaiting the final verdict of the Royal Council, and he returned to his duties. On 1 August 1579, Philip II entrusted him with investigating the damage caused by hunting at El Pardo to the properties located in the city of Madrid, from the beginning of 1573 to the end of 1579,[35] without reducing his commissional activity at all in the following years. On 8 October 1581, the king issued an order in Lisbon for Ortega to investigate the damage caused by royal hunting in the area around Aranjuez since late 1572 until then, and the following year, on 1 April, he received an allowance of 375,000 mrs. to investigate the harm caused by poaching in the forest of El Pardo, as well as for the commission he was about to start on in Aranjuez and another carried out the previous year in El Bosque de Segovia.[36] In July 1582 he was in Carabanchel de Abajo dealing with the damage caused by hunting in the forest of El Pardo area during the previous seven years.[37]

This sequence indicates that, during this period, *obras y bosques* affairs were more effectively articulated through the institutional channel of the board, but continued to be the dealt with through commissions. On 5 August 1582, when Ortega's enquiries were completed and various poachers had been charged, the king commissioned the *alcaldes de Casa y Corte* to hear their appeals, by means of a Cédula Real issued in Lisbon and signed by the Count of Barajas, *licenciado* Fuenmayor and the accountant Garnica,[38] who was then handling

33 Félix Labrador Arroyo (ed.), *Diario de Hans Khevenhüller: embajador imperial en la Corte de Felipe II* (Madrid: Sociedad Estatal para la Conmemoración de los Centenarios de Felipe II y Carlos V, 2001), 120–21.

34 AGP, SR, libros de cédulas, V, fol. 188v.

35 AGP, SR, libros de cédulas, V, fols. 238r–239r.

36 AGP, SR, libros de cédulas, VI, fol. 72r.

37 IVDJ, e. 99, caja 140, licenciado Ortega to Mateo Vázquez, 13 July 1582.

38 AGP, SR, libros de cédulas, VI, fol. 191r.

the affairs of the so-called 'Consejo de Obras y Bosques' (Council of Works and Woodlands). However, this course of appeal did not receive unanimous agreement from the injured parties. The question was discussed in a *memorial* (report) submitted by Fray Lorenzo de Villavicencio, the royal confessor, to Mateo Vázquez in Madrid on 20 January 1581. The *memorial* placed various aspects of the management of *obras y bosques* under the scrutiny of the royal conscience.[39]

This document was linked to the administrative upheaval at the Court caused by Philip II's stay in Portugal. The confessor considered the monarch's obligations as *paterfamilias* to be inconsistent with an excessive defence of hunting at the royal sites, particularly given the evident inadequacy of the compensation mechanism for appraising the damage caused by royal hunting in the areas surrounding the royal sites. For this reason, he suggested that Philip II should confine hunting to a smaller area, as he and Ortega had done at Valsaín, where the boundaries had been reduced. He also proposed that the hunting of royal deer that invaded private estates should be tolerated, and that the owners of such properties be given licences to possess and use hunting equipment. Interestingly, the ethical justification for this proposal had a very practical reason, since it aimed to reduce the hefty expenses that resulted from the compensation paid out by the Crown.

Regarding the appeals of these cases before the *alcaldes de Casa y Corte*, the city of Segovia and surrounding area demanded that they be handled instead by the *chancillería* of Valladolid. They argued that the *chancillería* was their ordinary court whose jurisdiction had been taken over by the *alcaldes*, resulting in a series of disadvantages: the local judges now needed the *alcaldes* to issue a judicial order (*Provisión Real*) in order to be able to proceed with cases. It is important to note that the jurisdiction of the *alcalde-juez de bosques* preempted that of the *corregidor* of Segovia, though both were equally competent. In most cases, the accused were unable to obtain this document, either because they were still in prison or because they lacked the means to travel to the Court. Even when they did manage to get it, Segovian justice was in the habit of releasing them only after they had paid the fine imposed on them, which meant that they remained in prison for months, with all the personal hardship that this entailed. Though defendants who did manage to appear before the *alcaldes de Casa y Corte* did not fare much better:

> Y si tienen hazienda para uenir a proseguir su causa aquí adelante los alcaldes como tienen tantos negocios no los despachan en mucho

39 IVDJ, e. 99, caja 140, fol. 376r.

> tiempo, cómense sus haziendas sin acabar nada no por más de hauerles hallado una red u otra cosa como esta.
>
> And if they have the wealth to pursue their cause here and now, the *alcaldes*, as they have so much business, do not dispatch them for a long time, and their wealth is used up without anything being finished, for no more than having been found with a net or something else like this.

Villavicencio also asserted that it was unacceptable for Court jurisdiction to be more influenced by the interests of the deer in the royal hunting grounds than by those of the royal subjects. *Alcalde* Ortega opposed the royal confessor's opinion, deeming that in Valladolid defendants were given lenient sentences for such offences.[40] Strictly speaking, the aforementioned unity of jurisdiction rendered the intervention of the *alcaldes de lo criminal* (criminal judges) of the *chancillerías* and the *alcaldes de Casa y Corte* indistinct and the former did indeed hear such appeals in the case of the royal sites near Valladolid (Abrojo) and Granada (Soto de Roma).

As for penal matters, the situation of *licenciado* Ortega improved with a gradual royal pardon; its first phase was in April and May 1583, when, with the help of Mateo Vázquez, he drew up a list of his services to the king, and others that he could perform thenceforth. Among the former, he cited His Majesty's lawsuit with Juan de Orbea, for which a sentence had been issued but not yet executed due to Ortega's imprisonment in Pinto, and it was expected to be very lucrative for the Crown. Another example was the revenues from the lands of Amaniel, which provided an annual income of 1,000 *ducados*, and another related to the Quejigal wood-pasture land (*dehesa*). He anticipated that this process would be advantageous for the monastery of El Escorial and would not require referral to ecclesiastical justice. From that point onwards, *licenciado* Ortega proposed intervening in two issues that called for exceptional consideration. The first was the deplorable state of the royal site of El Pardo.[41] The second was a point of particular importance for the Castilianist party within the framework of the extended royal domestic governance: the enforcement of social discipline in the area of the royal sites.[42]

40 IVDJ, e. 99, caja 140, fols. 379v–380r.

41 IVDJ, e. 99, caja 140, fols. 116r–117v, Madrid, 16 April 1583, 'Las cosas en que el licdo. Ortega diçe podría seruir a Su Mag[esta]d de presente son las siguientes, y en las que particularmente a seruido' ('The things in which licenciado Ortega says he could serve His Majesty in the present are the following, and in which he has particularly served').

42 IVDJ, e. 99, caja 140, fols. 116r–117, Madrid, 16 April 1583.

Having resumed his regular duties, Ortega expressed his gratitude to the secretary and the king for his return to more important matters. And in the same letter, he also reported on the inappropriate behaviour of certain servants of the Marquis of Aguilar in Carabanchel, who had undermined the reputation of several local ladies with the sole intention of marrying advantageously.[43] Ortega's collaboration with Rodrigo Vázquez de Arce on the Quejigal issue was also noted at the time.[44] But the *alcalde-juez's* main concern was that his partial pardon by the king should be completed and acquire legal force.[45]

This moment was still some time in coming, and meanwhile he took on new commissions. On 7 March 1584, he was entrusted with inspecting the officials of the Alcázar and Bosque de Segovia.[46] It took a year to perform the task, which coincided with the continued investigations into the harm caused by hunting at El Pardo in 1583 and 1584.[47] Likewise, a report by Mateo Vázquez on the Count of Santisteban, who was caught red-handed hunting in El Pardo, enables us to assess Ortega's intervention.[48] This marked the beginning of his full reinstatement to his duties by means of a royal decree issued in Monzón on 19 July 1585, which was undoubtedly aided by his zeal in prosecuting, in accordance with his attributions, a person as important as the count, as well as the protection offered by the Castilianist group, especially its secretary, Mateo Vázquez. It is possible that Ortega's expertise in *obras y bosques* affairs was a valuable aid for the secretary when he denounced the activity of the secretary Martín de Gaztelu and advocated the reform of his office and the management of *obras y bosques* by a specialised secretary. The *alcalde's* zealous defence of the royal patrimony, which led him to consider a simple cart and its mules to be proper hunting equipment, prompted him to denounce the inactivity of the *Junta de Obras y Bosques*. The cancellation of one of its sessions had prevented it from dealing with certain cases, such as that of Diego López, the priest of Carabanchel for whom holy orders were no obstacle to poaching. In this case, Ortega requested an order from the *Junta* making it possible to take action against López, and in doing so, despite the clergyman's excesses, he showed

43 IVDJ, e. 99, caja 140, fols. 111r–113v, licenciado Ortega to Mateo Vázquez, 21 May 1583.

44 IVDJ, e. 99, caja 140, fol. 119r, licenciado Ortega to Mateo Vázquez, Carabanchel, 3 August 1583; *ibidem*, fol. 118r–v, Mateo Vázquez to Rodrigo Vázquez and his answer, 10 August 1583.

45 IVDJ, e. 99, caja 140, fol. 119r, licenciado Ortega to Mateo Vázquez, Carabanchel, 10 August 1583.

46 AGP, SR, libros de cédulas, v, fols. 309r, and 324v–325v.

47 AGP, SR, libros de cédulas, v, fol. 397r–v.

48 Carlos Riba García, *Correspondencia Privada de Felipe II con su Secretario Mateo Vázquez, 1567–1591* (Madrid: CSIC, 1959), 335, Philip II to Mateo Vázquez, San Lorenzo de El Escorial, 11 September 1584.

that he was not entirely neutral in his role as *alcalde-juez*, as this case appeared to be tinged with political interest.

The priest was a first cousin of Jerónimo Pablo, Cardinal Quiroga's secretary, and Mateo Vázquez's support for Ortega in this matter was fuelled by this fact. The secretary and the *alcalde* aimed to highlight the poor governance and the lack of control among the prelate's clients by emphasizing that this poaching case, given its specific circumstances, was the only one that could go unpunished. In the case of Ortega, after his rehabilitation, it may be possible to discern the determination of someone who had himself experienced harsh judicial scrutiny that was not impartial and had suffered as a result. In any case, he took advantage of his involvement in the affair to request complete exoneration from Philip II. However, the king was concerned with avoiding mounting political tensions at the Court and instead referred the matter to Quiroga.[49] At the insistence of Ortega, Vázquez recommended that the king also seek the advice of Fray Diego de Chaves.[50] However, at this point, the confessor and the secretary were losing sway to a king who, as early as April 1586, complained about the large number of papers and petitions sent to him by the confessor.[51] This political context undoubtedly contributed to Ortega's request for an early departure from the Court and the subsequent withdrawal of his duties related to *obras y bosques*.[52]

3.2 *Licenciado Belvis Galarza*

The tasks of *alcalde* Ortega reveal the lack of involvement of the *Junta* during this period. The evidence indicates that, as with many other matters, the monarch reserved for himself the right to deal with affairs relating to *Obras y Bosques*, in accordance with their patrimonial nature. In this regard, it is worth mentioning that these matters were discussed with the confessor, Chaves, rather than involving the *Junta de Obras y Bosques* to a greater extent. The monarch's interest in maintaining direct and immediate control over his assets helps to understand the continuation of the commissionial system in this area of administration, and the consequent difficulty of perceiving the *Junta* beyond the documentary trace of its agreements validating this commissional functioning. However, the administrative reform dynamics being implemented

49 IVDJ, e. 90, caja 129, no. 609. On Quiroga's courtly and political role, Henar Pizarro Llorente, *Un gran patrón en la Corte de Felipe II: Don Gaspar de Quiroga* (Madrid: Universidad Pontificia de Comillas, 2004).

50 IVDJ, e. 90, caja 129, nº 608, licenciado Ortega to Mateo Vázquez and his reply, Carabanchel, 28 September 1586.

51 Riba García, *Correspondencia Privada*, 379, Philip II to Mateo Vázquez, 12 April 1586.

52 Díaz González, *La Real Junta*, 109–10.

by the monarchy since Philip returned from his Aragonese journey of 1585–86 suggest a shift towards the institutionalisation of the *Junta*. This tendency is reflected in the appointment of Francisco de Belvis Galarza as *juez de bosques* and the specification of his duties, issued on 30 June 1588. Belvis, previously *corregidor* of Oropesa and its territory, was made *alcalde mayor* of El Escorial at the suggestion of the prior Fray Miguel de Alaejos, and as a result of certain conflicts triggered by his predecessor, Doctor Camargo.[53]

However, the method of jurisdictional application within the 'territorial dominion of the royal household' continued to be commissional. Belvis was tasked with investigating the damage caused by royal hunting in the estates around Madrid from 1586 to 1588. In addition, he was also instructed to assess damage occurring around El Pardo (excluding some villages), Aranjuez, Valsaín and El Escorial. The judge was to examine their boundaries twice a year and his reports would be sent to the secretary, Juan de Ibarra, for the purpose of processing compensation. At the same time, he was to investigate and punish cases of poaching, carrying the rod of royal justice. For this, he would receive an annual salary of 300,000 mrs. with an allowance (*ayuda de costa*) of 75,000 mrs.[54] The appointment as *alguacil* of Juan de Gorbalán, a royal servant, to implement the orders of Belvis Galarza as judge of the woodlands of El Pardo by means of a Cédula Real of 12 December 1589, was evidence of the desire to reinforce the authority of the Court jurisdiction in the royal sites. This decree also designated Gorbalán as the public prosecutor in legal cases brought before the judge of *bosques* and in appeals filed with the *alcaldes de Casa y Corte* against sentences pronounced by the governor of Aranjuez and the *corregidor* of Segovia.[55]

These series of documents refined and made permanent the commissions described above. And the task of the *juez de bosques* also depended on the issuance of a wide-ranging and preliminary commission at a high level. As far as is known, Belvis Galarza received the first one on 24 September 1588, which enabled him to take action against poachers in Aranjuez.[56] The commissional game played by the *alcaldes* was varied: another decree, dated 19 April 1589, appointed Martín de Espinosa as the first instance judge for offences relating to the surveillance of hunting, fishing and firewood at El Pardo and the Casa de Campo, when Belvis had to attend to other commissions by royal order,

53 IVDJ, e. 99, caja 140, fol. 174r–v, Fray Miguel de Alaejos to Mateo Vázquez, San Lorenzo de El Escorial, 14 April 1583.

54 AGP, SR, libros de cédulas, VII, fols. 43r–v and 213r–v.

55 AGP, SR, libros de cédulas, VII, fols. 203v–204v.

56 AGP, SR, libros de cédulas, VII, fols. 97r–98r.

as had previously occurred during Ortega's time in office.[57] This order can be explained by the fact that on 8 March of 1588, another royal order was issued tasking Belvis Galarza with investigating an attack on Juan Fernández de Villarreal, a priest from Mocejón. Some poachers' relatives had accused the priest of being responsible for their imprisonment, and he sought the jurisdiction of the *corregidor* of Toledo or any other judge to plead his case.[58]

The administrative changes that took place around 1588 and 1592 coincided with a personal change in royal grace. This situation led to confusion and delays in the functioning of important institutions, such as the Royal Council and the Chamber, which extended to the management of *Obras y Bosques*. These circumstances highlighted the need to reinforce royal jurisdiction, and the appointment of Belvis Galarza as *alcalde de Casa y Corte* on 4 November 1591 was part of this context. He retained the post of *juez de Obras y Bosques*, the obligation to reside in El Escorial whenever the king was in San Lorenzo, and a condition that reflected the administrative lessons learned from years of accumulating commissions: to abstain from hearing civil and criminal cases related to the office of *alcalde*.[59] In a similar manner, his jurisdiction was completed by a royal decree signed in Valladolid on 20 July 1592, which, as in previous cases, required his presence at the hearing by the *alcaldes de Casa y Corte* of appeals against sentences that he had not pronounced at first instance, that is, those of the governor of Aranjuez and the *corregidor* of Segovia.[60] Therefore, appeals against sentences he had originally handed down at the first instance regarding the district of El Pardo would not be heard by him.

As had been the case with Ortega, this decision demonstrated the suitability of the jurisdictional supplement the post of *alcalde* entailed for the area of *Obras y Bosques*, which was characterised by its courtly nature. Notably, Galarza's promotion followed his assignment to inspect the royal sites under his jurisdiction, revealing the need for jurisdictional reinforcement.[61] To this task was added a new commission of inspecting the officers of the forest and Alcázar of Segovia[62] and the order given by the king on 29 June 1592 to attend to the needs of the *obras reales* (royal works), so that on the same day a commission was issued to *alcalde* Ayala to handle the matters in his charge during

57 AGP, SR, libros de cédulas, VII, fols. 155v–156r.

58 AGP, SR, libros de cédulas, VII, fols. 138r–139v, *Cédula Real* in Madrid, 8 March 1589.

59 AGS, Contaduría Mayor de Cuentas (CMC), 1ª época, leg. 1688, s.n.; Archivo Histórico Nacional (AHN), Consejos, lib. 707e, fol. 211r.

60 AGP, SR, libros de cédulas, VIII, fol. 156r.

61 AGP, SR, libros de cédulas, VIII, fols. 25r–26v.

62 AGP, SR, libros de cédulas, VIII, fols. 74v–75v.

Galarza's absence.[63] It is obvious from the simultaneous issuance of these last two decrees that the *alcalde* Pedro Bravo de Sotomayor had previously encountered difficulties in reconciling the latter task with searching for materials for the royal works. At that time, as a testament to his growing involvement, the king signed an order in Aranjuez on 15 May 1593 for paymaster Diego Lacorzana to indemnify the claimants for the damage caused to neighbouring estates by hunting at El Pardo in 1592.[64]

Despite the increasing responsibilities, in the case of Belvis Galarza the Court's jurisdiction showed its functionality and effectiveness in a wider sphere than that of the royal sites, since he was granted the title of *juez conservador de montes* (judge responsible for woodland conservation) on 29 September 1593.[65] It was thought that the limited effectiveness of Roque de Huerta since 1574 as *guarda mayor de los montes* (chief warden of the woodlands) in promoting forestry could be corrected by implementing greater royal domestic governance through the commissional route. The domestic and courtly nature of his position enabled Belvis Galarza to articulate a spatial cohesion with a wider radius.

On 2 October 1593, a commission was granted to Belvis Galarza to judge those guilty of stealing gold from the site of the Alcázar of Segovia,[66] and on 11 February 1594, he was appointed to judge the case of certain inhabitants of Yepes, Borox, Añover and other towns around Aranjuez, who had fought their guards with arquebuses when they were caught trespassing.[67] Once he had been granted complementary powers, the addition of new commissions would not have helped to improve the exercise of jurisdiction at the royal sites. Among the occupations that Galarza had to combine was taking possession of places incorporated into the king's patrimony. On 11 March 1595, together with Bernabé de Ávila, *alcalde mayor* of El Escorial and judge of the *fábrica* of San Lorenzo, he received a commission to take possession of the villages of Campillo, Monasterio and their annexes, until then owned by the Duke of Maqueda, Don Bernardino de Cárdenas.[68] Shortly afterwards, *licenciado* Belvis Galarza was commissioned to oversee the lawsuits arising between the two villages and the towns of Madrid, Guadarrama, Galapagar, and the places

63 AGP, SR, libros de cédulas, VIII, fols. 138r–139r.
64 AGP, SR, libros de cédulas, VIII, fols. 311r–312r.
65 AGP, SR, libros de cédulas, VIII, fols. 368v–370r.
66 AGP, SR, libros de cédulas, VIII, fols. 374v–375v.
67 AGP, SR, libros de cédulas, VIII, fols. 403v–404r, 412v–413r and 414v–415r.
68 AGP, SR, libros de cédulas, VIII, fols. 535v–536r.

of Navalquexigo, Collado Villalba and others over pastures, tree felling, boundaries, and jurisdiction.[69]

Despite his jurisdictional role and higher remuneration,[70] Belvis Galarza's struggles in fulfilling all his commissions meant, as on previous occasions, that the king had to simultaneously rely on his colleagues. On 2 March 1596, a Cédula Real appointed Francisco Mena de Barnuevo as a substitute for matters concerning *Obras y Bosques*, in the event that Belvis was absent from Court on royal commission. This was done in the same manner as with *alcalde* Ayala on previous occasions. Although it was a temporary appointment, subject to the presence of Belvis Galarza at the Court, it had a clear institutional profile, as Mena was entrusted with the task of replacing him in all the commissions he was leading.[71]

4 Conclusion: the Contribution of the *Alcalde-Juez de Bosques* to the Institutionalisation of the *Junta de Obras y Bosques*

The *Junta de Obras y Bosques*, which took on institutional shape during the reign of Philip II,[72] supported the empowering of the jurisdiction of the *alcalde-juez de bosques* in a commissional manner. Its indisputable permanence from that time onwards was unwittingly anticipated by several authors much earlier. The practice of hunting was supported by Charles V through the consolidation of royal forests in El Pardo and Aranjuez, among others, with different regulations to those previously used. A special board was set up in this context to organize and protect the forests, plantations, and *Obras Reales*. It was attended by the main ministers in royal service (*mayordomo mayor* [lord steward] *caballerizo mayor* [master of the horse], *montero mayor* [master of huntsmen]), the presidents of Castile and *Hacienda* (the Treasury) and two councillors of the *Cámara de Castilla*, and must have taken place in 1545, as stated by numerous credible authors.[73] However, Álvarez de Quindós pointed

69 AGP, SR, libros de cédulas, VIII, fols. 542r–543v, *Cédula Real*, Madrid, 28 April 1595.

70 AGP, SR, libros de cédulas, IX, fols. 10v–11r.

71 AGP, SR, libros de cédulas, IX, fol. 25r–v.

72 While I was making the last correction to this chapter, a very pertinent article came out on this subject, Félix Labrador Arroyo, "La administración de los sitios reales castellanos en la década de 1580: la institucionalización de la Junta de Obras y Bosques", *e-Spania* [*online*] 47 (2024) http://journals.openedition.org/e-spania/49512; DOI: https://doi.org/10.4000/e-spania.49512 (consulted on 5 March 2024).

73 Gil González Dávila, *Teatro de las Grandezas de la Villa de Madrid Corte de los Reyes Católicos de España* (Valladolid: Maxtor, 2003), 521; Alonso Núñez de Castro, *Libro Histórico*

out that this was the date of the oldest surviving documentation, according to a paper drawn up by the *Junta* itself during the reign of Philip III.[74] It appears that the initial request lacked institutional significance or continuity. Consequently, the issues of poaching and the susceptibility of the royal hunting grounds were left unaddressed. This is evident from Juan de Castilla's report to the emperor on hunting at Aranjuez and the management of the hunting grounds and royal sites, as well as Charles V's directive to regents Maximilian and Maria in 1548.[75] These indications support Díaz González's view that the *Junta* was no more than an *ad hoc* and informal convocation of the members of the king's Chamber. It lacked an institutional profile until very late in Philip II's reign, despite its domestic importance.

This evolution followed a logical progression: the shaping of the *Junta de Obras y Bosques* was gradual and dependent on the definition and consolidation of the royal sites. These sites were embedded in a dense network of jurisdictions and added a 'household state,' a royal territorial domain, to the pre-existing jurisdictional space. As a result, more direct forms of patrimonial dependence were imposed, in contrast to extended royal domestic governance. Thus, for instance, following the Instructions of 1543 and 1563, the *Junta* had issued the Instruction for the governance of Aranjuez on 7 June 1582. The board was the result of a process that had started at the beginning of the reign, with Philip II directing one of his secretaries to handle these affairs. In 1561, he assigned this duty to Pedro del Hoyo, who was subsequently replaced by Martín de Gaztelu after the latter's passing.[76] They served as royal secretaries with a clear mandate from the king to work within their respective

Político Sólo Madrid es Corte y el Cortesano en Madrid (Madrid: Roque Rico de Miranda, 1675), 111.

74 In addition to those already mentioned, the author included the governors of Casa de Campo and El Pardo, by virtue of their offices, the royal confessor, the dean of the Royal Council and a member of the chamber, the latter two acting as legal advisers. Later, the warden of the Buen Retiro and the governor of Aranjuez were added, both by special grant. Officers included the public prosecutor, the secretary, the notary of the chamber, the reporter, the procurator, the accountant and the gatekeeper, but this composition corresponded to a later phase, when the Junta was fully active at the institutional level, Álvarez de Quindós, *Descripción Aranjuez*, 411–12.

75 Report in Aranjuez, 11 April 1549, and Instruction in Brussels, 4 July 1549, *Corpus Documental de Carlos V*, III (Salamanca: Universidad, 1977), 116 and 137.

76 José Antonio Escudero, *Los secretarios de Estado y del Despacho*, (Madrid: Instituto de Estudios Administrativos, 1976), vol. 1, 182–83; José Martínez Millán, "Gaztelu, Martín de," in José Martínez Millán and Carlos Javier de Carlos Morales (coords.), *Felipe II (1527–1598). La configuración de la Monarquía Hispana* (Valladolid: Junta de Castilla y León, 1998), 383–84.

functional areas. It is unclear whether any additional duties were attributed to the position of 'secretary of *Obras y Bosques*' beyond that commission. That tasks, positions and negotiations were concentrated in Gaztelu's hands is a well-known fact, and after Francisco de Eraso died his duties were further extended to 'Orders, *Encomiendas*, religious houses and Churches [...] and he shall enter the Chamber'.[77]

When Gaztelu passed away in 1580, it was decided to distribute his extensive responsibilities among several secretaries. However, they were initially taken on by Mateo Vázquez in addition to his existing duties, despite his request to the king that Antonio Gómez de Eraso be appointed as as *secretario de Obras y Bosques*. In 1581, a 'Council of *Obras y Bosques*' was formed, which later evolved into the *Junta*. It is unclear how independent this 'council' was from the Chamber. Mateo Vázquez was appointed as its secretary.[78] The desire of the Castilianist group to ensure the continuity of this committee was a clear indication of its incipient and discontinuous nature, as can be seen from the statement made in 1580 to Mateo Vázquez by one of its members, the Count of Barajas, president of *Órdenes*: 'in this matter of woodlands and works I will need the commissaries to truly believe that the day of the week that is designated will not be missed'.[79] This approach, following the administrative restructuring, was upheld by political associates such as Rodrigo Vázquez de Arce, who as the president of the *Consejo de Hacienda* (Treasury Council) participated in guiding the panel.[80] It is worth considering whether the institutionalization of the administrative field of *Obras y Bosques* occurred simultaneously with that of the Chamber, as separate entities within a shared domestic context.

Although a precise date cannot be given, it appears that over the last two decades of Philip II's reign, the *Junta de Obras y Bosques* consisted, more out of custom than a result of precise regulations, of three permanent members (an *alcalde*, a public prosecutor and a secretary) who met with the ministers and officials of the *Casa y Corte* appointed for this purpose, namely the *mayordomo mayor*, the *cazador mayor*, the presidents of the Royal and Treasury Councils and two chamber councillors. However, the composition of its minimum of three members could vary greatly. In any case, the institutional consolidation

77 'lo de Órdenes, Encomiendas, háuitos y Iglesias [...] y que entre en la Cámara', Real Academia de la Historia (RAH), Salazar y Castro, A-67, fol. 221v, published by Santiago Fernández Conti, *Los Consejos de Estado y Guerra de la Monarquía Hispana (1548–1598)* (Valladolid: Junta de Castilla y León, 1998), 111.

78 Díaz González, *La Real Junta*, 70–74.

79 'en esto de bosques y obras tendré neçesidad que los comisarios crean de veras que el día que se señalare en la semana no se a de faltar', IVDJ, e. 56, caja 74, n.n.

80 IVDJ, e. 100, caja 141, fols. 161r and 162r.

of the *Junta* meant that it was finally staffed by an accountant, a notary, a fiscal agent, two doorkeepers and an *alguacil*.[81] Its purpose was to manage the royal residences, estates, sites, and woodlands, including selecting candidates to fill staff vacancies.[82] But this function extended beyond administrative tasks through its gradual jurisdictional consolidation. It served as a management body, as well as exercising governmental and judicial prerogatives. As a body, it held authority over the governance, justice, grace, and finances of the royal sites.[83] The larger the *Junta* was as an institution, the more evident its multiple roles were: it was not only responsible for the upkeep of the residences, palaces and royal forests, but also for the personnel under its jurisdiction, including the appointed judges, who were responsible for enforcing the decrees for the preservation of the forests and the obligations of the inhabitants of the surrounding areas. It also awarded titles such as *cazador mayor*, *montero mayor*, *marcador mayor*, and others, thus shaping the state of the royal sites. The *Junta de Obras y Bosques* provided clear evidence of the government's close ties to the administration of the household, making the two entities practically indistinguishable.

Among the activities that highlight this are the handling of exceptions to general laws on royal properties, as evidenced in a Cédula Real dated 19 January 1594, signed in Madrid and granting permission to the governor of Aranjuez, Vázquez de Cepeda, to use a carriage pulled by two mules for visiting the estate and travelling to Ocaña or other surrounding areas.[84] On occasions, the *Junta de Obras y Bosques* was even perceived to increase its involvement in a purely jurisdictional sense with respect to the *juez de bosques* and the *alcaldes de Casa y Corte*.

After consulting with the *Junta*, the king reduced the penalties imposed by the former on those who violated the royal hunting regulations, as proven by the case of Juan de Colmenarejo, a resident of Colmenar Viejo, who was caught hunting three times within the protected perimeter of El Pardo. As a result, Francisco Mena de Barnuevo sentenced him to a fine of 10,000 mrs, a four-year banishment from Colmenar and the limits of the forest of El Pardo and a three-league space. The convicted man appealed on the grounds of poverty to have

81 Cos-Gayón, *Historia Jurídica*, 77.

82 IVDJ, e. 99, caja 140, fols. 152r–153v.

83 Núñez de Castro, *Libro Histórico-Político*, 111–13.

84 AGP, SR, libros de cédulas, VIII, fol. 400r–v.

his fine reduced, and Philip II commuted it for a further two years' exile, to be spent with his entire family.[85]

However, it can be argued that in the field considered here, at the time the *Junta de Obras y Bosques* did not extend beyond the agreements signed by its secretary Juan de Ibarra. It played a coordinating role that was confined to the exercise of jurisdiction, which was carried out by the commissaries at first instance. Though there are significant exceptions to this perception. First, the drafting and processing of the regulatory instruments that governed the functioning of the different royal sites. Second, the issuing of the mandates for the creation and daily maintenance of the forests, gardens and royal palaces. And, third, the organization, dispatch and review of the inspections (*visitas*) to which they were subject.[86] But solely on a jurisdictional level, this restricted responsibility resulted in an inability to fulfil the *Junta de Gobierno's* call for efficiency, diligence, and coordination within its area of competence. Indeed, it is difficult to take stock of the *Junta de Obras y Bosques* during the last part of Philip II's reign, and the situation does not seem to have changed during that of his son.

In addition, there is evidence of Belvis Galarza's activity up to at least 1597.[87] However, it is clear that action in this field had been weakened for some time, as indicated by the fact that in 1599, Philip III instructed his confessor, *alcalde* Francisco Mena de Barnuevo, and the secretary Juan de Ibarra to assemble in Santo Tomás (which may represent the initial formation of the *Junta* during the new king's reign), to entrust Gil Negrete, *alcalde* of El Escorial, Francisco Gómez, notary of the *bosques*, Juan de Gorbalán, his public prosecutor, and Francisco del Basto with investigating hunting damages that had remained unpaid in the area of the forests of El Pardo, Escorial, Valsaín and Aranjuez since the time of the emperor.[88]

Likewise, the reports that Philip III requested from the start of his reign enable us to deduce a subtle involvement of the *Junta* since his father's reign. They indicate that he not only resolved such matters with the aid of a solitary councillor, but also unequivocally affirmed that 'I have not seen that a warrant or any other measure has been issued to enter this *Junta*'.[89] They did so in

85 AGP, SR, libros de cédulas, IX, fol. 312r, Madrid, 20 February 1598, signed by the prince Philip, future Philip III.

86 IVDJ, e. 99, caja 140, Doctor Pérez Manuel, Aranjuez visitor, to Mateo Vázquez, Ocaña, November 4 1583.

87 AGS, CMC, 1ª época, leg. 1688, s. n.

88 AGP, Sección Administrativa (SA), leg. 344.

89 'no he visto que se haya despachado cédula ni otro recaudo para entrar en esta junta', AGP, SA, leg. 853, year 1610, cited by María Victoria García Morales, "Los artistas que trabajan

the same manner as the lawyers Pedro and Manuel Antonio de Cervantes did during the time of Philip IV, when they cautioned against transforming the position of *juez de bosques* into that of *alcalde de Casa y Corte* in their renowned treatise.[90] In their argument, they demonstrated a lack of understanding of the jurisdictional development thus far, consequently leading astray many historians who have carried out research on the *Junta de Obras y Bosques*. The *Junta* appears to have been little more than the commissions that derived from it, whose execution was advisable, given the courtly nature of its administrative space, to complement the jurisdiction of the commissary judges with the title of *alcalde de Casa y Corte*.

However, it has been highlighted that although this was a significant aspect of the jurisdictional activity hosted by the royal sites, it did not account for it entirely. During the reign of Philip IV, the *Junta de Obras y Bosques* gradually evolved into a body responsible for governance, jurisdiction and administration within the royal sites, resulting in modifications to the aforementioned appeal process. The 1639 edition of the *Autos y Acuerdos* summarised it as follows: "The appeals pertaining to what is provided for in matters and damages relating to hunting at El Pardo and Aranjuez shall be submitted to the *alcaldes*" ("Que las apelaciones de lo que se proveyere en las cosas, i daños de la caza del Pardo, i Aranjuez, vengan a los alcaldes"). However, it added a very important point, which indicated that the provision was revoked: "This ceased with the *Junta de Obras y Bosques*" ("Esto cessó con la Iunta de Obras i Bosques").[91] And it achieved this through the final formalisation of the *Junta de Obras y Bosques* during the reign of Philip IV. The account of the authority exerted by the *Junta*, as presented by Álvarez de Quindós, exemplified this consequence. With regard to jurisprudence, the *Junta* acted punitively, imposing penalties on the officials and servants of the royal sites and forests who did not comply with their duties, both through the ordinary courts and through *visitas* to their offices. And when deemed appropriate, it pursued legal action against individuals who violated the royal regulations regarding the preservation of wildlife, fishing, vegetation and firewood resources within the royal woods.[92] This was the result of an institutionalization process based on a system of commissions.

para el rey: la Junta de Obras y Bosques", *Espacio, Tiempo y Forma. Serie VII. Historia del Arte*, 3 (1990): 131.

90 AGP, Expedientes Personales (EP), caja 226/11.

91 *Autos i acuerdos del Consejo*, f. 5.

92 Álvarez de Quindós, *Descripción Aranjuez*, 415.

Abbreviations

AGP: Archivo General de Palacio
  SA: Sección Administrativa
  SR: Sección Registros
  EP: Expedientes Personales
AGS: Archivo General de Simancas
  CMC: Contaduría Mayor de Cuentas
  E: Estado
IVDJ: Instituto Valencia de Don Juan
  E: Envío
  C: Caja
RAH: Real Academia de la Historia
  SC: Colección Salazar y Castro

Acknowledgements

This chapter is part of the Postdoctoral Project 'La Corte difusa. La articulación territorial de la jurisdicción real (Siglos XVI–XVIII) – The Diffuse Court. The territorial articulation of royal jurisdiction (16th–18th centuries),' under the 'Margarita Salas-María Zambrano' Programme, developed at the Universidad Rey Juan Carlos and funded by the Spanish Ministry of Universities and the European Union. The author is assistant professor doctorate in the Universidad de Cantabria and a collaborating researcher in the CINTER Group (Court, Image, Nobility and Territory) at the URJC, as well as a researcher at IULCE (UAM) and a collaborating researcher at CEDIS (NOVA School of Law, Lisbon).

Bibliography

Álvarez de Quindós, Juan Antonio. *Descripción Histórica del Real Bosque y Casa de Aranjuez.* Madrid: Doce Calles, 1993.

Autos y acuerdos del Consejo, de que se halla memoria en los libros, desde el año de 1532, hasta el presente de 618. Mandados imprimir por el ilustríssimo señor don Fernando de Azevedo, Arçobispo de Burgos, Presidente del Consejo, y señores dél. Madrid: Luis Sánchez, 1618.

Ayala Martínez, Carlos de. "Pérez, Arias". In *Diccionario Biográfico Español*, XL, 716–717. Madrid: Real Academia de la Historia.

Bauer Manderscheid, Erich. *Los montes de España en la Historia.* Madrid: Servicio de Publicaciones Agrarias, 1980.

Branchat, Vicente. *Tratado de los derechos y regalías que corresponden al Real Patrimonio en el reyno de Valencia y de la jurisdicción del intendente como subrogado en lugar del antiguo Bayle General ...* Tomo III, Valencia: Imprenta de Joseph y Tomás de Orga, 1786.

Cervantes, Pedro de and Manuel Antonio de Cervantes. *Recopilación de las Reales Ordenanzas de los Bosques del Pardo, Aranjuez, Escorial y Balsaín ...* Madrid: Melchor Álvarez, 1687.

Cos-Gayón, Fernando. *Historia jurídica del Patrimonio Real.* Madrid: Enrique de la Riva, 1881.

Covarrubias, Sebastián de. *Tesoro de la Lengua Castellana o Española según la impresión de 1611, con las adiciones de Benito Remigio Noydens publicadas en la de 1674.* Barcelona: Alta Fulla, 1998.

Díaz González, Francisco Javier. *La Real Junta de Obras y Bosques en la época de los Austrias.* Madrid: Dykinson, 2002.

Escudero, José Antonio. *Los secretarios de Estado y del Despacho.* Vol. 1, Madrid: Instituto de Estudios Administrativos, 1976.

Ezquerra Revilla, Ignacio, "La integración de la Casa en la Corte: los alcaldes de Casa y Corte." In *La Monarquía de Felipe II: la Casa del Rey,* coordinated by José Martínez Millan and Manuel Rivero Rodríguez, 697–800. Madrid: Fundación Mapfre-Tavera, 2005.

Fernández Álvarez, Manuel, ed. *Corpus Documental de Carlos V.* Vol. 3, Salamanca: Universidad de Salamanca, 1977.

Fernández Conti, Santiago. *Los Consejos de Estado y Guerra de la Monarquía Hispana (1548–1598).* Valladolid: Junta de Castilla y León, 1998.

García Morales, María Victoria. "Los artistas que trabajan para el rey: la Junta de Obras y Bosques," *Espacio, Tiempo y Forma. Serie VII. Historia del Arte,* 3 (1990): 121–1¡36. https://doi.org/10.5944/etfvii.3.1990.2161.

González Dávila, Gil. *Teatro de las Grandezas de la Villa de Madrid Corte de los Reyes Católicos de España.* Valladolid: Maxtor, 2003.

Hernando Ortego, Francisco Javier. "La lucha por el Monte de El Pardo. Rey, municipio y uso del espacio en el Madrid del Antiguo Régimen," *Cuadernos de Investigación Histórica* 12 (1989): 169–96.

Hespanha, António Manuel, *Vísperas del Leviatán: instituciones y poder político (Portugal siglo XVII).* Madrid: Taurus, 1989.

Hespanha, António Manuel, "El espacio político." In *La Gracia del Derecho,* edited by Antonio Manuel Hespanha, 85–121. Madrid: Centro de Estudios Constitucionales, 1994.

Labrador Arroyo, Félix, ed. *Diario de Hans Khevenhüller: embajador imperial en la Corte de Felipe II*. Madrid: Sociedad Estatal para la Conmemoración de los Centenarios de Felipe II y Carlos V, 2001.

Labrador Arroyo, Félix, "La administración de los sitios reales castellanos en la década de 1580: la institucionalización de la Junta de Obras y Bosques", *e-Spania* [*online*] 47 (2024) http://journals.openedition.org/e-spania/49512; DOI: https://doi.org/10.4000/e-spania.49512 (consulted on 5 March 2024).

Las Siete Partidas, del Sabio Rey don Alonso el nono, nuevamente glosadas por el Licenciado Gregorio López del Consejo Real de las Indias de Su Magestad. Salamanca: Andrea de Portonariis, 1555.

Manuel Valdés, Carlos. *Estudio Histórico-Selvícola del Monte de Valsaín (Siglos XVI–XX)*. Madrid: Organismo Autónomo Parques Nacionales, 1997.

Mariana, Juan de. "Del Rey y de la Institución Real." In *Obras del padre Juan de Mariana*, Vol. II, 463–576. Madrid: Rivadeneyra, 1854.

Marichalar, Antonio. "Tres figuras del XVI. Hernán Suárez de Toledo, Felipe de Borgoña y Briviesca Muñatones," *Escorial* 17 (1944): 9–67.

Martínez Millán, José. "Gaztelu, Martín de." In *Felipe II (1527–1598). La configuración de la Monarquía Hispana*, coordinated by José Martínez Millán and Carlos Javier de Carlos Morales, 383–84. Valladolid: Junta de Castilla y León, 1998.

Merlos Romero, Magdalena. *Aranjuez y Felipe II. Idea y forma de un Real Sitio*- Madrid: Comunidad de Madrid-Ayuntamiento de Aranjuez, 1998.

Núñez de Castro, Alonso. *Libro Histórico Político Sólo Madrid es Corte y el Cortesano en Madrid*. Madrid: Roque Rico de Miranda, 1675.

Ortega de la Torre, Eduardo. "Carlos III y la lucha por el Real Patrimonio en el País Valenciano: el informe Branchat (1784)." In *De la Ilustración al Romanticismo: IV Encuentro, Carlos III, dos siglos después. Cádiz, 7–9 de abril de 1988*, coordinated by Mariano Peñalves, vol. 2, 111–18. Cádiz: Universidad de Cádiz, 1994.

Pizarro Llorente, Henar. *Un gran patrón en la Corte de Felipe II: Don Gaspar de Quiroga*. Madrid; Universidad Pontificia de Comillas, 2004.

Riba García, Carlos. *Correspondencia Privada de Felipe II con su Secretario Mateo Vázquez, 1567–1591*. Madrid: CSIC, 1959.

Ruiz Ayúcar, Eduardo. *El alcalde Ronquillo. Su época. Su falsa leyenda negra*. Ávila: Institución Gran Duque de Alba, 1997.

Sánchez Meco, Gregorio. *El Escorial: de Comunidad de aldea a villa de Realengo*. Madrid: Ayuntamiento de El Escorial, 1995.

CHAPTER 2

The Royal Forests in the Spanish Monarchy's System of Royal Geographies: Offices and Social Integration during the Seventeenth Century

José Eloy Hortal Muñoz

1 The Royal Forests and Woodlands in the System of Royal Geographies: Representation, Provisioning and Territorial and Social Integration

In the past three decades, a new approach has been developed to studying the European courts of the Middle Ages and early modern period by applying several new methodologies. No doubt, one of the aspects that is receiving the most attention is court space, entailing an analysis of the existing bonds between space and power.

Space, taken as a construct, undoubtedly helps us understand how society has been dominated and organised, as well as the power relationships of individuals and groups of people with specific spaces. It is therefore appropriate to study its role in historical processes and conflicts.[1] In this sense, territory constituted a conglomerate stemming from the relationship between inhabitants, power and space, where dynasties of the late medieval and early modern times, both sovereign and noble,[2] were perceived and acted as the bond that connected the land with the people who populated it.[3]

Indeed, Stuart Elden stated that the term territory is at once a word, a concept, and a practice.[4] The so-called territory must therefore be taken to encompass several spaces – physical (land), political, jurisdictional, spatial (terrain) and living spaces – that could be interpreted differently by the various actors. Thus, space must be studied in accordance with how actors deal with it, paying

1 Paloma Bravo and Juan Carlos D'Amico, eds, *Territoires, lieux et espaces de la révolte. XVI–XVIII siècles* (Dijon: Éditions Universitaires, 2017).

2 Jeroen Duindam, *Dynasties: A Global History of Power, 1300–1800* (Cambridge: CUP, 2015).

3 Mario Damen and Kim Overlaet, eds, *Constructing and Representing Territory in Late Medieval and Early Modern Europe* (Amsterdam: Amsterdam University Press, 2022).

4 Stuart Elden, *The Birth of Territory* (Chicago: University, 2013).

special attention to its symbolism and the interaction of people with spaces and places.

In this connection, space also played an essential part in shaping the court system. Nevertheless, in studies conducted up until only a few years ago this system was held to encompass solely the palaces and cities where the Court sojourned frequently, and importance was attached to the impact of the latter on the cities which housed it.[5] Today, however, historians have begun to take into account the political interactions, ceremonial practices, cultural symbols, and institutional structures of the Court.[6]

Analysing these questions, it becomes clear that the Court generated its own geography, a term that here expresses our interest in the way the physicality of spaces and landscapes was acted upon and produced through cultural practices.[7] This interweaving of physical and human agency is naturally wide-ranging and encompasses image-making, architectural, agricultural, and administrative processes.

Of course, a significant element of that Court geography was what we can call "Royal Geographies", whose relevance to the configuration of the monarchies of the period has so far gone practically unnoticed. From the Middle Ages onwards, these networks of sites became increasingly significant means of consolidating sovereigns' power and key instruments for promoting their rule, as well as for extending their control over specific parts of the territory through kings' progresses.[8]

Taking all this into account, it can be said that in the early modern period the term Royal Geographies referred to properties belonging to the ruling

5 Robert Ashton, *The City and the Court, 1603–1643* (Cambridge: CUP, 1979); John P. Spielman, *The City and the Crown: Vienna and the Imperial Court 1600–1740* (West Lafayette: Purdue University Press, 1993); Léonard Courbon and Denis Menjot, eds, *La cour et la villa dans l'Europe du Moyen âge et des Temps Modernes*. Studies in European Urban History (1100–1800), 35 (Turnhout: Brepols, 2015); Eric Hassler and Anne Motta, eds, *Noblesses et villes de cour en Europe (XVIIe–XVIIIe): La ville de résidence princiere* (Rennes: Presses Universitaires Rennes, 2022).

6 The concept of *spatial turn* can be considered to have first been applied to studies on the Court in Marcello Fantoni, Giuseppe Gorse and Malcolm Smuts, eds, *The Politics of Space: European Courts ca. 1500–1750* (Rome: Bulzoni, 2009), where its parameters are explained in the introduction.

7 A very interesting example of this approach is Elisabeth Narkin, "Architectural Network as Dynastic Strategy in Fontainebleau's Galerie des Cerfs," *Renaissance Quarterly* 74:2 (2021): 454–97.

8 See José Eloy Hortal Muñoz and Merlijn Hurx, eds, *Building the Presence of the Prince. The Institutions Responsible for the Construction and Management of the Buildings of European Courts (14th–17th centuries)* (Turnhout: Brepols, 2024).

dynasty that were used as residences by the ruler and other members of the dynasty or were occasionally occupied by them, as well as places where other activities associated with the running of the monarchy and the Court were performed. Therefore, these Royal Geographies were not solely palaces or royal residences but complex spaces that also encompassed other types of constructions and lands.[9]

For example, in 1623, the *Junta de Obras y Bosques* (Board of Works and Woodlands) of the Spanish Monarchy had to take care of:

> Los Alcaçares, Casas y Bosques Reales que comprehende la Junta son, el Alcaçar Palacio Real de Madrid, quarto del Monasterio Real de San Jerónimo, Casa Real del Campo, Castillo y Monte del Pardo Casa de Vaziamadrid, Alcazares de Sevilla y los Palacios y Bosque del Lomo del Grullo; los Alcaçares de Toledo y Segovia, Casa Real y Bosque de Valsaín, Casa Real de Fuenfría, Casa de la Moneda del Ingenio de Segovia, Casas Reales de Valladolid, su huerta y su ribera, Casa Real y Bosque del Abrojo, Casa Real de Aondeguilla, Casa y Bosque de la Quemada, heredamiento de Aranjuez con su Casa Real y la de Aceca y el quarto de nuestra señora de la Esperanza, los bosques y dehesas deste heredamiento, la fabrica y patronazgo de San Lorenço el Real y conservación de sus bosques; la Alhambra de Granada y Soto de Roma, Archivo Real de Segovia y cavalleriza de Cordova.
>
> The Royal Fortresses [*Alcázares*], Residences and Woodlands covered by the *Junta* are the Alcázar Royal Palace of Madrid, the apartment in the Royal Monastery of San Jerónimo, the Royal Country Residence, Castle and Woodlands of El Pardo, the Residence of Vaciamadrid, the Alcázares of Seville and the Palaces and Woodlands of Lomo del Grullo; the Alcázares of Toledo and Segovia, the Royal Residence and Woodlands of Valsaín, the Royal Residence of Fuenfría, the Mint of Segovia, the Royal Residences of Valladolid, their vegetable garden and riverside land, the Royal Residence and Woodlands of El Abrojo, the Royal Residence of Aondeguilla, the Residence and Woodlands of La Quemada, which formed part of the estate of Aranjuez along with its Royal Residence, as well as the residence of Aceca and the apartments of Nuestra Señora de la Esperanza,

9 On the typology of Royal Geographies, see José Eloy Hortal Muñoz, "Royal Properties to Maintain: the Royal Geographies," in José Eloy Hortal Muñoz, Merlijn Hurx and Benjamin Ringot, eds, *Maintaining the Presence of the Prince. Management of Royal Geographies* (*XIVth–XIXth Centuries*) (Turnhout: Brepols, forthcoming).

> the woodlands and pastureland of its estate, the *fabrica ecclesiae* and patronage of San Lorenzo el Real and conservation of its woodlands; the Alhambra of Granada and Soto de Roma, the Royal Archive of Segovia and the stables of Cordoba.[10]

The type of Royal Geography most relevant to this chapter is the royal forests and woodlands, which constituted a step towards establishing the presence of the sovereign in the territory. The shaping of the royal forests and woodlands all over Europe progressed in parallel with that of the residences and systems of Royal Geographies, lasting from the late thirteenth to the seventeenth century.[11]

Generally speaking, once the monarchs decided to grant these forests and woodlands royal status, it was necessary to establish an administrative framework to mark out their limits and protect them jurisdictionally. There were two ways of doing this: by entrusting the upkeep of these wooded areas to their previous owners under the supervision of a royal officer, or by setting up a network of forestry agents devoted exclusively to preserving, exploiting and safeguarding the forests. This second option was unquestionably the most widely used in Europe. Places like Portugal, Venice and Wittenberg were pioneers in the process, though significant progress in the organisation of those spaces was made in others too, such as the modifications carried out in the forest of Soignies (Brabant) by Philip the Good around 1430.[12]

We find three organisational models in the process of defining the jurisdictional structure for these spaces. The first and the most common model was to attach these royal forests to the Royal Household, specifically the Hunt section, as occurred in Portugal with the *montaria-mor*, Denmark, England, Lithuania, and even the Papacy, among many other places.[13] The second model bound the

10 Gil González Dávila, *Teatro de las grandezas de la villa de Madrid. Corte de los reyes católicos de España* (Madrid: Thomas Iunti, 1623): 521–22.

11 See Koldo Trápaga-Monchet, "Power, Environment and Territory: the Creation and Implementation of Royal Forestry Legislation and Administration in Portugal in a Comparative Perspective (14th–17th centuries)," in Hortal Muñoz and Hurx, *Building the Presence of the Prince*: 130–47.

12 Sander Pierron, *Histoire illustrée de la forêt de Soignies*, 3 vols. (Brussels: Éditions Hansa, 1935–38).

13 There are numerous examples in Andrea Merlotti, ed., *Le Cacce Reali nell'Europa dei Principi* (Turin: Leo S. Olschki, 2017), in the chapters by John Robert Christianson on Denmark (3–20), Simon Adams on England (21–36), Austeja Braisunaite on Lithuania (37–44), and Franco Pignatti on the Papacy (161–94). Another source is the special monographic issue "Royal Hunts Issue" of *The Court Historian*, 18:2 (2013). For the Spanish Monarchy, see the chaptes devoted to the Royal Hunt by several authors in José Martínez Millán and

forests and woodlands and the Hunt-related offices to the territory, as in the Habsburg Netherlands, for example.[14] The third and last model involved making these spaces organisationally dependent on the institution in charge of managing the Royal Geographies, as in the case of the *Junta de Obras y Bosques* in Castile, though it also had significant ties to the Hunt section of the Royal Household. The *Junta de Obras y Bosques* itself specified about the Royal Forests in 1613:

> Los sotos propios que su Magestad tiene en estos reinos y se guardan conservan y administran por su real quenta como son los de Aranjuez, en el contorno del Escurial, los de San Lorenzo el Real en Madrid, los del Pardo, en Segovia los de Valsaín, en Valladolid los de la Quemada, en Sevilla los de los palacios del Lomo del Grullo y en Granada el Soto de Roma, consisten en dehesas particulares anexa a estos mismos sotos.[15]
>
> The wooded areas that His Majesty has in these kingdoms and which are guarded, preserved and administered at his royal expense, such as those of Aranjuez, in the surroundings of the Escorial, those of San Lorenzo el Real in Madrid, those of El Pardo, in Segovia those of Valsaín, in Valladolid those of La Quemada, in Seville those of the palaces of Lomo del Grullo and in Granada Soto de Roma, consist of private pastureland adjacent to those very woodlands.

The above organisation indicates that many of these forests were linked to the monarchs owing to their use as hunting grounds. Indeed, hunting was not simply a pastime but served a political and representative purpose as sovereigns met the most important members of their society while engaging in this activity. It likewise provided physical training for war, both for the various nobles and for the royal family, though royals stayed away from battlefields from the sixteenth century onwards.[16] Finally, it should be pointed out that experiments

---

Santiago Fernández Conti, eds. *La Monarquía de Felipe II: la Casa del Rey*, 2 vols (Madrid: Fundación Mapfre-Tavera, 2005); José Martínez Millán and Maria Antonietta Visceglia, eds. *La Monarquía de Felipe III: la Casa del rey*, 4 vols (Madrid: Fundación Mapfre, 2008); José Martínez Millán and José Eloy Hortal Muñoz, eds. *La Corte de Felipe IV (1621–65). Reconfiguración de la Monarquía Católica*. 4 vols. (Madrid: Polifemo, 2015).

14 Luc Duerloo, "The Hunt in the Performance of Archducal Rule", *Renaissance Quarterly* 69:1 (2016): 116–54.

15 Archivo General de Simancas (AGS), Casas y Sitios Reales (CSR), leg. 303, fol. 659.

16 The importance of the royal hunt is explored in Thomas T. Allsen, *The Royal Hunt in Eurasian History* (Philadelphia: University of Pennsylvania Press, 2006). More details in

and tests were also carried out on breeding different animals at the various hunting grounds with a view to improving the species besides guaranteeing the sustainability of these places and obtaining economic yields from them. It is therefore hardly surprising that we should come across posts related to catching vultures and foxes, as well as head goatherds, cattle herders, and shepherds.

The foregoing shows that although hunting was one of the chief reasons for making forests and woodlands royal property, it was not the only one. In fact, hunting lost part of its symbolic importance at the end of the sixteenth century and forests gained even greater economic significance after it was decided to register and fence them in to ensure their exploitation by the monarchs. For example, in Portugal, the royal forests were used to produce wood, especially for shipbuilding and, to a lesser extent, for firewood, charcoal, and construction. The same was true of Venice and Wittenberg and, to a lesser degree, of the Spanish Monarchy, where Soto de Roma was the only royal forest devoted exclusively from the reign of Philip II onwards to producing the wood used to craft cannon carriages for the Spanish Armada.[17]

Similarly, resources from the royal forests – revenues from the rental of land or from firewood and agricultural produce – contributed to the upkeep of the royal residences and provided income and allowances for the people who worked there. This has been very thoroughly studied in some cases, such as Versailles.[18] In the Iberian Peninsula, the king's possessions were used on several occasions to reward servants who did not receive regular wages for their duties in the royal households due to a shortage of funds in the royal treasury. In fact, there were moments when the king decided to prioritise these payments over performing certain necessary construction work on royal buildings.[19] Sites like

---

connection with the European kingdoms are provided in Wolfram Martini *et al*, eds, *Die Jagd der Eliten in den Erinnerungskulturen von der Antike bis in die frühe Neuzeit* (Gottingen: Vandenhoeck & Ruprecht, 2000); Claude d'Anthenaise and Monique Chatenet, eds, *Chasses princières dans l'Europe de la Renaissance* (Arles: Centre André Chastel, 2007), as well as the abovementioned Merlotti, *Le Cacce Reali*. Lastly, for Denmark, Jette Baagoe, *Kongens Skov-Verdens Arv. Parforcejagtlandskabet I Nordsjaelland / The King's Forest – The World's Heritage. The Par Force Hunting Landscape in North Zealand* (Rosendahls: The Danish Museum for Hunting and Forestry, 2016).

17 Felix Labrador Arroyo and Koldo Trapaga-Monchet. "Forestry, Territorial Organization, and Military Struggle in the Early Modern Spanish Monarchy", *Environmental History* 23:2 (2018): 318–41.

18 Vincent Maroteaux, *Versailles. Le Roi et son Domaine* (Paris: Editions Picard, 2000).

19 Archivo General de Palacio (AGP), Administraciones Patrimoniales (AP), Aranjuez, box 14131.

Lomo del Grullo in Seville, the Ingenio de la Moneda (mint) in Segovia,[20] and San Lorenzo de El Escorial and all its surrounding properties[21] were used for this purpose. However, there were three sites that enjoyed special economic importance as part of this strategy because they were used to pay rewards and sinecures to their own servants as well as to others without specific links to those sites.[22]

The first of the three was Aranjuez,[23] where wheat and barley were granted as payment in kind to individuals such as Antonia de Marquana y Alviz, daughter of the royal secretary Francisco Alviz, who received a yearly income of 400 ducats for her entire life,[24] or to royal convents such as that of the Encarnación, which received 300 *fanegas*[25] every year by standing order from Queen Margaret of Austria-Styria.[26] In addition, wood from its forests was given as charity for carrying out work on convents or schools.[27] The second site, the already mentioned and lesser-known Soto de Roma in Granada, was mainly employed as a royal hunting ground until the reign of Philip II. Thereafter the forests were used for their timber, which was necessary for shipbuilding and, on many occasions, as charity for convents.[28] Sustainable exploitation of the woodlands had to be implemented to prevent deforestation.[29] The third site, El Pardo, was where the king sourced the firewood he used to reward his officials,[30] as well

20 For coin minting since Philip II's reign see Victoria Soto Caba, "La primera fábrica de monedas: El Real Ingenio de Segovia", *Espacio, Tiempo y Forma, Serie VII, Historia del Arte* 4 (1991): 95–120.

21 Regarding the economic use of its meadows, José Antonio Martínez Bara, "Noticias sobre las dehesas del Monasterio de San Lorenzo del Escorial," *Anales del Instituto de Estudios Madrileños* 5 (1970): 109–19.

22 The same occurred in other places, such as the Royal Geographies of Valladolid where, for example, Doña Isabel de Mercado received an allowance as well as the salary of her husband, the royal physician Doctor Ruíz, although to a much lesser extent (AGP, Personal, boxes 160/12, 933/42, 1063/48, 1111/65).

23 There are excellent studies on economic exploitation, such as Carmen Magán Merchán and Jesús Espinosa Romero, "La evolución económica de un Real Sitio: Aranjuez en tiempos de Felipe II", *Reales Sitios* 153 (2002): 2–13.

24 AGP, AP, Aranjuez, box 14131.

25 This Castilian measurement of capacity is equivalent, in most cases, to 55.5 litres.

26 AGS, CSR, leg. 302/2 fol. 150.

27 There are numerous cases, for example, AGS, CSR, leg. 302/1, fols. 32 and 34.

28 There are several examples, such as the convent of the barefoot Carmelites of the Holy Martyrs of Granada that received 1,000 loads of firewood from Soto de Roma (AGS, CSR, leg. 303, fol. 506).

29 Labrador Arroyo and Trápaga Monchet, "Forestry, Territorial Organization", 320–29.

30 The importance that was attached to cutting firewood and the care it required can be seen in the *Instrucción para la corta de leña de El Pardo* (*n.d.*) (AGP, Registros (Reg) 25, fols. 351v–355v).

as the religious houses of the Capuchins, Trinitarians, Carmelites, Santa Isabel, and the Barefoot Franciscans of Barajas, a tradition initiated by Philip III.[31]

Of course, the wood and other raw materials produced at the Royal Geographies were used constantly for royal works, such as the wood from the woodlands of Aranjuez for the dome of the Monastery of San Lorenzo de El Escorial, or for other major constructions, such as the wood harvested from the forests of the Alcázar of Toledo that went into crafting the vault of that city's cathedral.[32] Finally, we find that wood was also employed in technological experiments, for instance that sent from Aranjuez to the Buen Retiro in 1639 with the idea of testing the design for a ship that the king wished to build to better navigate the Atlantic Ocean.[33]

Bearing all this in mind, we will now analyse the different offices entrusted with taking care of the royal forests and woodlands linked to the *Junta de Obras y Bosques*: their nature, the tasks they involved, who held them, and how the social status of the incumbents evolved.

2 The Offices Linked to Royal Forests and Woodlands in the Spanish Monarchy

No doubt many of the initiatives carried out in the Spanish Royal Geographies sought to preserve the plant and animal species that populated those places, albeit with the idea of using them for economic gain or for hunting. For this purpose, as well as to ensure the effective exploitation of the abovementioned resources, it was necessary to assign a number of people to the maintenance and care of the royal woodlands throughout the year.

The first point to stress is that in the Iberian Peninsula, except Portugal, these people were directly responsible to the Royal Geographies to which those woodlands belonged and were not necessarily members of the Royal Household. If there was a connection with the Royal Household it was due to other factors, but not to their relationship with the royal woodlands. This is the main reason why this study does not include the officials linked to the Royal Hunt. Many of the latter held posts that were mainly concerned with ensuring that hunting was carried out under the best possible conditions but also helped maintain the royal woodlands, such as the mounted falconers (*catarriberas*), and most of the huntsmen lived in villages (for example Fuencarral or San

31 Some examples can be found in AGS, CSR, leg. 302/2, no fol.

32 There are numerous examples in AGS, CSR.

33 AGS, CSR, leg. 310, fol. 12.

Sebastián de los Reyes) belonging to the royal residence with the largest forest, El Pardo. In any case, it should be pointed out that these posts have been profusely studied in the past few years.[34]

Similarly, nor will we examine here the officials related to the jurisdiction of the royal woodlands, as this group of people requires a specific study owing to their particular characteristics.[35] They were linked to the *Junta de Obras y Bosques* itself – for example the magistrates of works and woodlands (*alcaldes jueces de obras y bosques*) at El Pardo, El Escorial and Valladolid – and heard offences committed within a five-league radius of those woodlands together with the ordinary judges of those places.[36] We also find other officials such as the prosecutor of the woodlands of Segovia, whose duty was to prosecute the offences reported to him by the chief custodian (*guardamayor*) of Valsaín and the deputy *alcaide* (governor) of the Alcázar in Segovia.[37]

Having defined the posts we are going to examine, it is necessary to list the royal woodlands included in this study. Specifically, the Royal Geographies linked to the *Junta de Obras y Bosques* that had this connection with the royal forests and to whose staff we will refer were Aranjuez, El Pardo, Soto de Roma, Lomo del Grullo – which was connected with the Reales Alcázares of Seville – San Lorenzo de El Escorial, the Royal Residence and Woodlands of Valsaín and the Royal Sites of Valladolid, especially the woodlands of El Abrojo and La Quemada.

Unfortunately there is insufficient surviving documentation to be able to conduct a comprehensive study of the staff of each Royal Site, though we can gain an idea of the posts and the types of officials who held them all. These records are mainly located in two archives. The first is the Archivo General del Palacio Real de Madrid, whose Registros section contains two subsections that are helpful and contain the first 30 entries. The first subsection is entitled

34 See, among others, Manuel Rivero Rodríguez and Ignacio Javier Ezquerra Revilla, "La Caza en la Casa y corte de Felipe II", in José Martínez Millán and Santiago Fernández Conti, eds., *La Monarquía de Felipe II: La Casa del rey* (Madrid: Fundación Mapfre-Tavera, 2005), vol. I: 377–430; Ignacio Javier Ezquerra Revilla and Rubén Mayoral López, "La Caza Real y su protección: la Junta de Obras y Bosques", in José Martínez Millán and Mª. Antonietta Visceglia, eds, *La monarquía de Felipe III: la Casa del Rey* (Madrid: Fundación Mapfre, 2008), vol. I, 811–86; and José Martínez Millán and Félix Labrador Arroyo, "La pervivencia de la Casa de Castilla: la Caza", in José Martínez Millán and José Eloy Hortal Muñoz, eds, *La Corte de Felipe IV (1621–65). Reconfiguración de la Monarquía Católica* (Madrid: Polifemo, 2015), vol. II: 1041–1134.

35 A first approach in Ignacio Javier Ezquerra Revilla, "Atribuciones de la Junta de Obras y Bosques", in Martínez Millán and Hortal Muñoz, *La Corte de Felipe IV*, vol. III, 2066–83.

36 AGP, Administración General (AG), El Pardo, caja 9388, expediente 1.

37 AGS, Tribunal Mayor de Cuentas (TMC), leg. 1568, no fol.

'Libros donde se asientan todos los despachos tocantes a obras y bosques' (Books where all the offices related to works and woodlands are recorded) and the second, 'Colección de reales órdenes y disposiciones adoptadas por la Junta de Obras y Bosques en relación con el mejor gobierno de los Reales Sitios' (Collection of royal orders and provisions adopted by the *Junta de Obras y Bosques* in relation to the better governance of the Royal Sites). They are both known in the inventories of the palace archives as Cédulas Reales, and are all microfilmed and correctly foliated.

At the Archivo General de Simancas we consulted the Casa y Sitios Reales section, specifically bundles (*legajos*) 302 to 323, which concerns the decisions made by the *Junta de Obras y Bosques* throughout the seventeenth century, complemented by the bundles on the *Tribunal Mayor de Cuentas* (court of audits) from 1569 to 1572, which refer to a few particular Royal Geographies. Naturally there are also specific documents on each Royal Site, from which we will quote when appropriate during the course of this essay. Lastly, we will also draw on the instructions and ordinances on these posts, which we have progressively published in recent years.[38]

Before going on to list the offices in order of importance, it is advisable to point out that in general the social significance of the incumbents increased during the seventeenth century,[39] especially from Philip IV's reign onwards, owing to the overall reconfiguration process being carried out within the Spanish Monarchy.

Indeed, the model of monarchy formed during the time of Charles V was beset with crisis during Philip IV's rule, especially after the decade 1640–50. Throughout his 44-year-long reign, Philip tried to reconfigure the administrative machinery of the monarchy.[40] The main cause of the decline of the system was that many of its subjects could no longer be integrated into the Royal Household and therefore remained outside the protective umbrella that the monarch, as *pater familias*, had provided up until then, because the constitution of the Monarchy itself prevented it from successfully absorbing the various social groups from the different kingdoms, as the Household had

38 See especially, Martínez Millán and Hortal Muñoz, *La Corte de Felipe IV*, vol. II, 620–740.

39 Regarding the evolution of the type of individuals who lived at the Royal Geographies in the sixteenth and seventeenth centuries, see José Eloy Hortal Muñoz, "El personal de los Sitios Reales desde los últimos Habsburgos hasta los primeros Borbones: de la vida en la periferia a la integración en la Corte," in Lucio D'Alessandro, Félix Labrador Arroyo and Pasquale Rossi, eds, *Siti Reali in Spagna e in Italia. Tra Madrid e Napoli: aspetti e temi di una storia del territorio* (Naples: Università Suor Orsola Benincasa, 2014): 75–95.

40 As studied at Martínez Millán and Hortal Muñoz, *La Corte de Felipe IV*.

done in the past.[41] It therefore became necessary to reshape the model of the monarchy, and the Royal Geographies were one of the tools he used.

Although the Duke of Lerma had realised the potential of the Royal Geographies in this sense, the Count-Duke of Olivares was the main instigator of this strategy, a practice in which successive royal favourites followed suit, such as Don Luis de Haro, the Marquis of Eliche and the Count of Monterrey. It is clear that the interest of the king's favourites in controlling the royal residencies was aimed at interweaving them more tightly with the Court. The residences attained their height of splendour as a result of the construction projects to bring them up to date with current artistic trends and the number of servants attached to them was likewise increased.

As for the personnel employed in these Royal Geographies, thanks to the presence of the great court patrons in charge of some of them, and to their closer links with the Court, one of the few new ways of entering the Royal Household for people not yet part of it was through the Royal Geographies. Likewise, since other means of granting favours had been exhausted, prominent figures already serving in the Royal Household were rewarded with various offices in the Royal Geographies and residences, enabling them to combine both posts on occasions and accordingly strengthening the aforementioned link between the Court and the Royal Geographies. Moreover, the salary of staff attached to the royal residences was increased to reflect their higher status under Philip IV.[42]

As a result, the Royal Geographies in the surroundings of Madrid, which formerly only came into activity periodically during the visits of the royal family, became dynamic centres which contributed to the increasing presence of the Court in society through the growing number of officials, the extension of the administrative institutions, and the spread of court culture throughout broader segments of society.

This higher social status of the people appointed to the offices linked to the royal woodlands is found, first and foremost, in the most important post at each royal residence, that of *alcaide* (governor). The *alcaide,* an office of Muslim origin and first regulated under Alfonso X 'The Wise', was the highest-ranking

41 For a study of this process of integration, see José Martínez Millán, ed., *La Corte de Carlos V* (Madrid: Sociedad Estatal para la Conmemoración de los Centenarios de Felipe II y Carlos V, 2000); Martínez Millán and Fernández Conti, *La Monarquía de Felipe II*; Martínez Millán and Visceglia, *La Corte de Felipe III*; and Martínez Millán and Hortal Muñoz, *La Corte de Felipe IV.*

42 Hortal Muñoz, "El personal de los Sitios Reales", 75–95.

officer in those places as he held jurisdictional, administrative, economic and staff responsibilities. For the above reasons, before the Duke of Lerma became *valido*, the *alcaides* of these royal residences had enjoyed scant importance at the court except in isolated cases such as Antoine de Cröy at El Pardo and Juan de Ayala at Aranjuez. In other cases the post was associated with a particular family, such as the Marquises of Mondéjar at the Alhambra, the Counts of Chinchón at the Alcázar of Segovia, and Olivares' family at the Alcázares of Seville.[43]

Following Lerma's term as *valido*, however, the monarchs' favourites secured governorships of these places for themselves or used their influence to ensure that they went to members of their family or people they trusted. For example Lerma's son, the Duke of Uceda, was *alcaide* of the Alhambra (1604–24), while Olivares influenced the appointment of governors at Aranjuez, such as Don Melchor de Alcaraz (1625–28), Don Juan de Toledo y Meneses (1628–31), Don Diego Fernández de Zárate (1632–36, 1637–46) and Sebastián Antonio de Contreras y Brizuela (1636–37, 1646–54), and of the *alcaides* at El Pardo, such as the Marquis of Flores Dávila (1623–31) and Don Diego Ramírez de Haro (1631–45).

In fact, the appointment of the Marquis of Flores Dávila marked a before and an after for the royal woodlands owing to his importance at the Court, as attested by the fact that he swore his oath of allegiance before Olivares himself as opposed to before the secretaries of works and woodlands as had previously been done. This new custom continued with the subsequent *alcaides* of El Pardo:

> Todos los alcaides del Pardo hasta don Francisco de Herencia juraron en manos de los secretarios de Obras y Bosques; y en esta conformidad don Diego Ramírez juró en las mías solo con el marqués de Flores Dávila no se observó este estilo porque lo hiço en las del Conde Duque en virtud de orden de V. Magestad y porque se pudo fundar en la diferencia de la persona y de los puestos que ocupaba y parece debe concurrir la misma consideración en Don Luis Méndez de Haro. He querido dar quenta dello a V. Magestad para que con esta notiçia pueda ordenarme lo que más

43 See, for example, the orders (*cédulas*) appointing *alcaides* and deputy *alcaides* at El Pardo in AGP, Reg., Cédulas, from Reg. 1, fols. 37–38 (Antonio de Guzmán, deputy alcaide, in 1548), to Reg. 17, fols. 356–59 (Fernando de Haro Guzmán y Toledo, in 1695). Before the Marquis of Flores Dávila, the only significant *alcaide* was Antoine de Cröy in 1555 (AGP, Reg. 1, fol. 228 v).

fuere de su Real Servicio. Jure en manos del Marqués del Carpio. Madrid, 17 de febrero de 1645.[44]

All the *alcaides* of El Pardo up until Don Francisco de Herencia took oath of allegiance before the secretaries of Works and Woodlands; and accordingly Don Diego Ramírez took oath before me; only with the Marquis of Flores Dávila was this practice not observed because he took it before the Count Duke by virtue of an order from Your Majesty and because it was justified by the distinction of the person and of the posts he held and it seems that the same consideration should be given to Don Luis Méndez de Haro. I wished to report this to Your Majesty so that with this news you may order me whatever is most appropriate for your Royal Service. Oath before the Marquis of Carpio. Madrid, 17 February 1645.

Olivares subsequently appointed himself directly as *alcaide* of various royal residences in addition to the post he held through family inheritance at the Reales Alcázares in Seville. The first of these appointments was at the royal residence of San Jerónimo on 27 July 1630 – which eventually became the Buen Retiro palace, where he was the first *alcaide* from 8 November 1633 onwards. He also became the first *alcaide* of La Zarzuela on 16 February 1636 and of Vaciamadrid on 29 July 1634, a property that he would later split off from the Crown and incorporate into his own family's possessions.[45] Evidently, he could not personally perform the duties of the many offices he held in the sovereign's entourage, which is why he appointed a series of prominent men that he could rely on to manage the sites on his behalf, such as his son-in-law, the Marquis of Leganés, whom he made deputy *alcaide* of Vaciamadrid (1636–45),[46] and the Count of Puñoenrostro, who served as interim *alcaide* of the Zarzuela (*c.* 1638–*c.* 1646).[47]

Like Olivares, his successor Don Luis de Haro was aware of the importance of controlling the Royal Geographies, and after the death of the Count-Duke in 1645 he accumulated even more titles than his predecessor.[48] Between 1645 and 1650 Haro obtained the governorships of Buen Retiro, El Pardo, Valsaín, and the Reales Alcázares of Seville as well as the office of Master of the Horse

44 AGS, CSR, leg. 310, fol. 476.

45 AGP, Personal, boxes 518/8, 754/49 and 955/29, and Reg. 13, fols. 30r–v, 98v, 137v–138r, 159v and 181v; Archivo Histórico Nacional (AHN), Consejos, leg. 4422, fol. 1.

46 AGP, Personal, box 543/19, and Reg. 14, fols. 402r and 407v.

47 AGP, Personal, box 857/15.

48 AGP, Personal, boxes 325/48, 496/36 and 668/2, and Reg. 14, fols. 96r–v, 213r and 267r–v; AGS, CSR, leg. 306, fol. 401.

(*caballerizo mayor*) of the Royal Stables in Cordoba, holding all of these offices until 1661. His son Gaspar de Haro y Guzmán[49] would act as interim *alcaide* at practically all of these sites and would become incumbent *alcaide* of each of them after a brief period from 1662 until he was exiled in 1665,[50] at which point the role passed to the Duke of Medina de las Torres (Buen Retiro) and the Count of Monterrey (Valsaín, El Pardo, and Zarzuela). Subsequently, by then under Charles II, Fernando de Valenzuela was also given the post of *alcaide* of several Royal Geographies such as El Pardo, as well as the superintendency of Royal Works (1674–77).[51] It should be stressed that, owing to the significance of the place and the need for the person in charge to be physically present, at Aranjuez it was not an *alcaide* that was appointed but a *gobernador* of noble rank but not as important as the *valido*. He also acted as chief custodian (*guardamayor*) of the forest.

At El Pardo and Valsaín, the *alcaides* and *gobernadores* were also appointed as *guardamayores* of that forest. This occurred until the term of Don Luis de Haro as *valido*, when, as the result of a personal decision of the favourite, the two posts became separate: that of *alcaide* continued to be held by grandees, whereas the *guardamayores* had a more technical profile; that is, they had previously held offices related to the Hunt or the guard of those royal forests. The type of care given to those forests therefore became even more specialised.

This occurred at El Pardo: Don Luis Méndez de Haro was appointed 'Alcaide de mi Castillo y casa real del Pardo y guardamayor de sus bosques' (*Alcaide* of my castle and royal residence of El Pardo and chief custodian of its forests) on 22 February 1645, but in 1649 the office of *guardamayor* became separate and was given to Juan de Gaona, with 100,000 *maravedís* in emoluments.[52] The same was true of the Royal Residence and Woodlands of Valsaín, where former members of the royal guard held the combined post of *alcaides* and *guardamayores* until 1645, such as Gaspar de Mollenghien (1607–25) and his son with the same name (1625–37), and a gentleman of Calatrava and perpetual alderman of Segovia, Don Antonio Fernández de Miñano (1637–45).[53] On

49 Regarding his connection with the Royal Geographies, see María Asunción Flórez Asensio, "El Marqués de Liche: Alcaide del Buen Retiro y "Superintendente" de los Festejos Reales," *Anales de Historia del Arte* 20 (2010): 145–82.

50 María Asunción Flórez Asensio. *La corte en llamas. Proceso al marqués de Heliche (1662–1663)* (Madrid: Marcial Pons, 2023).

51 Antonio Álvarez-Ossorio Alvariño, "El espacio de la privanza. Fernando de Valenzuela y los Reales Sitios", in Marina Mestre-Zaragoza, ed., *L'Espagne de Charles II, une modernité paradoxale: 1665–1700* (Paris: Classiques Garnier, 2019): 47–68.

52 AGP, Personal, box 390/20.

53 AGP, Personal, box 16904/8 and Reg. 13, fol. 230 r.

the advice of the *Junta* of 22 November 1645, the king appointed Don Pedro de la Canal as *guardamayor* of Valsaín, separating the office from that of *alcaide* of the Royal Site with which it had previously been joined, and in 1650 it was given to Don Luis de Haro 'con atenzión a lo que conbenía a que tubiese aquella Alcaidía persona de autoridad para que estuviesen aquella Real Casa y Bosques con mejor cobro y cuidado' (with attention to the appropriateness of that office of *alcaide* having a person of authority in order for that Royal Residence and Forests to be better guarded and tended to).[54] At any rate, it is worth pointing out that in those places a *guardamayor* had always been previously appointed to cover the absences and illnesses of the *alcaide*, so that in practice the office had been separate on several occasions. La Zarzuela, a property that became linked to the Buen Retiro in 1639 and to El Pardo in 1662, always had an interim *alcaide* and a separate *guardamayor* of the forests, the first being the archier Miguel Carlier, who was appointed in 1638.

Throughout this process, it is worth highlighting the peculiarity of the case of the Monastery of San Lorenzo de El Escorial, where an *alcaide* was never appointed because of the privileges already enjoyed by the Hieronymite monks in the monastery.[55] Therefore, the office of *guardamayor* of the royal woodlands of El Escorial depended to an extent on the Hieronymite monks.[56] As an exceptional measure, there were two *guardamayores* there owing to the huge expanse of land at that Royal Site, in order to be able to feed the Hieronymites and the workers and residents who were dependent on the place;[57] specifically, one received the name of *guardamayor* of the royal woodlands of the San Lorenzo el Real area, and the other of *guardamayor* of El Espadañal, the first being higher up in the hierarchy. The woodlands of the Lomo del Grullo were run by the *alcaide* and perpetual *guardamayor* of the Reales Alcázares of Seville, though it was the deputy *alcaide* who was in charge of day-to-day affairs owing above all to the many tasks that busied the *alcaides* outside Seville because of their importance at the Court. Lastly, the Soto de Roma had an *alcaide* and *guardamayor* who was appointed by the *Junta de Obras y Bosques* with a salary of 30,000 *maravedís*.[58]

54 AGP, Personal, box 16904/9 and Reg. 14, fols. 125 r–v and 363 r.

55 Discussions about this topic can be found in AGP, Personal, box 738/10.

56 In AGS, CSR, leg. 302/3, fol. 323 and onwards illustrate the profile of the candidates for the post of *guardamayor* of the Royal Woodlands of San Lorenzo El Real, which was vacant in August 1621.

57 Martínez Bara, 'Noticias sobre las dehesas'.

58 AGS, CSR, leg. 303, fol. 513.

A document addressed to the *Junta de Obras y Bosques* regarding the *guardamayor* of the Soto de Roma[59] illustrates the main duties inherent in the post:

> lo que he podido entender de los papeles que he visto el que quando este reino se ganó no avía mas personas para la conservación del Soto y sus vedados que la guarda maior y las menores que el nombraba que con solo su poder podrían hacer causas y denunciaciones a los que cazaban cortaban leña o madera y pescaban en el dicho Soto y sus términos y estas causas las sentenciaban las justicias ordinarias y por no aver juez particular de las causas de Soto y las condenaciones eran enteramente para Juan de Guzmán criado de los reyes católicos por merced particular que dellas le hicieron y fue el primer guarda maior que se nombró que todo esto consta de la primera comisión y nombramiento que se hiço el rey don Fernando en Écija el 3 de diciembre de 1499 (…) El Emperador nombró en el mismo oficio a Hernando de Luzena con poder de nombrar guardas con treinta mil maravedís de salario aplicando las penas de condenación de las causas por terçias partes a la cámara juez denunciados y que los sentenciasen los jueces y ordinarios. (…) y que se vendiese la leña seca, los árboles caídos y arrancados y la yerba.

> What I have been able to understand from the papers I have seen is that when this kingdom was won there were no more people for the conservation of the Soto and its game preserves than the chief custodian [*guardamayor*] and the lesser guards whom he appointed, who alone had the powers to bring charges and complaints against those who hunted, cut wood and firewood and fished in the said Soto and its districts and these charges were ruled on by the ordinary magistrates and as there was no specific judge of the charges relating to the Soto and all rulings were issued by Juan de Guzmán, servant of the Catholic Monarchs by a particular favour that they granted to him and he was the first chief custodian who was appointed. All of this is recorded in the first commission and appointment made by King Ferdinand at Écija on 3 December 1499 (…) The Emperor appointed to the same office Hernando de Luzena with powers to appoint guards with a salary of thirty thousand *maravedís* and to enforce the penalties resulting from the charges brought against third parties, to be paid through the chamber of justice, ruled on by ordinary

59 AGS, CSR, leg. 303, fol. 539.

> magistrates (…) and to sell dry firewood, fallen and uprooted trees, and grass.

Indeed, the *guardamayor* was the officer mainly in charge of each woodland, and organised the service of the guards. He therefore needed to be familiar with the post of guard; in fact, most of the *guardamayores* had previously served as guards, such as Juan López, a mounted boundary guard at El Pardo (1653–63) and *guardamayor* of that place (1663), and Mateo Moreno, mounted guard of the Royal Residence and Woodlands of Valsaín (1651–53), and *guardamayor* of that place (from 1653 onwards), as well as serving as a guard and warden (*conserje*) at Fuenfría (1653–55).

The candidates for the post hailed from very different social backgrounds, as can be seen, for example, in the appointment of Antonio de Alosa Rodarte as *guardamayor* of the Royal Residence and Woodlands of Valsaín after serving as groom of the chandlery (*mozo de la cerería*) to Margaret of Austria-Styria; that of Miguel Carlier as *guardamayor* of the forests of the Zarzuela after serving as an archier; and, in particular, the proposal put forward on 7 November 1636 for endowing the post of *guardamayor* at Valsaín with a salary of 150,000 *maravedís* provided from the income from the land owned by the military orders (*mesa maestral*).[60] The candidates listed were Juan de Molina, who had been employed for 7 years in the king's tapestry department, Don Pedro de Barreda Yebra, who had spent '60 [sic] years' serving the king in the war as a soldier and captain, Gerard Goethals, archier to the king, Joan de Villarroel, the brother of Bernardo Ruiz de Valdivieso, who was aid of the king's oratory for 28 years; Juan de Medina, a soldier of the Spanish guard for more than 13 years; and Matías del Castillo, falconer for more than 15 years. However, the man finally chosen for the post was Don Juan de Castro Villafañe, a knight of the Order of Santiago who had been *aposentador* (in charge of maintenance, furnishing, and supply) of the Household of Burgundy, gentleman of the household, and clerk and accountant of works at the Alcázar of Segovia and the Royal Residences of Valsaín and Fuenfría.[61] It should be remembered that the social significance of the chosen candidate was linked to the fact that he was still appointed as both *alcaide* and *guardamayor*, though this changed when Don Luis de Haro became *alcaide* of that place.

Similarly, it should be pointed out that there was a constant movement of officers who were transferred between the sites with royal woodlands located around Madrid – such as El Pardo, Valsaín and Aranjuez – to be appointed

60 AGP, AP, San Ildefonso, box 13536, file 1, n.f.

61 AGP, Personal, boxes 463/20 and Reg. 13, fols. 169 r, 232 r–v and 277v.

as *guardamayores*. Some examples are Miguel Muñoz, who was a mounted guard at the boundaries of El Pardo (1652–63) until becoming *guardamayor* of the Royal Residence and Woodlands of Valsain (1663); Pedro de Tovar, senior guard (*sobreguarda*) at El Pardo (1651–58) and subsequently *guardamayor* of the Royal Residence and Woodlands of Valsaín (1658–59); and Don Andrés de la Plaza Eguiluz, senior guard of the forests of Aranjuez (1640–53) and *guardamayor* of the Royal Residence and Woodlands of Valsaín (1653–58).

The *guardamayor* was in charge of a number of guards, mounted or on foot, whose number depended on the size and importance of the royal forest in question and its needs. We know how many there were specifically at El Pardo thanks to the instructions that were issued in 1563–64 and extended in 1614.[62] In 1625 there were a chief boundary guard, a senior guard and two mounted guards of the boundaries marked by the related *prágmatica* (a type of royal law), who received their title from the *Junta de Obras y Bosques*.[63] Other boundary guards on foot and on horseback were appointed by the *alcaide* or his deputy;[64] in 1593 they numbered four on horseback and eight on foot.[65] From 1650 onwards we also find a few assistant guards. As for Aranjuez, also in 1625,[66] there were a chief guard and a senior guard appointed by the *Junta de Obras y Bosques*, together with various guards whose number varied over the years, fluctuating between four and six; they served either on foot or on horseback and were appointed directly by the governor. Meanwhile, at San Lorenzo de El Escorial the mounted guards of the royal forests, whose number varied between five and six, were appointed by the prior and the *guardamayor* and their wages were paid by the monastery. Valsaín had six guards, two of whom were on horseback and four on foot, all appointed by the *guardamayor*.[67] At Valladolid there were three guards stationed in different locations, namely the park and forest of Ribera, the forest of Abrojo and the forest of Quemada.[68] We occasionally come across guards for hunting, fishing, firewood and grass in the woodlands around Valladolid. They were all appointed directly by the *alcaide*. At Lomo del Grullo there were two guards who were paid two *reales* daily and were duty bound to each maintain a horse until in 1623 when, thanks to the influence of the *alcaide*, the Count-Duke of Olivares, the number of guards

62 *Instrucciones* studied in Virginia Tovar Martín, *El Real Sitio de El Pardo* (Madrid: Patrimonio Nacional, 1995), chapter II, which partially publishes those of 1614 (43).

63 AGP, AG, leg. 853, no. fol.

64 AGS, CSR, leg. 302/1, fol. 109.

65 AGP, AP, El Pardo, box 9383, file 7.

66 AGP, AG, leg. 853, no. fol.

67 AGS, TMC, leg. 1569, n.f.

68 AGS, CSR, Valladolid, leg. 8–1, n.f.

increased to eight, and they each earned four *reales* daily.[69] At Soto de Roma we find one mounted guard and four foot guards, who were appointed and dismissed by the *alcaide* and *guardamayor*.[70] Lastly, at the Zarzuela there were mounted guards; it is not known how many but the number must have been very small. As a rule, they were all provided with accommodation near their workplace for themselves and their families.

The scarcity of the existing biographical information about the guards indicates that their social status and mobility were not very high, though they sometimes held guard posts in other royal forests around Madrid, such as Agustín García, who was a mounted guard of the Royal Residence and Woodlands of Valsaín (1657–61) and subsequently a mounted boundary guard at El Pardo (1661–65).[71] Most of them had family ties to a specific royal forest and the post was even handed down from fathers to sons. The family remained linked to the place after the guards died, as their widows or children received various pensions and alms that enabled them to subsist. Similarly, a few guards succeeded in becoming *reservados* after holding the post for many years, meaning that they received all or part of their salary without needing to perform the work.[72]

In nearly all cases, their only possibilities of promotion were within the body of guards of the relevant royal forest. Indeed, also depending on the importance of each royal forest, the *guardamayor* was assisted by a few subordinates in organising the service, which was different in each place.[73] Similarly, the duties of these officers with intermediate posts between the *guardamayores* and the guards depended on the customs at each place. For example, at Aranjuez we know from the title of Alonso Carrizo as chief guard of that forest as of 31 December 1658 that:

> es superior a los otros guardas que ay y hubiere y como tal visite de ordinario todos los dichos montes sotos y bosques y vea como y de qué manera cada uno de los dichos guardas sirven en sus oficios y tienen guardado y conservado el quartel que fuere a su cargo y de razón de todo lo que hubiere y se ofreciere al mi gobernador del dicho heredamiento

69 AGS, TMC, leg. 1571, n.f.

70 AGS, CSR, leg. 303, fol. 513.

71 AGP, Personal, box 391/1.

72 On these pensions, see José Eloy Hortal Muñoz, "Reservados y pensionistas: Una nueva vía de integración de los reinos en la Casa Real", in Martínez Millán y Hortal Muñoz, *La Corte de Felipe IV*, vol. III, 2327–41, especially the charts on 2330–35.

73 At Aranjuez there were a chief guard and a senior guard and at El Pardo there were a principal boundary guard and two senior guards. There is no record of intermediate guard posts either at El Escorial or at Valsaín, or at Soto de Roma, Valladolid or Lomo del Grullo.

para que se provea lo que cerca de cada cosa conviniere, por tanto por la buena relación que tengo de la suficiencia, diligencia y cuidado y a los servicios de vos Alonso Carrizo es mi voluntad que en execución dello seáis tal guarda principal de la caza, montes y sotos del dicho Aranjuez, Otos y Aceca y de todo lo demás incorporado en el dicho heredamiento para que residiendo de ordinario en él y en la casa y estancia de Otos, os ocupéis en todo lo susodicho y visitéis por vuestra persona los quarteles y estancias donde los otros guardas residen y quando el caso requiere y no hubiere lugar de comunicarlo con el dicho gobernador les ordenaréis lo que debieren hazer para que ellas los obedezcan y cumplan dando vos siempre noticia al dicho gobernador de lo que se ofreciere (…) y mando a los dichos guardas que os obedezcan y respeten como a superior suyo y de noche y de día os acojan en sus estancias que tubieren en los quarteles donde residieren y salgan con vos, adonde y en las partes que les ordenáredes y fuere necesario para la buena guarda de la caza, montes y sotos y guarden y executen las órdenes que les diéredes, no siendo contrarias a las del dicho gobernador y es mi voluntad que por vuestra ocupación y trabajo y para vuestro sustentamiento hayáis y gocéis de salario a razón de ochenta mil maravedís en dinero, cien fanegas de trigo y otras ciento de cebada en cada un año.[74]

He is the superior of the other guards there are and may be and as such he shall visit ordinarily all the said woodlands, copses and forests and oversee how and in what way each of the said guards perform their duties and guard and preserve the district for which they are responsible, and report on everything there is and that occurs to my governor of the said estate so that appropriate provision can be made for each thing; therefore, owing to the good report I have of the competence, diligence and care and the service provided by you, Alonso Carrizo, it is my wish that in execution thereof you be the principal guard of the game, woodlands and copses of the said Aranjuez, Otos and Aceca and of everything else that is part of the said estate so that, residing ordinarily on it and in the house and dwelling in Otos, you take charge of all the aforesaid and visit personally the districts and houses where the other guards reside and, when necessary and if it were not appropriate to report it to the said governor, you shall give orders as to what they must do so that they obey and fulfil them, with you always reporting to the governor on what arises […] and

74 Title given to Alonso Carrizo as main guard of the estate of Aranjuez following the retirement of Juan Martínez de la Higuera, 31 December 1658 (AGS, TMC, leg. 1568, no. fol.).

> I order the said guards to obey and respect you as their superior and by night and by day welcome you to the dwellings they have in the districts in which they reside and accompany you to wherever and to whatever places you order them to go and which are necessary to properly guard the game, woodlands and copses and carry out the orders you give them, which shall not go against those of the said governor, and it is my wish that for your duties and work and for your sustenance you have and enjoy a salary of eighty thousand *maravedís* in money, one hundred *fanegas* of wheat and another one hundred of barley each year.

The senior guard of that site was to assist him in these tasks, as is indicated in the title of this post, which was granted to Don Andrés de la Plaza Eguiluz at Aranjuez following the death of Juan Moreno, on 25 June 1641:

> Para que sirvais el dicho oficio según y de la manera que lo han hecho los demás vuestros antecesores hicieron y debieron hacer ayudando a la guarda principal a quien haveis de estar subordinado y guardando la orden que os diere assimismos el gobernador del dicho heredamiento, teniendo quenta con el quarto de Villamejor y con las guardas de aquel Partido.
>
> In order that you serve in this post in accordance with, and in same manner, as the rest, your predecessors did and were duty bound to, assisting the main guard to whom you shall be subordinate and following the orders given to you likewise by the governor of said estate, watching over the district of Villamejor and the guards of that area.[75]

The same was true of the royal forest of El Pardo.[76] The senior guards were also to be responsible for the wellbeing of the guards and for seeing to it that their houses and workplaces were in suitable condition. There was even a barber and surgeon specifically for the guards at El Pardo: Blas de Almorox, who handed over to Juan de Villoslada in 1654.[77]

The guards' main duty involved keeping watch over the royal woodlands. Naturally the process of separating these spaces from their surroundings triggered numerous disputes and issues with the surrounding villages involving

75 AGP, Reg. 6513, fols. 13 v–14r.

76 *Nuevas órdenes que se dieron en tiempos del conde de Monterrey para El Pardo*, 2 June 1662 (AGP, Reg. 15, fols. 158v–160r).

77 AGP, Personal, box 1105/29.

damage caused by royal hunting parties or game, as well as clashes with the poachers who took advantage of the small number of guards and the vast expanse of land to catch game in those places. All this forced the monarchy to create numerous jurisdictions establishing the boundaries of the various royal woodlands.[78]

Accordingly the 'Instruççiones en lo tocante a la cassa real, Monte y Bosque del Pardo' (Instructions concerning the royal residence, woodland and forest of El Pardo) stated that:

> 8. Los dichos guardas no salgan del dicho monte y bosque ni le dexen de guardar de ordinario y prendan y denuncien qualquiera persona que hallaren cazando, pescando, cortando leña o madera o haziendo algún daño o otra cossa alguna de las prohibidas por provisiones y cédulas de V. Majestad y no hagan conçiertos con las partes sobre los daños y denunçiaciones que hubieren hecho o tuvieren obligación de hazer ni cobren las penas ni parte de ellas hasta estar juzgadas y sentençiadas y lo cumplan sin remisión ni disimulación alguna so pena de privación de offiçio. […].
>
> 10. A dicho alcayde procure con diligencia y cuidado que se maten las zorras, aves de rapiña y otros animales que hizieren daño en la caza y pesca del dicho monte y bosque y para este efecto se les permita a los guardas que puedan traer arcabuz, tirar y matar con él los dichos animales dañosos, aves de rapiña, lobos y perros que entraren en el dicho monte y bosque con que luego manifiesten lo que ubiere muerto y si los dichos guardas u otra persona hallaren caza mayor o menor, muerta o herida lo manifiesten a el dicho alcayde y al juez de obras y bosques para que procure saber quien la mató o de que murió y se haga de ella lo que el ordenare con que no se dé a la persona que la ubiere hallado.[79]

> 8. The said guards shall not leave the said woodland and forest nor cease to guard it ordinarily and shall arrest and report anyone found hunting, fishing, cutting firewood or wood or doing harm or anything else prohibited by the provisions and orders [*cédulas*] of Your Majesty and shall not reach agreements with the parties on the damages and complaints they

78 Compiled in Pedro de Cervantes and Miguel Ángel de Cervantes, *Recopilación de las Reales Ordenanzas y Cédulas de los Bosques Reales del Pardo, Aranjuez, Escorial, Balsaín y otros* (Madrid: Oficina de Melchor Álvarez, 1687).

79 AGP, AG, El Pardo, box 9386/7.

> may have brought nor have the duty to impose or collect penalties or part of them until they are judged and sentenced and shall comply with these duties unremittingly and without dissimulation under penalty of removal from office. [...].
>
> 10. The said *alcaide* shall diligently and carefully ensure that foxes, birds of prey and other animals harmful to hunting and fishing in that woodland and forest are killed and for this purpose the guards are permitted to carry a harquebus and shoot and kill with it the said harmful animals, birds of prey, wolves and dogs that enter that woodland and forest and shall subsequently report that they have been killed and if the guards or anyone else should find large or small game dead or wounded they shall report this to the said *alcaide* and to the magistrate of works and woodlands in order to try to find out who killed it or of what it died, and shall do with it whatever is ordered and not give it to the person who found it.

Similarly, the guards of the royal woodlands played a very important role in ceremonially separating the royal residences from their surroundings, as during the seventeenth century the boundaries of these Royal Geographies became more defined.[80]

This newly delineated space, reserved only for courtiers, firstly needed an increasing number of personnel to demarcate its boundaries. For example, royal guard units were specifically assigned to protect the Royal Geographies where ceremonies and other royal functions would take place, clearly marking the separation of these spaces from the rest of society. In this sense, a paradigmatic example was set by France, a kingdom that had enormously increased the number and quality of its guard units due to the unstable political situation following the death of King Henry II in 1559 and the beginning of the Wars of Religion. While the number of guards and units increased, a division appeared between the guards *du dedans* and *du dehors* – inside and outside – of the Louvre. The latter watched over the exteriors of royal palaces, as well as escorting the king at public functions outside the palace walls.

This division, which became the model for numerous European kingdoms, could not be fully adapted for the royal guards of the Spanish Monarchy, who were more like a parade and ceremonial body than an actual military force

80 Analysed in depth in José Eloy Hortal Muñoz, "Courtly and Ceremonial Spaces in Spanish Royal Sites: An Evolution from the Renaissance to the Baroque", in Anna Kalinowska and Jonathan Spangler, eds, *Power and Ceremony in European History Rituals, Practices and Representative Bodies since the Late Middle Ages* (London: Bloomsbury, 2021), 87–104.

equipped to guard the Royal Geographies.[81] Indeed, specific and effective units responsible for these tasks were not created until the eighteenth century.[82] Meanwhile, our mounted and dismounted guards were the ones in charge of differentiating the royal spaces from their surroundings.

Below the guards was a group of officials whose main duty was to maintain the royal woodlands in suitable condition for hunting or other purposes. It should be added that these posts did not belong specifically to the section of the Hunt but were attached to the relevant royal woodland and involved maintaining and preparing the woods to be used for royal hunts. This sometimes led to clashes with the guards of those places.[83]

Among these posts, at Aranjuez we find several that did not belong directly to the Royal Hunt; rather, they were only involved in the hunting parties in which the monarch engaged in the woodlands of that royal residence owing to the difficulty of transferring all the officers of the Royal Hunt who lived in Fuencarral. They all resided in Aranjuez or the nearby villages of Esquivias, Ocaña and Ciempozuelos, and were commonly promoted to the above posts or to others related to the Royal Site of Aranjuez, especially the guards. Among them we find as many as ten huntsmen, as well as harquebusiers (*arcabuceros*) and fox hunters (approximately three of each), and a ferret hunter[84] when necessary.

Owing to the particular nature of the type of hunting engaged in at El Pardo, in addition to fox hunters we find *ordinarios de las buitreras*, orderlies entrusted with ensuring that the vultures' nests and chicks were in good condition so

81 As discussed in José Eloy Hortal Muñoz, "Las guardias de los Austrias, ¿cuerpo militar o de parada?", in *Perspectivas jurídicas e institucionales sobre guerra y ejército en la Monarquía Hispánica*, eds. Sara Granda Lorenzo, Leandro Martínez Peñas and Manuela Fernández Rodríguez (Madrid: Dykinson, 2011): 119–51.

82 Enrique Martínez Ruiz and Magdalena de Pazzis Pi Corrales, *Proteccion y seguridad en los Sitios Reales desde la Ilustracion al Liberalismo* (Alicante: Universidad de Alicante, 2010).

83 'Que a los dos Çorreros que ay se les paguen con puntualidad las muertes de alimañas y aves de rapiña, conforme lo que está dispuesto por lo pasado, o poniéndoles tasa de nuevo para que con más cuidado asistan a su ocupación, sin que los guardas se entrometan a matar ni tirar a las dichas alimañas y aves, ni se ocupen en esto por ningún casso' (The two fox catchers there are shall be paid punctually for the deaths of vermin and birds of prey, in accordance with what was established in the past or setting a new rate so that they perform their duties with greater care, without the guards interfering by killing or shotting the said vermin and birds, or taking charge of this in any case) *Nuevas órdenes que se dieron en tiempos del conde de Monterrey para El Pardo*, 2 June 1662 (AGP, Reg. 15, fols. 159v–160r).

84 Because they harmed game. In AGS, CSR, leg. 303, fol. 225, we find an instruction from the *Junta de Obras y Bosques* for them to make sure that there were no ferrets within a radius of eight leagues of the royal woodlands of Valladolid.

that these birds could perform their task of cleaning up animal remains in the forest. The social class, background and mobility of the people who held the posts at El Pardo were based on the same parameters as those at Aranjuez.

Lastly, it should be pointed out that officers were often hired temporarily to perform specific duties, such as to clear the trees of caterpillars, a task that needed to be done regularly,[85] and labourers were recruited temporarily to cut wood. It was sometimes attempted to avoid hiring these temporary workers at some of the royal residences by renting the land to individuals, but the experience did not turn out well in most cases and this ended up becoming an infrequent practice.[86]

3 Conclusion

Until a few years ago, certain elements that made up the Court of the Early Modern times were hardly studied, or their relevance was not taken into account, as each discipline focused its work on its own fields of study and interdisciplinary studies were not very abundant. But the historiographical progresses made by researchers in various fields over the past decades has enabled us to advance beyond the traditional view, and we can assert that the holistic studies conducted from various approaches in the lasts decades have shown the relevance of forgotten spaces. In this sense, historians have begun to take into account the relevance of the space at the political interactions, ceremonial practices, cultural symbols, or institutional structures of the Court. These new needs have motivated us to work on the concept of Royal Geographies that, no doubt, amplify the relevance of these royal spaces, and where we can face and study multiple aspects: religious, political, geographies of power, organisation of the royal works, etc.

In this chapter, for example, we have studied how royal forests and woodlands were an essential element in shaping the court space and increasing the sovereign's presence in the territory. Their connection with the Hunt, which was unquestionable, has often led all the other roles these places played

85 For example, in AGP, AP, El Pardo, box 9382, file 4, we find a 'Nómina de la gente que ha andado a desorugar y limpiar las encinas en este monte del pardo por mandado del rey nuestro señor desde lunes cinco de diciembre 1588 hasta savado diez del dicho mes y año' (List of people who have removed caterpillars and cleaned the holm oaks in this woodland of El Pardo on the orders of our lord the king from Monday December the fifth 1588 to Saturday the tenth of that month and year).

86 As pointed out by the *Junta de Obras y Bosques* on 11 July 1598 in a document it sent to the palace of Lomo del Grullo, but extended to other Royal Geographies (AGS, CSR, leg. 303, fol. 739).

within the court system of the period, namely representation, provisioning, and territorial and social integration, to be overlooked. In order for these duties to be correctly performed, it was necessary to have a number of officials entrusted with maintaining the woodlands and separating them both physically and ceremonially from what were sometimes aggressive surroundings – the inhabitants of the nearby villages sought to benefit from the resources of the royal forests and woodlands – as well as endeavouring to minimise the damage caused by royal hunting parties.

So, this chapter studies the officers who were in charge of the main royal forests and woodlands of the Spanish Monarchy, such as the governors, chief custodians and other posts related to their maintenance, as well as the duties and social background of the officers who held them. All this provides us with a clearer picture of the role of the royal forests and woodlands in the configuration of the courtly space of the Spanish Monarchy, especially that of the kingdom of Castile.

In conclusion, I would like to point out that this chapter is just a part of the multiple studies related with these Royal Geographies that we are developing in the lasts years; so, more novelties to come.

Abbreviations

AGP: Archivo General del Palacio Real, Madrid (Spain)
 AG: Administración General
 AP: Administraciones Patrimoniales
 Reg.: Registros
AGS: Archivo General de Simancas, Valladolid (Spain)
 CSR: Casa y Sitios Reales
 TMC: Tribunal Mayor de Cuentas
AHN: Archivo Histórico Nacional, Madrid (Spain)

Acknowledgements

This chapter was funded as part of the projects 'Protection, Production and Environmental Change: The Roots of Modern Environmentalism in the Iberian Peninsula (XVI–XVIIIth centuries)' (AZ 60/V/19); 'Madrid, Sociedad y Patrimonio: pasado y turismo cultural' (H2019/HUM-5898); and Proyecto Puente 'Conservar es Innovar. El patrimonio de lo Real a lo Rural: Madrid y su entorno', URJC (ConInMad/V1254).

Bibliography

Allsen, Thomas T. *The Royal Hunt in Eurasian History.* Philadelphia: University of Pennsylvania Press, 2006.

Álvarez-Ossorio Alvariño, Antonio. "El espacio de la privanza. Fernando de Valenzuela y los Reales Sitios". In *L'Espagne de Charles II, une modernité paradoxale: 1665–1700*, edited by Marina Mestre-Zaragoza, 47–68. Paris: Classiques Garnier, 2019.

Anthenaise, Claude d' and Monique Chatenet, eds. *Chasses princières dans l'Europe de la Renaissance.* Arles: Centre André Chastel, 2007.

Ashton, Robert. *The City and the Court, 1603–1643.* Cambridge: CUP, 1979.

Baagoe, Jette. *Kongens Skov-Verdens Arv. Parforcejagtlandskabet I Nordsjaelland / The King's Forest – The World's Heritage. The par force hunting landscape in North Zealand.* Rosendahls: The Danish Museum for Hunting and Forestry, 2016.

Bravo, Paloma and Juan Carlos D'Amico, eds. *Territoires, lieux et espaces de la révolte. XVI–XVIII siècles.* Dijon: Éditions Universitaires, 2017.

Cervantes, Pedro de and Miguel Ángel de Cervantes. *Recopilación de las Reales Ordenanzas y Cédulas de los Bosques Reales del Pardo, Aranjuez, Escorial, Balsaín y otros.* Madrid: Oficina de Melchor Álvarez, 1687.

Courbon, Léonard and Denis Menjot, eds. *La cour et la villa dans l'Europe du Moyen âge et des Temps Modernes.* Studies in European Urban History (1100–1800), 35. Turnhout: Brepols, 2015.

Damen, Mario and Kim Overlaet, eds. *Constructing and Representing Territory in Late Medieval and Early Modern Europe.* Amsterdam: Amsterdam University Press, 2022. DOI: https://doi.org/10.1515/9789048551804.

Duerloo, Luc. "The Hunt in the Performance of Archducal Rule". *Renaissance Quarterly* 69:1 (2016): 116–54.

Duindam, Jeroen. *Dynasties: A Global History of Power, 1300–1800.* Cambridge: CUP, 2015. DOI: https://doi.org/10.1017/CBO9781107447554.

Elden, Stuart. *The Birth of Territory.* Chicago: University, 2013.

Espinosa Romero, Jesús. "La evolución económica de un Real Sitio: Aranjuez en tiempos de Felipe II". *Reales Sitios* 153 (2002): 2–13.

Ezquerra Revilla, Ignacio Javier. "Atribuciones de la Junta de Obras y Bosques". In *La Corte de Felipe IV. Reconfiguración de la Monarquía Católica*, edited by José Martínez Millán and José Eloy Hortal Muñoz, vol. III: 2066–83. Madrid: Polifemo, 2015.

Ezquerra Revilla, Ignacio Javier and Rubén Mayoral López. "La Caza Real y su protección: la Junta de Obras y Bosques". In *La monarquía de Felipe III: la Casa del Rey*, edited by José Martínez Millán and Mª. Antonietta Visceglia, vol. I, 811–86. Madrid: Fundación Mapfre, 2008.

Fantoni, Marcello, Giuseppe Gorse and Malcolm Smuts, eds. *The Politics of Space: European Courts ca. 1500–1750.* Rome: Bulzoni, 2009.

Flórez Asensio, María Asunción. "El Marqués de Liche: Alcaide del Buen Retiro y "Superintendente" de los Festejos Reales." *Anales de Historia del Arte* 20 (2010): 145–82.

Flórez Asensio, María Asunción. *La corte en llamas. Proceso al marqués de Heliche (1662–1663)*. Madrid: Marcial Pons, 2023.

González Dávila, Gil. *Teatro de las grandezas de la villa de Madrid. Corte de los reyes católicos de España*. Madrid: Thomas Iunti, 1623.

Hassler, Eric and Anne Motta, eds. *Noblesses et villes de cour en Europe (XVIIe–XVIIIe): La ville de résidence princiere*. Rennes: Presses Universitaires Rennes, 2022.

Hortal Muñoz, José Eloy. "Las guardias de los Austrias, ¿cuerpo militar o de parada?". In *Perspectivas jurídicas e institucionales sobre guerra y ejército en la Monarquía Hispánica*, edited by Sara Granda Lorenzo, Leandro Martínez Peñas and Manuela Fernández Rodríguez: 119–51. Madrid: Dykinson, 2011.

Hortal Muñoz, José Eloy. "El personal de los Sitios Reales desde los últimos Habsburgos hasta los primeros Borbones: de la vida en la periferia a la integración en la Corte". In *Siti Reali in Spagna e in Italia. Tra Madrid e Napoli: aspetti e temi di una storia del territorio*, edited by Lucio D'Alessandro, Félix Labrador Arroyo and Pasquale Rossi, 75–95. Naples: Università Suor Orsola Benincasa, 2014.

Hortal Muñoz, José Eloy. "Reservados y pensionistas: Una nueva vía de integración de los reinos en la Casa Real". In *La Corte de Felipe IV. Reconfiguración de la Monarquía Católica*, edited by José Martínez Millán and José Eloy Hortal Muñoz, vol. III, 2283–341. Madrid: Polifemo, 2015.

Hortal Muñoz, José Eloy. "Courtly and Ceremonial Spaces in Spanish Royal Sites: An Evolution from the Renaissance to the Baroque." In *Power and Ceremony in European History Rituals, Practices and Representative Bodies since the Late Middle Ages*, edited by Anna Kalinowska and Jonathan Spangler, 87–104. London: Bloomsbury, 2021. DOI: 10.5040/9781350152212.ch-005.

Hortal Muñoz, José Eloy. "Royal Properties to Maintain: the Royal Geographies". In *Maintaining the Presence of the Prince. Management of Royal Geographies (XIVth–XIXth Centuries)*, edited by José Eloy Hortal Muñoz, Merlijn Hurx and Benjamin Ringot. Turnhout: Brepols, forthcoming.

Hortal Muñoz, José Eloy and Merlijn Hurx, eds. *Building the Presence of the Prince. The Institutions Responsible for the Construction and Management of the Buildings of European Courts (14th–17th centuries)*. Turnhout: Brepols, 2024.

Labrador Arroyo, Félix and Koldo Trapaga-Monchet. "Forestry, Territorial Organization, and Military Struggle in the Early Modern Spanish Monarchy." *Environmental History* 23:2 (2018): 318–41. DOI: 10.1093/envhis/emx133.

Maroteaux, Vincent. *Versailles. Le Roi et son Domaine*. Paris: Editions Picard, 2000.

Martínez Bara, José Antonio. "Noticias sobre las dehesas del Monasterio de San Lorenzo del Escorial". *Anales del Instituto de Estudios Madrileños* 5 (1970): 109–19.

Martínez Millán, José, ed. *La Corte de Carlos V*. 5 vols. Madrid: Sociedad Estatal para la Conmemoración de los Centenarios de Felipe II y Carlos V, 2000.

Martínez Millán, José and Santiago Fernández Conti, eds. *La Monarquía de Felipe II: La Casa del rey*. Madrid: Fundación Mapfre-Tavera, 2005, 2 vols.

Martínez Millán, José and Mª. Antonietta Visceglia, eds. *La monarquía de Felipe III: la Casa del Rey*. Madrid: Fundación Mapfre, 2008, 4 vols.

Martínez Millán, José and José Eloy Hortal Muñoz, eds. *La Corte de Felipe IV (1621–65). Reconfiguración de la Monarquía Católica*. Madrid: Polifemo, 2015, 4 vols.

Martínez Millán, José and Félix Labrador Arroyo. "La pervivencia de la Casa de Castilla: la Caza". In *La Corte de Felipe IV (1621–65). Reconfiguración de la Monarquía Católica*, edited by José Martínez Millán and José Eloy Hortal Muñoz, vol. 2, 1041–1134. Madrid: Polifemo, 2015.

Martínez Ruiz, Enrique and Magdalena de Pazzis Pi Corrales. *Proteccion y seguridad en los Sitios Reales desde la Ilustracion al Liberalismo*. Alicante: Universidad de Alicante, 2010.

Martini, Wolfram et al. *Die Jagd der Eliten in den Erinnerungskulturen von der Antike bis in die frühe Neuzeit*. Gottingen: Vandenhoeck & Ruprecht, 2000.

Merlotti, Andrea, ed. *Le Cacce Reale nell'Europa deis Principi*. Turin: Leo S. Olschki, 2017.

Narkin, Elisabeth. "Architectural Network as Dynastic Strategy in Fontainebleau's Galerie des Cerfs." *Renaissance Quarterly* 74:2 (2021): 454–97. DOI: https://doi.org/10.1017/rqx.2021.3.

Pierron, Sander. *Histoire illustrée de la forêt de Soignies*. Brussels: Éditions Hansa, 1935–38, 3 vols.

Rivero Rodríguez, Manuel and Ignacio Javier Ezquerra Revilla. "La Caza en la Casa y corte de Felipe II." In *La Monarquía de Felipe II: la casa del rey*, edited by José Martínez Millán and Santiago Fernández Conti, vol. 1, 377–430. Madrid: Fundación Mapfre-Tavera, 2005.

Soto Caba, Victoria. "La primera fábrica de monedas: El Real Ingenio de Segovia." *Espacio, Tiempo y Forma, Serie VII, Historia del Arte* 4 (1991): 95–120.

Spielman, John P. *The City and the Crown: Vienna and the Imperial Court 1600–1740*. West Lafayette: Purdue University Press, 1993.

Tovar Martín, Virginia. *El Real Sitio de El Pardo*. Madrid: Patrimonio Nacional, 1995.

Trápaga Monchet, Koldo. "Power, Environment and Territory: The Creation and Implementation of Royal Forestry Legislation and Administration in Portugal in a Comparative Perspective (14th–17th centuries)." In *Building the Presence of the Prince. The Institutions Responsible for the Construction and Management of the Buildings of European Courts (14th–17th centuries)*, edited by José Eloy Hortal Muñoz and Merlijn Hurx. Turnhout: Brepols, 2024, 130–147.

CHAPTER 3

Institutionalization of Forestry Management for Naval Construction in the Spanish Monarchy (16th–19th Centuries)

Alfredo José Martínez González

> To demonstrate physically and civilly the need and usefulness of the woodlands in a kingdom would amount to wanting to persuade people of an eternal truth which nobody has doubted since the creation of the universe and which originates from God himself, for He ordered the Earth to produce them. Its certainty is undeniable, reason endorses it, the sanctions of any cultivated and uncultivated society confirm it; and experience over the centuries has strengthened the infallibility of this principle.[1]

1 Introduction

This sentence, penned at the Naval Department of Cádiz in 1806 by a jurist of the Spanish Armada following the defeat by the Nelson's British fleet at the battle of Trafalgar, sums up the core issue of this essay: the Spanish monarchy's need to safeguard forest masses over the centuries as strategic assets through the triple combination of God's work, technical expertise gained through reason, and empirical experience. This protection spanned the period from the second half of the 1500s to the beginnings of the Liberal State and led to the

1 'Demostrar física, y civilmente la necesidad, y utilidad de los Arbolados en un Reino, sería querer persuadir una verdad eterna, que desde la creación del Universo nadie ha dudado, y que trae su origen del mismo Dios, pues mandó a la Tierra los produjese. Su certeza es innegable, la razón la aprueba, las sanciones en toda sociedad culta e inculta la confirma; y la experiencia por todos los siglos ha afianzado lo infalible de este principio.' Sentence transcribed from Alfredo José Martínez González, *Francisco Mier y Terán: vida y perspectiva de un jurista de la Marina y de la Real Audiencia de Canarias en la transición del Antiguo al Nuevo Régimen*; (Granada: Comares, 2021), 210.

DOI:10.1163/9789004745698_005

adoption of a plethora of institutional and legislative measures of huge significance designed to give impetus to wooded areas for naval and war purposes. From Philip II onwards, the Spanish monarchy took charge of managing large forest masses, almost invariably to cater to shipping needs, not through local bodies but using delegate agents who were granted the title of *superintendente de montes y plantíos* (superintendent of woodlands and plantations) or *conservadores de bosques* (forest conservators) in Mediterranean Catalonia. Although they received various other names following the advent of the Bourbon dynasty, they all had in common the political and economic significance they attached to expanses of woodlands, as well as the need for them to be sustainable.

The reasons for this institutionalization were clear. Forest masses were part of the daily sustenance of the rural societies that were widespread during the Early Modern Age. However, they competed directly with the naval aspirations of the Crown, which required logs and planks to maintain the integrity of its territories and connections between them, since, as historians have aptly summarized, 'everything is wood, from the peasant's chair to the King's throne'.[2] All the silvicultural management tasks involving various Crown delegates and their related rules underpinned the entire naval framework of a monarchy which, on account of its universality, was forced to protect its peninsular coastline as well as to take measures to preserve its huge territories in Europe, Africa, America and the Pacific.[3]

2 The Pursuit of Harmony above Silvicultural Exploitation: between the Royal Right and the Cantabrian Local Institutions for Atlantic Naval Policy (16th–17th Centuries)

Following the abdication of Charles I of Spain (1556), his son Philip II (1556–1598) came to rule over vast territories in Europe, the Americas and Asia, and Spain became the most powerful monarchy on earth. However, as his empire spanned several continents, it clashed with the expansionist interests of other European powers.[4] Under these circumstances, in order to maintain the territories he had inherited, the Prudent King – as Philip II was known – decided

2 'de la silla del labriego al trono del Rey todo es madera', José Luis Urteaga González, *La tierra esquilmada. Ideas sobre la conservación de la naturaleza en la cultura española del siglo XVIII* (Barcelona: Ediciones del Serval S.A. – Madrid: C.S.I.C., 1987), 115.

3 Esteban Mira Caballos, *Las Armadas imperiales. La guerra en el mar en tiempos de Carlos V y Felipe II* (Madrid: La Esfera de los Libros, 2005), 173.

4 Magdalena de Pazzis Pi Corrales, *Felipe II y la lucha por el dominio del mar* (Madrid: Editorial San Martín, 1989), 18.

to make full use of all possible raw materials along with other economic and human resources.[5] From 1562 onwards, he therefore reinforced the activity of his governing bodies, both local and central, and set in motion a programme of regulatory and institutional initiatives to guarantee the integrity of his territories, which were so far apart that they could only be connected by sea.[6]

Such a programme inevitably involved stepping up shipyard activity and a consequent increase in the stockpiling of timber. For this purpose, a whole forestry management system was put in place in an attempt to harmonize the felling of trees with the planting of new ones throughout the north of the Iberian Peninsula. A prominent role in this system was played by Cristóbal de Barros y Peralta, who brought with him a wealth of experience in nautical matters and who has come to be compared in Spanish historiography with figures such as the Frenchman Jean-Baptiste Colbert or the Englishman Samuel Pepys, except that he was active a century before both.[7]

Barros was summoned to the Court at the beginning of February 1562 and received instructions to inspect the entire Cantabrian coast from the French border to the Portuguese frontier, since that area was regarded as 'the national shipyard.'[8] The records still preserved in the Archivo General de Simancas indicate that he was entrusted with analysing the situation of the existing ships on the Spanish coast – their units, dimensions and suitability for war – as well as enquiring about potential manufacturers of ships useful for royal service, possible ship designs, the economic aspects of the whole operation and the accounting records of all of them. Furthermore – and this is relevant to our study – all the inspections that Cristóbal de Barros carried out were to take into account 'what concerns the planting of woodlands',[9] and not as a minor matter but 'because of its great importance' to Spain's policy.[10] For this reason, Philip II ordered all the local Cantabrian institutions located less than two leagues from the coast or from navigable rivers – whose situation allowed logs to be floated to the shipyards – to collaborate with this royal delegate to give a boost

5 Juan A. Ortega y Medina, *El conflicto anglo-español por el dominio oceánico (siglos XVI y XVII)* (Málaga: Editorial Algazara, 1992), 148; Ricardo Cerezo Martínez, *Las Armadas de Felipe II* (Madrid: Editorial San Martín, 1988), 45.

6 Pazzis Pi Corrales, *Felipe II y la lucha*, 81–82.

7 José Luis Casado Soto, "Barcos para la guerra. Soporte de la Monarquía Hispánica," *Cuadernos de Historia Moderna. Anejos* 5 (2006), 31.

8 José Luis Casado Soto, "Los barcos de la expansión oceánica española," in *Estudios del Mar (XVIII Semana de Estudios del Mar)*, (Pontevedra: Fundación ASESMAR, 2000), 199–222. http://www.asesmar.org/conferencias/semana18/semana18.htm.

9 'lo que toca a las plantaciones de los montes'.

10 'por lo mucho que importa'.

to the naval construction industry and to create a whole forestry management network in the woodlands.[11]

More specifically, the monarch ordered the municipal institutions – *corregidores* (royal representatives), *jueces de residencia* (special judges entrusted with reviewing the conduct of officials) and local justices – to establish and distribute annually to the residents of their municipalities the quantity of oaks that they deemed appropriate to plant in view of the relief, size and quality of their land. These operations were to be paid for out of each council's assets and everything was to be recorded in writing and a report submitted 'so that we [Philip II] may know how our orders are being fulfilled'.[12] In the event of non-compliance, both the disobedient resident and the negligent local authority would be liable to pay a fine of one *real* for each tree they failed to plant. One-third of the amount was to be given to the person who had reported it, another to the judge in charge of hearing the case and the last to the royal chamber. The locals were also duty bound to preserve the land given over to planting from the harmful effects of livestock during the early years because, the monarch stressed, it would benefit the 'public good' (*bien público*). This shows that his measures were designed to reconcile the monarchy's foreign policy with the needs of the inhabitants of those woodlands.[13]

At the end of the 1560s a royal councillor, Hernán Suárez de Toledo, joined in Cristóbal de Barros's forestry work. The Royal Council of Castile instructed him to take advantage of a reforming visit around the north of the peninsula, where he had been entrusted with inspecting the University of Oñate, to gather as much information as possible about the nautical situation on the coast of the province of Guipúzcoa, as well as on the state of the forest masses best suited to shipbuilding in that enclave in the east Cantabrian region.[14] For this purpose, he was ordered to take statements in various municipalities from the most prominent residents of Guipúzcoa about the woodlands in their area and the condition they were in. He was also to ask the residents to give witness

---

11 All this information was recorded three months later in an "Ynstruizion que se dio a Xrobal de Barros sobre lo de los navios; Madrid, 6 de mayo de 1563." Archivo General de Simancas (AGS), Guerra y Marina (GyM.), leg. 347, doc. 15.

12 'para que nos [Felipe II] sepamos cómo se cumple nuestro mandato.

13 David Goodman, *Poder y penuria. Gobierno, tecnología y ciencia en la España de Felipe II* (Madrid: Alianza Universidad, 1990), 110; AGS, GyM, leg. 347, doc. 19.

14 María Rosa Ayerbe Iríbar, "Universidad de *Sancti Spiritus* de Oñate. Fuentes y líneas de investigación," in *Universidades Hispánicas. Modelos territoriales en la Edad Moderna (II): Valencia, Valladolid, Oñate, Oviedo y Granada*, ed. by Luis Enrique Rodríguez San Pedro Bezares and Juan Luis Polo Rodríguez (Salamanca: Ediciones Universidad de Salamanca, 2007), 108.

statements on what condition their ports were in, what ships were moored there and who had crafted them.[15] He was likewise entrusted with enquiring about whether each locality already had any municipal legislation on managing its forests and to find out about other naval matters of use to the monarchy. For this reason, he was also to ask each interviewee what other potential rules they might suggest to help boost Spain's naval power. All this information was to be subsequently sent to the Royal Council of Castile.[16]

To comply with this royally commissioned task, between 20 and 21 June 1569 Suárez de Toledo drew up two questionnaires (one on forestry matters and the other on naval issues) for interviewing the chosen residents of the places with closest links to the sea and the eastern Basque woodlands. The first contained four questions: what residents knew about the existence of woodlands and, above all, those closest to the coast; in what condition these forests were in; if the interviewee was aware of the existence of forestry measures that could be employed by the Crown; and, lastly, what rules might be drawn up and put in place to increase the supply of wood useful for constructing ships fit for taking on other maritime powers. The second questionnaire had three parts: the number and size of existing ships in their ports; who the shipbuilders in Guipúzcoa were; and, lastly, what treatment they had traditionally received from the monarchy and whether there were any naval construction regulations that might also be useful.[17] The answers to these questions were sent to the Madrid Court and all of them reflected the concerns both of the woodland populations and of Philip II about reconciling the protection of forest masses with the need to supply timber to the shipyards.

Following on from Hernán Suárez de Toledo's inspections, in the early 1570s Cristóbal de Barros set about combining the results of that earlier visit with the experience gained since 1562. He penned his own report (*Memorial*) to the king summing up the approach that needed to be taken to naval policy and administration in Philip's Spain.[18] His text described in detail the common mentality of the coastal and woodland localities in the north of the peninsula, how the municipal institutions acted, the successes achieved and mistakes made in the

15 Xabier Alberdi Lonbide, "Conflictos de intereses en la economía marítima guipuzcoana. Siglos XVI–XVIII" (PhD diss., Universidad del País Vasco, 2012), 438.

16 Alfredo José Martínez González, "Bosques guipuzcoanos para la construcción de navíos y recopilación normativa por el Real Consejo de Castilla (1569)," *Tiempos Modernos. Revista Electrónica de Historia Moderna* 39, no. 2 (2019), 330.

17 Both questionnaires are housed in Madrid, specifically in the Archivo Histórico Nacional (AHN), Consejos, 15651, exp. 1.

18 Cerezo, *Las Armadas de Felipe II*, 106. The author dates the document to 1573, but does not cite the source or explain how this date was deduced.

construction of vessels in their shipyards and the need to provide a legal basis for all naval matters, among other matters. In short, Barros conveyed to the court the need for the monarch, through his delegates, to control and supervise all aspects of maritime affairs, calling for the creation of a new institution known as the *Superintendencia de Montes y Plantíos* (Superintendency of Woodlands and Plantations), as follows:[19]

> Que en la Provincia de Guipúzcoa y Señorío de Vizcaya se mande a los Corregidores y Justicias que guarden una Provisión y mandato que en esta conformidad se ordenaría, y que ellos sean comisarios de estos Plantíos, y que en cada Lugar nombren un Comisario por lo que toca à la correspondencia y aviso de lo que se hiciere, y estos serán más a propósito los Escribanos en cuyo poder quedare el registro del repartimiento, y en los Lugares, Puertos de mar los Escribanos de los registros de las Naos, pues es todo una materia y la han oído platicar muchas veces, y porque como hasta aquí las Justicias no han hecho lo que se les ha ordenado como conviene, y de la mesma manera se ha de tener por cierto que lo harían de aquí adelante, y ya esta materia no sufra dilación que nombre V. Majestad un Comisario General y Superintendente como se hizo para levantar gente de las Marinas y Sierras del Reyno de Granada, el cual pueda tomar cuenta à los susodichos de como cumplen la dicha Provisión, y ordenarles que repartan más número de Plantíos en la parte que pareciere han repartido pocos, y que les ordene y haga guardar lo que en esto convenga, y que avise à V. Mag. de lo que cumplen los dichos Corregidores, y dejan de cumplir para que V.M. envíe ejecutor que les ejecute en la cantidad de dineros que à V. Mag. parezca, y de esta manera pienso que se enmendará mucho el plantío de los Montes en estas dos Provincias [Vizcaya y Guipúzcoa], en las cuales no se advierte ni apunta lo que toca a la conservación y corta de ellos, ni à la prohibición del pasto, porque como los plantíos han de ser gruesos, y altos los ganados no los pueden comer, y porque como es de tanta sustancia el tener montes cada uno los gasta y usa de ellos muy convenientemente, y contra quien se los cortare, según su costumbre pueden proceder como contra ladrón que hurtase otra cosa. En estas partes casi todos los montes son particulares ò Concejiles, à lo menos están en esta posesión, y si à los dichos Corregidores se les diesen cuarenta mil maravedís de salario librados en gastos de Justicia por veinte días o pocos más que se han de ocupar en esto, como

19 Archivo del Museo Naval de Madrid (AMN), Colección Navarrete nº XXII, doc. 28; fols. 86 rº and vº.

tuviesen que gastar harían lo mejor, y esto habían de cobrar por libranza del comisario general que se la daría cuando tuviese satisfacción de lo que habían cumplido y hecho, descontándoles lo que de las condenaciones de sus tercias partes hubiesen llevado, con este salario holgarían ellos. V. Mag. mandara lo que sea servido.

That in the Province of Guipúzcoa and Lordship of Biscay, the *Corregidores* and Justices be ordered to observe a Provision and mandate that would accordingly be ordered, and that they be commissaries [*comisarios*] of these Plantations, and that in each Place they appoint a Commissary for the correspondence and notification of what is done. These tasks are most suited to clerks, who shall be in charge of keeping a register of the distribution, and in the coastal places and ports the clerks of the ships' registers, since it is one and the same matter and they have heard it discussed many times, and because up until now the Justices have not done properly what they have been ordered to do, and it should likewise be taken for granted that they will do so from now on. And so that this matter is not delayed, Your Majesty should appoint a Commissary General and Superintendent, as he did to remove people from the coasts and mountains of the kingdom of Granada,[20] who shall keep check of how the abovementioned comply with the said Provision, and order them to distribute a greater number of seedlings in the part where few appear to have been distributed, and order them and have them observe what is appropriate, and notify Your Majesty of what the said *Corregidores* comply and fail to comply with so that Your Majesty may appoint an executor to carry them out for the sum of money that Your Majesty deems fit. And I think that that way the planting of the woodlands will be much improved in these two provinces [Biscay and Guipúzcoa] in which there are no reports or records of matters relating to their conservation and cutting, or to the prohibition of grazing, because as the planted trees must be thick and tall livestock cannot eat them, and because, since having woodlands

20 This is an allusion to the population depletion in the kingdom of Granada following the expulsion of the *moriscos* (Muslim converts) as a consequence of the *Rebellion of the Alpujarras* against Philip II's policy. The revolt began in 1568 and led to the deportation of tens of thousands of *moriscos* to other parts of Spain in late 1570. Their expulsion triggered a huge economic decline in that area in the southeast Iberian Peninsula and led the monarch to seek new settlers. There is evidence that Cristóbal de Barros was involved in recruiting Basque families willing to move to this formerly *morisco* area. AGS, Cámara de Castilla (CCA), Libro de Registro de cédulas (CED) 259, doc. 75 Letter of Philip II, Madrid, 24 March 1571.

> is so important, everyone uses them very appropriately, and according to custom they can proceed against anyone who were to cut them as against a thief who were to steal something else. In these parts nearly all the woodlands are private or common, having this ownership at the very least, and if the said *Corregidores* were given a wage of forty thousand *maravedís* paid in legal expenses for twenty days or slightly more, the time they need to spend on this, having money to spend they would do what was best. And they should be paid this on the orders of the commissary general, who would issue payment orders when satisfied with what they had fulfilled and done, discounting the one-third parts of the penalties they had collected, and they would be content with this wage. May Your Majesty give orders for what he deems fit.

This text, together with the urgent need to give impetus to the building of warships, proved effective, and on 7 December 1574 Cristóbal de Barros was officially chosen as the first superintendent ('*super yntendente*' or more commonly '*superintendente*') of woodlands and plantations for the whole Cantabrian area.[21] This post involved preserving the forests in the north of the Iberian Peninsula and fostering their growth, as well as attending to all the operations needed to supply the shipbuilding industry with materials, it being crucial to secure stocks of trunks and timbers.[22] The first words of his letter of appointment made it clear that the deforestation of the area assigned to him (owing, among other reasons, to the existence of forges that consumed enormous quantities of wood) had led to a sharp increase in the prices of those raw

21 ASG, GyM, leg. 347, doc. 57. Real Cédula de Felipe II (*Tanto del titulo de xptobal de Varros. Para los Plantios de rrobles su besita y de montes y dehesas*). Madrid, 7 September 1574.

David Goodman, *El Poderío naval español. Historia de la Armada del siglo XVII* (Barcelona: Ediciones Península, 2001), 109 and 111.

Gómez Rivero states that "the word superintendent arose in France. The first superintendent who appears in the French central administration is that of finance. From the fourteenth century in France royal finances were divided into ordinary and extraordinary, which were administered by the treasurers and the generals. […] From 1552 one of those intendents came to be called superintendent of finances: the actual head of the financial administration who was in charge of all the services and *ordonnance toutes de dépenses de l'Etat.*" See Ricardo Gómez Rivero, "Aproximación a la Superintendencia en Guipúzcoa (s. XVII)," in *Centralismo y Autonomismo en los siglos XVI–XVII*, ed. by Aquilino Iglesia Ferreirós and Sixto Sánchez-Lauro (Barcelona: Servicio de Publicaciones de la Universidad de Barcelona, 1989), 161 (footnote 2).

22 Ricardo Gómez Rivero, "La Superintendencia de construcción naval y fomento forestal en Guipúzcoa (1598–1611)," *Anuario de Historia del Derecho Español* 56 (1986), 594–95. Álvaro Aragón Ruano, *El bosque guipuzcoano en la Edad Moderna: aprovechamiento, ordenamiento legal y conflictividad* (San Sebastián: Sociedad de Ciencias Aranzadi, 2001), 47.

materials on account of their scarcity, which had a direct effect on the rate of production in the shipyards. In order to be able to proceed more effectively, Philip II granted Barros full powers of action, while at the same time compelling him to obey the specific mandates that were added to his appointment.[23] As we shall see in the following pages, this aspect was crucial for the future, since the measures established by Barros, in his executive capacity as the first superintendent of woodlands and plantations, to solve the shortage of useful trees had far-reaching consequences.[24]

Firstly, he was empowered to analyse the accounting records of the Cantabrian localities that usually claimed to be in economic straits to avoid having to plant new trees. If they were proven to be lying, Barros could impose a specific tax on them in order to raise the capital he needed for his forestry work. In addition, he was authorized to appoint forest guards and to dismiss the local justices if they failed to cooperate. He could also examine cases of infringement of the rules governing the preservation of forests and impose financial penalties in the event that local authorities acted negligently in both the prevention and regeneration of forest stands. He was likewise granted the necessary authority to supervise the *corregidores* (the monarch's regional delegates), reporting any wilful or negligent actions to Philip II. In order to be able to perform all these duties, the superintendent was not answerable to any authority other than the king, and both the *corregidores* and justices were expressly forbidden to obstruct his tasks. Finally, among other powers, he was authorized to make decisions on everything he considered necessary: 'regarding whatever you deem fit, you can do and order everything'.[25] Thanks to this royal delegation of powers in 1575, Cristóbal de Barros developed a set of regulations, which he called an *Instrucción*. Its tenets laid the foundations for all subsequent forestry organization and remained in force, directly or indirectly, for centuries, regardless of the circumstances and dynasties reigning in Spain. Due to their importance, we will describe them below.[26]

These regulations established that the plantings to be carried out by each municipality, depending on the number of active inhabitants, should take

---

23 AGS, GyM, leg. 78, docs. 108 and 377. Madrid, 7 December 1574. Casado, *Los barcos españoles del siglo XVI*, 113.

24 Goodman, *El Poderío naval español*, 109.

25 en lo que os pareciere convenir podais vos hacer y ordenar todo', ASG, GyM, leg. 347, doc. 57, edict (*cédula*) issued by Philip II, 7 September 1574, Madrid.

26 We do not know whether the original *Instrucción* document has survived to the present day because, despite being searched for, it has yet to be found. We have succeeded in reconstructing part of its contents thanks to two manuscripts written nearly twenty years later, to which we will return in due course. These are:

place annually between the months of November and January, except in low-lying, very wet areas in which case they would be performed between February and March to prevent the roots from rotting due to the winter rains. This point shows that the first superintendent was familiar with forestry science, the rural customs of the Cantabrian settlers and naval needs, since, depending on how trees were planted, the timber could be used in the shipyards for different parts of vessels, in accordance with the modus operandi employed since ancient times by the inhabitants of the villages where the forests useful for royal service were located:

> Los plantíos que se hacen en las partes bajas y húmedas son los más espesos con que se suben en alto y sus ramas no se dilatan y estos son para tabla y sus dueños los guardan como cosa de mucho valor, y de suyo suben sin que haya necesidad de limpiarlos. Los que se plantan en tierras altas y notales [sic.] se hace en más distancia, unos de otros y estos se hacen más parrados y extienden sus Ramas y son a propósito para maderas tuertas y no se han de podar ni limpiar porque por allí les entra el agua y pudre todo el árbol y esto es ciertísimo y se ve en Vizcaya donde hacen esto.

> The trees in the low-lying and humid parts are those planted most densely and therefore grow upwards and their branches do not spread out, and these are for planks and their owners guard them as something of great value, and of their own accord they grow upwards without there being any need to pollard them. Those that are planted on high ground and new ground are spaced further apart from each other, and these are more spread out and their branches grow outwards and they are suitable for twisted timbers and must not be pruned or pollarded because water gets in and rots the whole tree and this is absolutely true and can be seen in Biscay where they do this.

Oak trees, for instance, a crucial species for shipbuilding, needed to be allowed to reach a minimum size so that certain ship parts could be crafted from them.

AGS, GyM, leg. 403, doc. 97; 9 July 1594. This is a copy made under the superintendent who was Cristóbal de Barros's successor, and is dated 25 October 1575 and specifically aimed at the municipality of Comillas.

AGS, GyM, leg. 403, doc. 102. Report by Cristóbal de Barros, dated 11 October 1594 in Seville, on his actions as superintendent of woodlands and plantations which, among other things, reflects much of what was mentioned in his *Instrucción* of 1575.

Specifically, according to the metric system of the time, they had to be 'three and a half *varas* high from the ground upwards and a short spear's [*venablo*] length thick, more or less'.[27] In addition, Barros established that monitoring and inspecting the plantings, as well as imposing of penalties on those who did not comply with these duties, were the responsibility of the authorities of each locality. They were likewise required to mark the areas to be used for planting, bearing in mind that the most important task was to ensure that the most fertile land was used and that the new trees were not far apart but placed all together. Municipalities made up of scattered population centres were allowed to plant them in the vicinity of 'each neighbourhood' ('cada barrio'), so that residents would have easy access to acorns and leaves, as well as being able to safeguard the plantations more comfortably. With regard to techniques, local residents were obliged not to plant seedlings in the ground until each had at least two leaves, and the trees had to be spaced two feet apart 'so that the grass and other weeds would not take away the goodness of the soil and the water would remain for longer when the soil was dug and it would be more advantageous as they would gain time to grow and thrive'.[28]

In addition, in order to ensure the growth of the plants, he gave orders for the residents to plant hawthorns and gorse trees as barriers at the boundaries of the plantations to prevent livestock from entering and eating the future trees destined to meet royal needs. He also tried to dissuade arsonists by establishing that 'burning and fires must be avoided and punished as Your Majesty has ordered'.[29] Finally, his *Instrucción* attached great importance to having an accurate record-keeping system, making it compulsory for the authorities of each municipality to continue to have 'a book in which the details of the said plantations are recorded',[30] which would also include the penalties imposed on anyone who breached his orders.

As we will see in the following pages, all the measures designed by Cristóbal de Barros designed were long lasting. However, this continuity was not only witnessed in the Spanish-speaking world: nearly a century later it was also reflected in *Sylva, or discourse of Forest-Trees and the propagation of Timber in his Majesties Dominions*, a treatise written by John Evelyn to structure the

27 'de alto del suelo para arriba de tres varas de media y del grosor de un venablo, poco más o menos'.

28 'para que la yerba y otras malezas no les llevasen la virtud de la tierra y el agua se detuviese más estando la tierra cavada y les fuese de más provecho con que ganen tiempo en el medrar y crecer.

29 'las quemas e incendios se han de evitar y castigar como Vuestra Majestad lo tiene mandado'.

30 'un libro en que se asiente lo tocante a los dichos plantíos'.

conservation and encouragement of timber-producing species to cater to the Royal Navy's growing needs.[31] Drafted following the presentation of a document to the Royal Society in 1662,[32] it addressed various issues, among them those previously developed by Cristóbal de Barros in the sixteenth century, such as penalties for anyone who damaged the woodlands and special protection for forest stands located near the coastlines or rivers and watercourses. This may have been the result of John Evelyn's study of the forest management systems of other European countries. Article 12 referred to practices relating to woodlands and ironworks along the Cantabrian coast of the Iberian Peninsula, specifically in Vizcaya (Biscay), an area which, as we have pointed out, was expressly quoted in the *Instrucción* of 1575:

> The King of Spain has neer *Bilbao* sixteen times as many Acres of *Copse-wood* as are fit to be cut for Coal in one year; so that when 'tis ready to he fell'd, an *Officer* first marks such as are like to prove *ship-timber*, which are let stand, as so many *sacred*, and *dedicate* Trees: But by this means the Iron-works are plentifully supplied in the same place, without at all diminishing the stock of *Timber*. Then in *Biscay* again, every proprietor, and other, plants *three* for *one* which he cuts down; and the *Law* obliging them is most severely executed. There indeed are few, or no *Copses*; but all are *Pollards*; and the very lopping (I am assur'd) does furnish the Iron-works with sufficient to support them.[33]

Cristóbal de Barros eventually left the post of superintendent of woodlands and plantations in 1592, after being appointed as the Seville-based purveyor of the Indies fleet by a royal decree of 10 December 1592.[34] This change led the territorial jurisdiction of the first forestry superintendency to be fragmented into other superintendencies that held the same powers but operated in a much smaller, and therefore more manageable, area. The aim was to improve efficiency, as the vast expanse of territory that Philip II had assigned to Barros in 1574, spanning from the French to the Portuguese border, made it very difficult

31 John Evelyn, *Sylva, or discourse of Forest-Trees and the propagation of Timber in his Majesties Dominions* (London: Jo. Martyn & Ja. Allestry, Printers to the Royal Society, 1664), http://www.archive.org/details/sylvaordiscourse00eveluoft.

32 John Nisbet, *Sylva or a discourse of Forest Trees: by John Evelyn F.R.S. with an essay on the life and works of the author by John Nisbet* (vol. 1) (London: Published by Arthur Doubleday & Company Limited at York Building Adelphi, 1908), http://www.archive.org/details/sylva00eveluoft.

33 Evelyn, *Sylva, or discourse*, 110.

34 Archivo General de Indias (AGI), Contratación 3263.

to carry out the duties attached to his position. In this respect, we know that the mayor of the Fort of Hano, in Santander, Fernando de la Riva Herrera, replaced him on 1 September 1593 for the area of Asturias and the Cuatro Villas de la Costa de la Mar.[35]

In Guipúzcoa, a crucial province for naval policy because it was home to several shipyards of vital importance to the Spanish monarchy, forestry matters were entrusted around 1597 to Antonio de Urquiola, who had years of experience in shipbuilding.[36] Finally, Captain Agustín de Ojeda, who is described by the Spanish Royal Academy of History as 'perhaps the most important Basque shipbuilder of the 1590s,'[37] was appointed in early 1598 as superintendent of woodlands and plantations for the Lordship of Biscay. Nevertheless, although this institution was fragmented into smaller ones and inherited by other superintendents over the decades, all of them continued to act in accordance with the criteria that Barros had established years earlier to meet the Catholic Monarchy's timber needs, though the intensity of their actions gradually dwindled.

In the mid-seventeenth century – specifically on 9 July 1649 – an individual named Toribio Pérez de Bustamante was appointed as 'superintendent of factories and plantations of the Cuatro Villas' (an area roughly equivalent to the present-day autonomous community of Cantabria).[38] After inspecting the woodlands in the Cantabrian area, he came to the conclusion that the preexisting rules from the previous century were in fact being obeyed less and less and therefore drew up a new set of forestry regulations to give fresh impetus to the care of the woodlands that served royal needs. This is how the so-called *Instrucción* of Toribio Pérez de Bustamante came into being.[39] However, Bustamante did not design anything truly new; rather, he largely repeated the main points that Cristóbal de Barros had established the previous century when reflecting on the need to promote the forests under his jurisdiction. Indeed, his regulations were based on a core idea that can be considered a precedent

35 AMN, Colecc. Guillén, Ms. 1294, fols. 1–2.

36 In relation to the role played by Antonio de Urquiola, see AGS, GyM., leg. 484, doc. 54, and 490, docs. 106, 123 and 127.

37 Lourdes Odriozola Oyarbide, "Agustín de Ojeda" (Madrid: Diccionario Biográfico de la Real Academia de la Historia), available at https://dbe.rah.es/biografias/72037/agustin-de-ojeda.

38 AMN, Col. Guillén LXXXVII, doc. 9, fols. 118–19, and Ms. 1294.

39 Its full title was *Instrucción que han de guardar los Lugares del Corregimiento de las Cuatro Villas de la Costa de la mar que estuvieren dos leguas distantes de ella y de los ríos navegables a ella, en el aumento y conservación de sus Montes y Plantíos desde ahora para adelante y mientras Su Majestad, que Dios guarde, no mandare otra cosa de la cual se da copia autentica a la Justicia y Procurador de cada Concejo Para que no puedan alegar ignorancia.* AGS, GyM, leg. 3309. Santander, February 1650.

for what came to be known as sustainable development centuries later: 'there can be no good place without woodlands; and if people of the past had not preserved them, those of the present will not enjoy them; and if those of the present do not preserve them, those of the future will not have them'.[40] The above concept enshrined in this set of regulations – which, it should be remembered, was based on Cristóbal de Barros' earlier principles – was not only territorially confined to the area of the Cuatro Villas de la Costa de la Mar but ended up extending its scope of action to Spain's entire northern coastline.[41]

3 Reconciliation Attempts in Timber Management in Eastern Spain to Address Mediterranean Threats during the Habsburg Rule

The Cantabrian region was not the sole target of the policies promoted by the monarchy to control the forests useful for meeting royal needs. The Principality of Catalonia was also heir to a Mediterranean political, legal and naval tradition that was not only necessary to maintain Spanish hegemony over what the Romans called the Mare Nostrum but also took into account the global interests of a Crown that needed to connect its various territories in the area.[42] Philip II, the architect of the great Spanish Armadas, was thus also forced to rely on Mediterranean galley warfare and the organization of naval units, including those of Catalonia, as one of the key aspects of his policy,[43] especially in the face of Turkish threats and Barbary corsairs. It was implemented through a government apparatus that sought to control the centralized construction of galleys between the second half of the 1500s and the first half of the 1600s. These vessels had been the main European warships since medieval times and were specifically designed for the intrinsic characteristics

---

40 'no puede haber lugar bueno sin montes; y si los pasados no los hubieran conservado, no los gozaran los presentes; y si los presentes no los conservaren, no los tendrán los venideros'.

41 Lourdes Soria Sesé, *Derecho municipal guipuzcoano. Categorías normativas y comportamientos sociales* (Bilbao: Administración de la Comunidad Autónoma de Euskadi, 1992), 260.

We also know that in 1675 an *Instrucción* was likewise issued for the kingdom of Galicia, which reiterated the forestry measures of Cristóbal de Barros and Toribio Pérez de Bustamante, AMN, Ms. 2137–4. *Copia Dela Rl. Ynstrucion dada pr. la Reyna Governadora para la conservacion y aumento de los Montes de este Reyno en 19 de Mayo de 1675.*

42 Miguel Ángel de Bunes Ibarra, "La defensa de la cristiandad; las Armadas en el Mediterráneo durante la Edad Moderna," *Cuadernos de Historia Moderna. Anejos* 5 (2006), 77.

43 Irving A. A. Thompson, *Gobierno y administración en la España de los Austrias, 1560–1620* (Barcelona: Crítica, 1981) 201.

of the Mediterranean, whose currents and swells differed from those of the Atlantic.[44] In other words, the Spanish Monarchy's theatre of forestry operations was not confined to the Cantabrian area but also extended to the eastern coastal region of the mainland Spain, where the Barcelona Atarazanas (shipyard played a key role.[45]

Timber and rigging from this region[46] had traditionally been used since the late Middle Ages to supply the naval militia of the Crown of Aragón (a supra-territorial entity to which the Principality of Catalonia belonged). It was stored in the grounds of the Atarazanas of Barcelona, since, in addition to the boatyard, its facilities had large warehouses for storing all kinds of naval equipment.[47] However, from the fifteenth century onwards, much of the naval power of the Crown of Aragón on the eastern coast of the Iberian Peninsula fell into decline and mercantile activity in the Mediterranean dwindled.[48] For that reason, by the time Charles I acceded to the throne, the Spanish monarchy's naval forces in the Mediterranean numbered no more than a dozen units, and attempts were consequently made to reactivate the Barcelona shipyard from time to time during the first half of the 1500s.[49] Later on, due to increased Turkish and Barbary threats, this issue was addressed at the Cortes of Monzón in 1552, presided over by the then Prince Philip (the future Philip II), and a set of regulations (*Capitol de Corts*) was passed prohibiting the removal of any kind of logs or planks except those destined for royal ships. In addition, in the event of infringement, the offender would incur a penalty of one thousand ducats and the loss of the ship and the timber.[50]

---

44 Bunes Ibarra, "La defensa de la cristiandad," 84.

45 The name *atarazanas* originates etymologically from the Arabic *dár-as-sána*, which strictly means "house of industry." See Pablo Emilio Pérez-Mallaína Bueno, *Andalucía y el dominio de los espacios oceánicos. La organización de la Carrera de Indias en el siglo XVI* (Seville: Fundación Corporación Tecnológica de Andalucía, 2010), 22; Geoffrey Parker, *La revolución militar. Innovación militar y apogeo de occidente (1500–1800)* (Madrid: Alianza, 2002), 122.

Ramón Carande, "La navegación y el comercio en el Mediterráneo durante el siglo XVI," *Estudios de Historia nº1, Temas de historia de España* (Barcelona: Crítica, 1989), 321.

46 Antonio Campmany de Montpalau, *Memorias históricas sobre la Marina, Comercio y Artes de la antigua Ciudad de Barcelona* (Madrid: Imprenta de D. Antonio de Sancha, 1779), 28 and 258.

47 *Ibid.*, p. 29.

48 Goodman, *Poder y penuria*, 117; John Lynch, *Los Austrias (1516–1700)* (Barcelona: Crítica, 2000), 149, 595.

49 Jaime Carrera i Pujal, *Historia política y económica de Cataluña. Siglos XVI al XVIII* (Barcelona: Bosch, 1947), vol. 2, 362.

50 AGS, GyM, leg. 953. *Relacion de lo q se reduce …*; Carrera i Pujal, *Historia política y económica*, part II, 362–63.

The reason was clear: interest in woodlands also lay at the heart of Mediterranean warfare, which was considered an instrument preserving the integrity of the Spanish monarchy that served to sustain its hegemony on the Italian Peninsula and periodically reaffirm its leadership position against Muslim threats from across the sea.[51] For all these reasons, at the beginning of the Philip II's reign, attempts were made to reactivate the shipbuilding industry in the Principality of Catalonia, with the royal authority intervening in the management of its pine forests. However, the main problem was that the decline was so great that no skilled craftsmen with the expertise to rig a galley were to be found in Barcelona, and in 1562 (the year Cristóbal de Barros was first summoned to inspect the Cantabrian woodlands) a shipbuilder, Captain Arana, had to be sent from Vizcaya with three hundred artillery arsenal and depot workers.[52] This fact highlights the interest of the Council of War, which was directly responsible for the Atarazanas, in revitalizing them during Philip's reign.[53] Indeed, shortly after acceding to the throne, the monarch had already reaffirmed his control of the naval network along the Mediterranean coast to address the threat of Turkish and Barbary corsairs. He also aimed for all the galleys in his service to belong to him and not to private owners.[54] For this reason, the Cortes of Monzón-Barcelona in 1564 discussed the 'continual' expenditure on pine for building galleys, which had caused its price to rise and even led it to become costlier than wood brought from other countries. Consequently, preparations were made to ensure timber supplies by once again banning the removal of pinewood from Catalonia unless it was used to build royal galleys and other ships destined for the navy.[55]

The situation in Catalonia did not improve throughout the sixteenth century, even though Philip II sent instructions from Madrid on 15 June 1574 (the same year that Barros was appointed superintendent of woodlands and plantations for the Cantabrian region), written in Catalan and published in Barcelona on 12 October of the same year.[56] They were based on the assumption

51 Irving A. A. Thompson "Las galeras en la política militar española en el Mediterráneo durante el siglo XVI," *Manuscrits: Revista d'història moderna* 24 (2006): 249.

52 Goodman, *Poder y penuria*, 117; Cesáreo Fernández Duro, *A la mar madera. Libro Quinto de las Disquisiciones Náuticas* (Madrid: Imprenta de Aribau, 1880), 23.

53 Juan Carlos Domínguez Nafría, *El Real y Supremo Consejo de Guerra (siglos XVI–XVIII)* (Madrid: Centro de Estudios Políticos y Constitucionales, 2001), 87.

54 Thompson, *Guerra y decadencia*, 204.

55 AGS, GyM, leg. 953, *Relación de lo que se reduce disponer para la conservación de los montes de Cataluña, con carta del obispo de Solsona.* 4 September 1627; Carrera i Pujal, *Historia política y económica*, part II, 363.
Capmany de Montpalau, *Memorias históricas*, 258.

56 AGS, GyM, leg. 399, doc. 79. Instrucciones al Maestre de Montesa. Madrid, 15 June 1574. AGS, GyM, leg. 953.

that there had previously been a great abundance of various species of trees in 'our principality of Catalonia and the counties of Rossello and Cerdanya',[57] which were very useful for the galleys and fleets that had traditionally served to 'defend against so many and such powerful enemies'.[58] However, the king considered that its inhabitants 'have stopped sowing and planting trees as they used to',[59] a circumstance that he did not wish to allow 'also considering that everything a galley shipyard has to do is of public and private benefit and of use to our vassals',[60] and he therefore gave orders for trees to be planted and their prices to be appraised.

In addition, the duty to visit all the useful Catalan woodlands was established, for which purpose the captain general of the principality was given the responsibility of appointing a 'doctor of the Royal Court' (*doctor de la Real Audiencia*). He and other people were 'to go to and travel along the eastern and western coasts of the said principality and counties, as far as three and four leagues inland',[61] noting not only the existing trees but also how many could be planted 'in accordance with the nature and location of the land'.[62] However, despite these intentions, it was clear that the Crown could not easily succeed in its attempt to increase its number of galleys and consequently use large quantities of trees for logistical reasons: since the second half of the sixteenth century there had been signs that the Barcelona Atarazanas did not have sufficient capacity to undertake the project on their own, and this was further aggravated by the depletion of timber stocks.[63] In addition, specialised naval carpenters were in short supply.[64]

Nor did Catalan feelings about the exploitation of their forests calm down after Philip II died. In fact, the captain-general of the principality, the Duke of Monteleon,[65] was even forced to publish a proclamation the year of the king's death (1598) declaring that it was forbidden to remove any kind of wood or timber from Catalonia under severe penalties. He considered that trees were

57 'nostre principat de Catalunya y condats de Rossello y Cerdanya'.

58 'defendarse de tants y tant poderosos enemichs'.

59 'han dexat de sembrar y plantar arbres com soliâ'.

60 'considerant tambe q quanta se ha de fer alguna fabrica de galeres redunda en benefici i utilitat publica i particular de dits nostres vassals'.

61 'vajen, y discorreguen per las costas de lleuant y ponent del dits principat y comtats, fins a tres y Quatre llegues dintre terra.'

62 'conforme a la naturala y dispositio de la terra.'

63 Thompson, *Guerra y decadencia*, 220.

64 Thompson, *Guerra y decadencia*, 220.

65 Héctor de Pignatelli y Colonna, Duke of Monteleón, was viceroy and captain-general of Catalonia from 19 May 1606 to 5 November 1610. See Jesús Lalinde Abadía, *La Institución Virreinal en Cataluña (1471–1716)* (Barcelona: Instituto Español de Estudios Mediterráneos, 1964), chronological chart of viceroys.

absolutely necessary for the country's strength and maintenance due to the importance of building ships for its defence and reckoned that many private individuals who were driven by greed were still extracting large quantities of them. As a principal measure, he made it compulsory for anyone wishing to cut and purchase trees for the king's service and the galley works to take with them the money needed to cover their living expenses and everything necessary for felling and transporting the wood by boat or wagon.

For his part, Philip III began his reign with the intention of continuing in the same direction as his father and addressing the war situation on the Mediterranean coast by attempting to carry on reactivating the galley fleet.[66] It was at this point that a forestry institution proper was created. We are referring to the appointment in Catalonia of Pedro de Montagut as the first 'conservator of forests of that principality' (*conservador de los bosques de aquel principado*), also known as 'commissary of plantations' (*comissario de plantios*), on 28 April 1606.[67] To fill this new post, the monarchy needed a suitable person who could safeguard, conserve and increase the number of Catalan trees – which were in a state of 'considerable damage and a torment to all these Kingdoms of Spain'[68] – available for building the Mediterranean squadrons. Shortly afterwards, on 20 June 1606, the viceroy of Catalonia published an *Instrucción* on the post of conservator of forests urging 'all and any whatsoever officials of justice and war, greater and lesser, royal and seignorial, and any other ministers to aid Montagut in his tasks.[69]

Later, from the 1620s onwards and during the reign of Philip IV, the king's favourite, the Count-Duke of Olivares, attempted to impose a common defence scheme known as the *Unión de Armas* on all the territories of the Spanish monarchy. However, according to Lynch, the diversity of the different territories, including Catalonia,[70] prevented the *valido* 'from taxing the periphery by executive means, and raised the issue of royal prerogative versus regional

66 Thompson, *Guerra y decadencia*, 220.

67 AGS, GyM, leg. 1088. Appointment of *Pedro de Montagudo* by Philip III. Ventosilla, 28 April 1606.

68 'en tan notable daño, y de tormento de todos estos Reinos de España'.

69 'a todos y quales quier officiales de Justicia y guerra mayores y menores Realengos y de Señorio y a otros qualesquier ministros'. AGS, GyM, leg. 1088. "Instrucción del Duque Ende Monteleon Virrey de Cataluña para el ejercicio de Conservador de los Bosques de aquel Principado." Barcelona, 20 June 1606.

70 Eulogio Zudaire Huarte, *El Conde-Duque y Cataluña* (Madrid: Escuela de Historia Moderna – C.S.I.C., 1964), 37 and 351.

privilege.'[71] In the case of Catalonia, this resulted in a certain reluctance on the part of the principality to impose a military union on the various territories of the monarchy. These misgivings stemmed both from an attempt to have a particular cultural-political-administrative idiosyncrasy recognised and, from a strictly legal point of view, from the absence of pacts between the king and the Cortes.[72] A *pact-based* parliamentary model was by no means new as it had emerged around the thirteenth century.[73] As in the other territories of the Crown of Aragón, these rules were legal provisions approved by the king and by the three branches of the realm – clergy, nobles and knights, and syndics of the cities and towns – in the form of independent bodies, which made up the Cortes.[74] Precisely because of this nature, in Catalonia laws agreed by pact were more valid than the king's own decisions, and monarchs themselves were obliged to respect them in an oath sworn when they acceded to the throne.[75] In short, from a purely regulatory point of view, it was unthinkable for there to have been a forestry administration in the government of the principality without the involvement of the Cortes.[76]

Nevertheless, despite the legal difficulties, 1627 proved to be a key year for Catalonia, which witnessed hectic activity in shipbuilding, the inspection of tree stands and the stockpiling and transportation of logs and planks, although these operations came up against problems. In addition to the shortage of funds and material, as pointed out above, the ban on exporting timber established in the constitutions of the principality was still not being respected. Miguel de Santos y de San Pedro,[77] bishop of Solsona and viceroy of Catalonia, even complained about this:

---

71 Lynch, *Los Austrias*, 523. Quoted from the original English edition *Spain Under the Habsburgs: Spain and America, 1598–1700* (Oxford: B. Blackwell, 1967), 95.

72 Mateo Ballester Rodríguez, *La identidad española en la Edad Moderna (1556–1665). Discursos, símbolos y mitos* (Madrid: Tecnos, 2010), 439.

73 On the beginnings of the Catlan parliamentary pacts from the reigns of Peter the Great to Alfonso the Benign, see Jaime Sobrequés Callicó, "La práctica política del pactismo en Cataluña," in *El pactismo en la Historia de España*, (Madrid: Instituto de España – Cátedra "Francisco de Vitoria," 1980), 55–57.

74 Juan Vallet de Goytisolo, "Valor jurídico de las Leyes Paccionadas," in *El pactismo en la Historia de España* (Madrid: Instituto de España – Cátedra "Francisco de Vitoria," 1980), 84–85.

75 Eva Serra i Puig, "La vida parlamentària a la Corona d'Aragó: segles XVI i XVII. Una aproximació comparativa," in *Actes del 53è Congrés de la Comissió Internacional per a l'Estudi de la Història de les Institucions Representatives i Parlamentàries*, coord. by Jaume Sobrequés i Callicó et al., vol. 1, 501–36, (Barcelona: Parlament de Catalunya-Museo d'Història de Catalunya, 2005), 518 and 525.

76 *Ibid.*, p. 504.

77 Zudaire, *El Conde-Duque y Cataluña*, 62.

> Habiendo tenido noticia que había grandes excesos en cortar los montes para sacar la madera de este Principado, hice recibir información de ello, y ha constado que sin embargo que por constituciones de este Reino está prohibido el cortar los montes para sacar la madera fuera de él, ha habido tan grande exceso que si no se pone remedio, no se ha de poder fabricar galeras ni otros bajeles del servicio de Vuestra Majestad y en beneficio de esta Provincia, cargando en las playas de estas Marinas muchos bajeles para reinos extraños con grandes cantidades de carbón, pez, y alquitrán, talando y quemando para esto los dichos montes, y bosques, y cortando los árboles a raíz que en muchos años no pueden volver a ser de provecho. Y así para reparo de esto he hecho publicar pregones prohibiendo so graves penas la dicha corta sino es para bajeles de Vuestra Majestad y fábrica de galeras.
>
> Having heard news that great excesses were being committed in cutting down trees to take timber out of this Principality, I asked for information about this, and it has been confirmed that although cutting down trees to take timber out of this Kingdom is prohibited by the constitutions, the excess has been so great that if it is not remedied, it will not be possible to make galleys or other vessels for Your Majesty's service and for the benefit of this Province. On the beaches along these coasts, many ships bound for foreign kingdoms are loaded with large quantities of charcoal, pitch and tar; the said woodlands and forests are felled and burned for this purpose and trees are cut at the roots so that they can be of no use again in many years. Therefore, in order to remedy this, I have made public announcements prohibiting such cutting, under severe penalties, unless it is for seafaring ships for Your Majesty and to build galleys.[78]

The situation of those woodlands called for a detailed examination of the causes that had led to their destruction and an analysis of the most effective remedies for promoting measures to restore forest stands to an acceptable level. The diagnosis was entitled 'Account of what needs to be done to conserve the forests of Catalonia, with a letter from the bishop of Solsosa of 4 September 1627'.[79]

---

78 AGS, GyM, leg. 953. Barcelona, 24 April 1627.

79 'Relación de lo que se reduce disponer para la conservación de los montes de Cataluña, con carta del obispo de Solsona de 4 de sept. de 1627'. AGS, GyM, leg. 953. Barcelona, 7 September 1627.

In this document the viceroy surveyed the regulatory history of forestry matters in the principality, going back even as far as

> the 3rd constitution, title on tributes, made by the Most Serene King Don Jaime the 2nd in the Cortes of the year 1299 and by others, with everything in the same constitution prohibiting and forbidding the extraction of wood, pitch, tar, and other things, and in the same title of the seventeenth constitution made by the then Prince Philip in the Cortes held in the town of Monzón in the year 1547.[80]

He also identified the reasons he considered harmful to the forests, pointing out that 'this damage has been a threat for many years',[81] and proposed measures to remedy the situation. Although he acknowledged that in much of Catalonia and its counties there were still many forest masses full of sufficient timber reserves for crafting galleys, it was no less true that the geographical conditions deriving from the fact that 'the terrain is so mountainous'[82] prevented them from being extracted; and so it once again became clear that it was in the most accessible areas, from which wood could be transported to the Atarazanas, that exploitation had been so huge that the situation had become a cause for concern. He acknowledged that since medieval times the regulations issued on the matter had been sufficient, although the problem lay not in the wording of those provisions but in the way they had been put into practice, as the constitutions contained

> prohibiciones y ordenaciones muy acertadas, sobre lo cual no se puede ordenar, ni disponer más de lo que por dichas constituciones y pragmática está dispuesto, pero el daño procede y resulta de la mala ejecución que han tenidovery appropriate prohibitions and regulations, on which no more can be ordered or provided for than what is provided for in the said constitutions and ordinance [*pragmática*], but the damage comes from and results from the poor implementation that they have had.[83]

---

80 'la constitución 3ª, titulo de vectigals hecha por el serenísimo Rey don Jayme el 2º en las Cortes del año 1299 y por otras, con todo en la misma constitución prohibe y veda la saca de la madera, pez, alquitran, y otras cosas, y en la constitución 17 del mismo titulo hecha por el inclito don Philipe entonces Principe en las Cortes celebró en la Villa de Monçon el año de 1547'.

81 'este daño viene ya amenazando de muchos años atrás'.

82 'la tierra es tan montuosa'.

83 *Ibid.* The *Relación* began with "la constitución 3ª, titulo de vectigals hecha por el sermo. Rey don Jayme el 2º. en las Cortes del año 1299," furthermore citing as precedents the various constitutions of 1547, 1564 and the *pragmática* of 1574.

He also reported that authorisations of tree cutting were in turn conducive to multiple frauds, as much more wood was actually extracted than was permitted. Furthermore, he criticised the actions of the chief royal official appointed to supervise and encourage forestry work:

> Su Majestad, que Dios guarde, tiene un oficial que le llaman conservador de los bosques, y plantíos, con muy buen sueldo, y no carece de mucha culpa viendo este daño, como no lo ha repuntado, y instado con veras su remedio, y según su omisión se puede creer que es la mayor parte de esta devastación y ruina.

> His Majesty, may God preserve him, has an official who is called the conservator of the forests and plantations, with a very good salary, and he is not lacking in much of the blame for seeing this damage and not reporting it, nor has he truly encouraged it to be remedied, and judging by his omission, he may be believed to be the main cause of this devastation and ruin.[84]

In view of the negative situation described above, the chief official of the principality proposed a series of "conservationist" remedies. Among others, he recommended inspecting other not yet depleted areas with abundant woodlands containing useful timber and communication routes for the Atarazanas; making repeated announcements declaring that the conservation of the Catalan woodlands was in the public interest and announcing 'severe financial and corporal penalties'[85] for those who infringed the law. He also stressed the need for a ban on selling timber or ships to foreigners under the threat of 'severe pecuniary and corporal penalties and the loss of the ships themselves'.[86] A further recommendation was to restrict the granting of permits for felling trees near the port of Tortosa (a traditional pine forest area) and, if any were granted, to control the felling, thus preventing the number of authorized trees from being exceeded. Regarding the proposals that the owners of woodlands should be paid a fair price for each tree that was cut down, the bishop of Solsona felt that this was impossible, given the Crown's financial straits: 'because the cost of a galley would be so expensive that if it currently costs four thousand to make,

---

84 AGS, GyM, leg. 953.

85 'graves penas pecuniarias, y corporales'.

86 'graves penas pecuniarias y corporales y perdición de los mismos navíos'.

it could not be made with thirty thousand'.[87] Nor did he support the idea of buying land. Instead, he was in favour of planting trees, which would have to be done by contract, and pointed out that 'there are always many inconveniences due to the negotiations and profits of private individuals'.[88] Finally, taking into account the different circumstances, he suggested a single solution consisting of drafting a new set of regulations reiterating 'that established by King Philip II in 1574 with the points referred to in this list'.[89]

Years later, during the second half of the 1630s, the viceroy and captain-general, Dalmau de Queralt y Codina, Count of Santa Coloma, also issued a series of regulations for the protection and growth of trees, adding 'the particularities which have been deemed necessary for the proper management and conservation of the Forests with the desire to fulfil the requirements of the Royal Orders of Your Majesty and my obligations'.[90] These "particularities" were set out in a list of regulations drawn up in Catalan and published on 26 October 1638 'in execution of the orders given by His Majesty'.[91] The text was drafted at the end of the 1630s, by which time the Count-Duke Olivares had firmly resolved that the Union of Arms should be consolidated.[92] Santa Coloma was fully compliant with Philip IV's favourite, being willing to promote the war industries in response to the Crown's requirements,[93] disregarding the principality's constitutions and institutions if necessary. These regulations included truly strict measures in their articles and were based on the idea that, despite the great abundance of forests in many areas of Catalonia, the captaincy general had been informed that excesses were still occurring that were depleting its forests, and there was no sign of forestry management. Furthermore, they stated that although the old constitutions prohibited the sale of timber and vessels to foreign countries, this practice continued and the

87 'porque sería tan caro el coste del buque de una galera, que si ahora se hace con cuatro mil entonces con treinta mil no se haría'.

88 'siempre suele haber muchos inconvenientes por las negociaciones y provechos de particulares'.

89 'la del Sr. Rey don Phelipe 2º. del año de 1574 con los apuntamientos referidos en esta relación'.

90 'las particularidades que han parecido necesarias a la buena dirección, y conservación de los Bosques con el deseo de acertar lo que piden las Reales Ordenes de V.M. y mis obligaciones'. AGS, GyM, leg. 1250. Letter from the Count of Santa Coloma to Philip IV. Barcelona, 11 November 1638.

91 'en executio dels ordens donats per sa Magestad', AGS, GyM, leg. 565, doc. 59; AGS, GyM, leg.1250. Announcement by the Count of Santa Coloma, Barcelona, 26 October 1638.

92 John. H. Elliot, *La rebelión de los catalanes (1598–1640)* (Madrid: Siglo XXI de España Editores, 1977), 454. Zudaire, *El Conde-Duque y Cataluña*, 147.

93 Zudaire, *El Conde-Duque y Cataluña*, 142–43. Elliot, *La rebelión*, 454.

matter was becoming increasingly worrying because the Spanish Monarchy was then engaged in a conflict with France in the context of the Thirty Years' War. In view of this situation, the terms of reference of Dalmau de Queralt's regulations, apparently issued 'to serve His Majesty and for the public good',[94] contained many prohibitions accompanied by harsh penalties that were particularly severe for anyone who cut all or part of a tree; indeed, they established not only fines but also deprivation of liberty and even banishment to remote islands or years' service as galley slaves.[95]

We do not know whether these regulations were implemented in full or in part, because they were issued at a very unstable time in Catalonia and may have been barely enforced, as the Count of Santa Coloma did not enjoy the sympathy of the populace of the principality. In fact, in 1640, he was eventually assassinated by mobs of *segadors* (insurgent casual labourers) on the day known in Catalan history as the *Corpus de Sang* (Blood Corpus).[96]

Lastly, we have found no further archival data on the principality's conservators of woodlands after the Catalan revolt against the Spanish Crown in the mid-seventeenth century. Perhaps their non-existence was partly due to the fact that, after that time, the Mediterranean lost importance to land warfare and privateering was promoted, sparing the Monarchy from having to charter squadrons itself, and leading to the evident decline of the Mediterranean fleet.[97]

4 Forest Management for Naval Construction following the Advent of the Bourbons

Charles II, Spain's last Habsburg king, died on 1 November 1700. Under his will, the French House of Bourbon came to power, giving rise to a gradual programme of institutional reforms.[98] Among the fronts this modifying process had to address were the decline of the Spanish fleets, which had been dragging on since the end of the seventeenth century, and the management of timber for shipbuilding.[99] As a result, the first decades of the eighteenth century

94 'per lo seruey de sa Magestad y benefici public'.

95 AGS, GyM, leg. 565, doc. 59.

96 Joan Reglà, *Els Virreis de Catalunya* (Barcelona: Vicens-Vives, 1980), 115–16; Elliot, *La rebelión*, 393–97.

97 Thompson, "Las galeras," 116 and 119.

98 Domínguez Nafría, *El Real y Supremo*, 161.

99 Emilio de Diego García, "Estructuras y Organización Naval: Departamentos y Arsenales Peninsulares," XXVI *Jornadas de Historia Marítima "Arsenales y Construcción naval en el siglo de la Ilustración"*, ed. by Instituto de Historia y Cultura Naval, 41 (2002), 15–17.

witnessed various changes aimed at developing a new war fleet made up of new ships of the line that was powerful enough to deter the ambitions of other European nations. This led to the strengthening of the old forestry measures, adapting them to the new institutions.[100]

One of the milestones of the new French-inspired Bourbon structure that was progressively put in place was the establishment of three naval divisions by means of a royal order of 5 July 1726, with arsenals in El Ferrol (in the north of the Iberian Peninsula), Cádiz (south) and Cartagena (east), as it can be seen in the Figure 3.1.[101] The whole naval strategy was organised around these divisions and their territorial jurisdictions spanned vast tracts of forest that were subject to the naval authorities.[102]

However, setting up these bodies led to a depletion of forests, spurring the naval secretary at the time, the Marquis of La Ensenada, to promulgate a royal ordinance on the *montes de marina* (woodlands designated for naval construction) on 31 January 1748, which became the Spanish Bourbons' key forestry legislation for supplying the arsenals of all three divisions with logs and planks.[103] Structured into 74 provisions, it was aimed at regulating all kinds of measures conducive to the promotion of forests for the construction of ships. Mid-eighteenth-century vessels required enormous quantities of timber that furthermore had to be of excellent quality. But the main issue that needed to be addressed in order to meet this supply target was the 'declining state which the forests are currently in, especially those close to the sea, due to the cutting,

---

100 Josep Juan Vidal and Enrique Martínez Ruiz, *Política interior y exterior de los Borbones* (Madrid: Ediciones Istmo, 2001), 173.

101 Ana Crespo Solana, "Las reformas del comercio gaditano a comienzos del siglo XVIII en el contexto europeo de las políticas navales y comerciales," in *La Casa de la Contratación y la navegación entre España y las Indias*, coord. by Enriqueta Vila Vilar, Antonio Acosta Rodríguez and Adolfo Luis González Rodríguez (Sevilla: Universidad de Sevilla-Consejo Superior de Investigaciones Científicas, 2004), 186.

102 Manuel Díaz Ordóñez, *Amarrados al negocio. Reformismo borbónico y suministro de jarcia para la Armada Real (1675–1751)* (Madrid: Ministerio de Defensa – Secretaría General Técnica, 2009), 129.

103 Its full title was *Ordenanza que su Magestad, (Dios le guarde) manda observar para la Cria, Conservacion, Plantios y Corta de los Montes, con especialidad los que estàn inmediatos à la Mar, y Rios Navegables. Metodo, y Reglas que en esta materia deben seguir los Intendentes de Marina, establecidos en los tres Departamentos de Cadiz, Ferrol y Cartagena*, of 31 January 1748, enacted by Ferdinand VI at the Buen Retiro. We examined the copy held in the AMN, F006–41.

Emilio de la Cruz Aguilar, *La destrucción de los montes (claves histórico-jurídicas)* (Madrid: Servicio de Publicaciones de la Facultad de Derecho de la Universidad Complutense, 1994), 84.

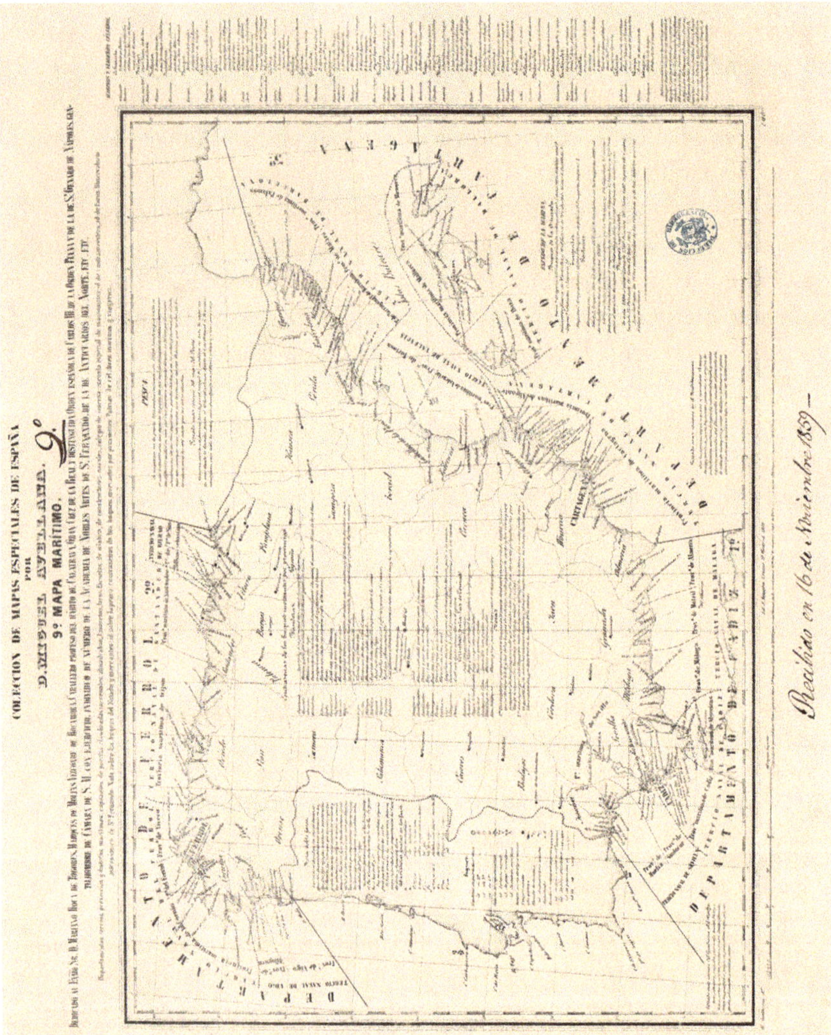

FIGURE 3.1 Map of the Spanish Maritime departments in early 19th century
SOURCE: AMN, AHA, E 14-10-9

felling and burning that has been carried out very frequently, and the lack of care that has been and still is taken'.[104]

However, although this royal ordinance was intended to cover all aspects of Enlightenment forestry knowledge, recent research has shown that the Bourbon administration was fully aware of the *Instrucción* issued by Toribio Pérez de Bustamante (1650), whose legislation has been proven to have largely constituted the backbone of the 1748 regulations. In turn, as we have already shown, in the mid-1600s Pérez de Bustamante had been inspired by the all the regulations designed almost a century earlier by Cristóbal de Barros. This leads us to conclude that over the centuries the legislation drawn up by the Habsburg Monarchy and the regulations developed by the Bourbon monarchy were linked by a common thread.[105]

There is insufficient space in this volume to expand on the substantial validity of this ordinance, but it should at least be pointed out that as the naval jurisdiction was progressively extended to a large part of the Iberian Peninsula (during the Habsburg period only certain forests in the Cantabrian Sea and Catalonia had been linked to shipbuilding), forestry casuistry became so diverse that there was a constant need to specify or interpret matters not envisaged in the 1748 legislation. This was despite the fact that the 1748 royal ordinance on the *montes de marina* had been very thoroughly drafted. The proliferation of specific cases that fell outside the scope of the legislation highlighted the need to regulate them, leading to an overregulation that affected compliance. We know this because at the beginning of the nineteenth century the naval authorities in Madrid asked the heads of the three naval divisions to compile and send to Madrid all the forestry regulations drawn up to improve or clarify the 1748 *montes de marina* ordinance. Once the compilation had been made, the lists of complementary regulations sent from the divisional arsenals were huge – so much so that the very naval officers who compiled them declared that they were unaware of many of them even though they could be found in their archives.[106]

104 'decadente estado en que están presentemente los Montes, con especialidad los inmediatos à la Mar, à causa de las cortas, que indebidamente se han hecho con mucha frecuencia, talas y quemas, y el ningún cuidado, que se ha tenido, y tiene'. Urteaga González, *La tierra esquilmada*, 126 and 129.

105 Alfredo José Martínez González, "La Ordenanza de Montes de Marina, de 31 de enero de 1748, y sus fuentes jurídicas durante los Austrias," in *Estudios jurídicos e históricos sobre navegación en la Edad Moderna: visiones cruzadas*, ed. by Alfredo José Martínez González (Granada: Comares, 2021), 102–32.

106 Archivo General de la Marina "Álvaro de Bazán" (AGMAB), Montes, 4222. *Índice que contiene un extracto de todas las reales órdenes adiciones a la ordenanza de montes del año de*

In the end, this 1748 ordinance did not disappear suddenly, though its last years in force were marked by ups and downs stemming from the period of upheaval experienced by the Spanish Monarchy. During the reign of Charles IV, on 27 August 1803 a royal ordinance was issued at San Ildefonso on the governance of the woodlands and forests under naval jurisdiction (Real Ordenanza para el gobierno de los montes y arbolados de la jurisdicción de Marina), designed to replace the 1748 *montes de marina* ordinance.[107] However, on 10 February 1805, its implementation was suspended and it was decided to return to the 1748 legislation, since at the beginning of the nineteenth century the navy did not have sufficient material resources to comply with the new 1803 ordinance until it had drawn up topographical plans of the forests under its jurisdiction.[108]

5 The Beginnings of the Liberal State in Spain and the End of the *Montes de Marina* System

At the beginning of the nineteenth century, Spain's social structure, like that of other neighbouring countries, was still based on the estates. This system had begun to be challenged by Enlightenment ideology, which advocated as the foundation of society a conception of the individual endowed with the qualities required to achieve happiness and prosperity on the basis of reason. Furthermore, according to this doctrine, the State was merely the consequence of a freely agreed contract established between the individuals who made up society. This idea of freedom spread to very diverse fields of thought, religion and economics.[109] And it was precisely this latter realm of new economic conceptions that ultimately had a negative impact on the forestry matters managed since centuries earlier during the ancien regime.

Spain could not have begun its transition to the new nineteenth-century Liberal State in a more dramatic and disastrous way. The Napoleonic invasion and the abduction of Charles IV's royal family in 1808 roused the masses in arms, and the notables of the provinces gradually formed Juntas that generally

1748 hasta 1 de mayo de 1802 que se remiten a la Dirección General de la Armada consecuente la orden de 13 de febrero último.

107 Biblioteca de la Universidad de Sevilla (BUS), B Res.002652.

108 Francisco Gil de Lemos reporting to the directorate general of the navy Manuel Godoy's decision on the return to the 1748 ordinance. Aranjuez, 10 February 1805. AGMAB, Montes, 4222.

109 José Sánchez-Arcilla Bernal, *Historia de las instituciones político-administrativas contemporáneas (1808–1875)* (Madrid: Dykinson, 1994), 1.

called for people to be granted the sovereignty which they were presumed to have ceded to the monarch in the past.[110] The reasoning was simple: since the king had been abducted, they understood that sovereignty had reverted to the people. However, the concurrence of various provincial Juntas led to a political fragmentation that hindered the taking of any coordinated action against the French armies. For this reason, on 25 September 1808, a new institution was set up, the supreme central governing Junta of Spain and the Indies (Junta Central Suprema y Gubernativa de España e Indias), which represented the various Spanish territories and became the highest governmental body.[111] Almost eight months later, on 22 May 1809, this institution agreed to promulgate a very important agreement entitled Decree on the reestablishment and convening of Cortes (Decreto sobre el restablecimiento y convocatoria de Cortes).[112] This document was published not in Madrid but in the Andalusian city of Seville, where the Junta was based, as much of the Iberian Peninsula was occupied by French troops. This text was also known as the 'nationwide consultation' decree (*Consulta al País*) insofar as one of its main purposes was to send a questionnaire to various institutions and prominent people in the kingdom requesting their opinions so that certain matters could be taken into account in possible debates of the future Cortes, which were to culminate in the adoption of a Constitution.[113]

Finally, early in 1810, following the entry of French troops into Andalusia, the Central Junta was forced to move further south to the Cádiz area as it was in a more sheltered geographical position than Seville. Months later, on 24 September, the so-called General and Extraordinary Cortes (*Cortes Generales y Extraordinarias*) were solemnly opened.[114] From that moment on, its parliamentarians began an ambitious programme of debates aimed at achieving reforms designed to demolish the foundations of the ancien regime, some of them far-reaching. These changes were not only political and legal but also social in nature, with land (and consequently also forests) being one of the major issues to be dealt with. The members of the Cortes of Cádiz tried to define a new regime in accordance with the demands of a revolutionary bourgeoisie who were socially *two sided*: on the one hand, they advocated freeing

110 Mario Onaindía, *La construcción de la nación española. Republicanismo y nacionalismo en la Ilustración* (Barcelona: Ediciones B, 2002), 305.

111 Sánchez-Arcilla Bernal, *Historia de las instituciones*, 9.

112 The word *Cortes* is the name by which the parliamentary institutions were historically known in the Spanish kingdoms.

113 José Manuel Nieto Soria, *Medievo constitucional. Historia y mito político en los orígenes de la España contemporánea (c. 1750–1814)* (Madrid: Akal, 2007), 149.

114 Sánchez-Arcilla Bernal, *Historia de las instituciones*, 13–14.

workers so that their strength and activity could be harnessed; on the other hand, they encouraged the expropriation of land rights in favour of private property.[115] In order to complete the transition to a market economy, the Cádiz parliamentarians needed to put an end to the property ownership of the privileged sectors. Therefore, according to this mentality, the aspiration of the farmers was none other than to become landowners and they justified their own interests with the statement that proper exploitation of the land was only feasible when there were sufficient guarantees that they would receive appropriate compensation for the improvements made by them – a guarantee which, in their opinion, only ownership could provide. In other words, at Cádiz the parliamentarians held that stepping up production inevitably involved access to individual property. And it was here that new approaches emerged through regulatory debates that led to an invalidation of much of the Spanish forestry legislation in force until then. Given the limited space available here, we will attempt to provide a brief overview of the debates, which dragged on for months in the Cortes.

As agreed at a secret session held the previous day,[116] on 5 October 1811 the interim naval minister, Cádiz-born José Vázquez Figueroa, read out a report that fully addressed the problems posed by the regulations in force until then on the *montes de marina* and their jurisdiction. The issue was presented in a simplified manner, without distinguishing between the forestry problems of territories not allocated to naval shipbuilding and the conflicts that could be triggered by forests that *were* linked to the navy:

> Primero: La Ordenanza de matrículas de mar ¿es útil o perjudicial?
> Segundo: En el caso de ser útil, ¿puede seguir del mismo modo que en la península en las provincias de Ultramar?
> Tercero: ¿Es útil o perjudicial el reglamento de montes?
> Cuarto: ¿Son las mismas circunstancias de los montes en las provincias de Ultramar que las de los de la península?

115 Bartolomé Clavero, Pedro Ruiz Torres and Francisco J. Hernández Montalbán, *Estudios sobre la Revolución Burguesa en España* (Madrid: Siglo XXI de España Editores, 1979), 35–6.

116 There is only one, brief reference on 4 October 1811, that 'the time of twelve noon was set in a public session, for the naval minister to inform HM on matters pertaining to his branch." Cortes de Cádiz, *Actas de las sesiones secretas de las Cortes Generales extraordinarias de la nación española, que se instalaron en la Isla de León el día 24 de setiembre de 1810 y cerraron sus sesiones en Cádiz el 14 de igual mes de 1813*, Madrid: Imprenta de J. Antonio García, 1874, 429.

> One: The Ordinance on naval registers: is it useful or harmful?
> Two: If it is useful, can it carry on in the same way as on the peninsula in the overseas provinces?
> Three: Are the regulations on woodlands useful or harmful?
> Four: Are the circumstances of the woodlands in the overseas provinces the same as those of the peninsula woodlands?[117]

This speech is of key importance to properly understanding the spirit that prompted the activation of all the liberal parliamentary mechanisms against the *montes de marina* regulations dating from the ancien regime during the first constituent assembly. After posing these four questions, Vázquez de Figueroa pointed out that the depleted state of the forests and the timber trade in early nineteenth-century Spain was due to the supposedly pernicious effects of the naval shipbuilding regulations. He stated that the interest that the navy had had until then 'was none other than to take advantage of the timber grown in the woodlands of the Crown to use it to build warships, persuading itself that the timber would be much cheaper than if it had been purchased; but how deluded it was'.[118] Thus, after various deliberations, on 14 January 1812, a decree was issued abolishing the laws and ordinances on woodlands and plantings and the related posts (Abolición de las leyes y ordenanzas de montes y plantíos, y extinción de su conservaduría, subdelegaciones, etc.). Article IV did away with the centuries-old system governing the woodlands and forests destined for shipbuilding as follows:

> Queda desde ahora extinguida la Conservaduría general de montes, y todas las Subdelegaciones y Juzgados particulares del mismo ramo, así en todas las provincias marítimas como en las demás, con todos los visitadores y sus tenientes, auditores, promotores fiscales, escribanos, guardas, zeladores, y finalmente todos los dependientes y subalternos de las mismas Subdelegaciones y Juzgados, qualquiera que sea su denominación. Las denuncias que se ofrezcan se pondrán ante las justicias de los pueblos respectivos, y en apelación entenderán las Audiencias territoriales, como en los demas asuntos contenciosos; pero los jueces que determinen

117 Cortes de Cádiz, *Diario de Sesiones de las Cortes Generales y Extraordinarias*, no. 368, Cádiz: Imprenta Real, 1811, 1994.

118 'la Marina no fue otro que el de aprovecharse de las maderas criadas en los montes de la corona para emplearlas en construir los navíos de guerra, persuadiéndose que así las maderas serian mucho más baratas que si se comprasen; ¡pero quanto engaño ha padecido'.

> las denuncias no continuarán recibiendo la parte que hasta ahora han recibido de las condenaciones, la qual se aplicará al fisco.[119]
>
> The post of conservator general of woodlands is abolished from now on, along with all the regional offices and special courts of the same branch, and in all the coastal provinces and all others, with all the inspectors and their deputies, auditors, prosecutors, clerks, guards, wardens and all the assistants and subordinates of these regional offices and courts, whatever they be called. Any complaints shall be filed with the justices of the respective villages and appeals shall be heard by the territorial courts, as in all other disputes; but the judges who rule on complaints shall not continue to receive the portion of penalties they have received until now, which shall be applied to the prosecutor.

By means of this decree, the liberalism of the Cádiz Cortes took another step forward in its attempt to enshrine its conception of private property as a natural right that constituted civil society and made owners the main guarantors of social order. The forestry regulations established centuries earlier during the Habsburg rule to support the shipyards and strengthened during the Bourbon dynasty were thus abolished in one fell swoop by a decree passed by a liberal parliamentary assembly that had set itself up as the nation's representative body.

However, these measures were in force for a very short period. After Ferdinand VII returned, having been released from arrest by Napoleon, his absolute rule was re-established in 1814 in a decree of 4 May. By means of this decree, the monarch revoked all the actions of the Cortes of Cádiz, declaring them null and void with no legal effect or value whatsoever 'as if such acts had never been passed and were erased from time, and with no obligation for my people and subjects of any class and condition to comply with them and to observe them'.[120] In other words, it was the starting point for a strategy that sought to delegitimize the work carried out since 1810 by the parliamentarians who had met at Cádiz. This process of delegitimization was based on branding these individuals' efforts as revolutionary and subversive activities that

119 Cortes de Cádiz, *Colección de Decretos y Órdenes de las Cortes de Cádiz* (Madrid: Publicaciones de las Cortes Generales, 1987), vol. 1, 344–45.

120 'como si no se hubiesen pasado jamás tales actos y se quitasen de en medio del tiempo, y sin obligación en mis pueblos y súbditos de cualquiera clase y condición a cumplirlos y a guardarlos', Sánchez-Arcilla Bernal, *Historia de las instituciones*, 28.

undermined the rights and historical traditions of the Spanish nation.[121] Its repercussions also extended to forestry: the Ancien regime forestry principles were restored in the country by means of a royal order of 13 September 1814, which referred to an allegedly calamitous situation assumed to be a consequence of 'the evils that the State is suffering from owing to the scandalous fellings, fires and destruction of all kinds that the Kingdom's woodlands are experiencing, and which are threatening to totally ruin them',[122] blaming all these alleged excesses on the measures approved by the liberal parliamentarians. For this reason, on 18 September 1814, by means of a royal order, Ferdinand VII declared the decree issued by the Cortes on 14 January 1812 to be defunct and stated that 'things should be restored to the condition and state they were in in 1808 with regard to the common and royal woodlands for naval use'.[123]

Nevertheless, the implementation of this royal order was very limited not so much for legal as for political reasons: as the Spanish American nations progressively gained their independence, Spain gradually found itself less and less compelled to connect a large part of its territories by sea. Therefore, although from a purely legal point of view the system of forests designated for shipbuilding was prolonged, in reality it no longer made as much sense as it had done in the past and was eventually dealt a coup de grâce by a royal decree of 9 December 1832 that permanently deprived the Spanish navy of its forestry powers and assigned those wooded areas to the ministry of development.[124]

6 Conclusions

Despite the limitations of space, this article sets out to show how for centuries the Spanish monarchy attempted to maintain a series of institutions and legislative measures in order to be able to establish a sustainable forestry system. The reason, as David Goodman pointed out, was that 'shipbuilding, the principal reason for the policy of conservation, was at the same time a

121 Nieto Soria; *Medievo constitucional*, 173.

122 'los males que está sufriendo el Estado por las escandalosas talas, incendios y destrozos de toda especie que experimentan los montes del Reyno, y amenazan su total ruina'.

123 'las cosas al ser y estado que tenían en el año 1808 en quanto á los montes comunes y realengos de la comprensión de la Marina', copy of the royal order enacted a few days later, on 27 September 1814. AGMAB, Montes, 4223.

124 Ana María Vigón Sánchez, *Guía del Archivo Museo "D. Álvaro de Bazán"* (Viso del Marqués: Instituto de Historia y Cultura Naval, 1985), 324.

principal cause of deforestation'[125] and that is what both the Hapsburg (on the Cantabrian and Catalan coasts) and Bourbon (extending these actions to the rest of the Iberian Peninsula) dynasties tried to prevent.

It would evidently be a mistake to address the forestry issues examined in this study from a present-day perspective based on parameters more in keeping with modern ecologism. On the contrary, the monarchy's agencies' attitude to forest masses was always utilitarianist, driven by the concern not to exhaust timber stocks in order to maintain a steady supply for its shipyards between the sixteenth and seventeenth centuries and for its arsenals throughout the 1700s and early 1800s. Nevertheless, despite this utilitarianism, it is impossible to deny the conservationist nature of those forestry policies.

In theory, this management system attempted to strike a balance between naval requirements, which were subject to international events triggered by other maritime powers, and the daily needs of the rural societies that relied on forests for their daily sustenance. In practice, however, agreements and disagreements between the Crown's agents and local communities were frequent. After all, the forestry policy that favoured shipbuilding for maritime warfare often attempted to encroach on the powers that local institutions had enjoyed since medieval times over the surrounding woodlands from the second half of the 1500s until the first third of the 1800s.

The period witnessed a succession of systems of plantings, interaction with the local communities, preservation of forest stands from the pernicious action of livestock, systems of pruning and felling in accordance with a calendar, and guided growth of particular trees to obtain suitable parts for shipbuilding, among other measures. In other words, it can be affirmed that there was a common thread running through all these forestry matters starting from the original system designed by Cristóbal de Barros, and that by and large its essence was maintained until 1832, when it was abolished as a result of conceptualisations of private property more in line with nineteenth-century liberalism.

Abbreviations

AGI: Archivo General de Indias
AGMAB: Archivo General de la Marina "Álvaro de Bazán"

125 Goodman, *El Poderío naval español*, 120. Quoted from the original English edition *Spanish Naval Power, 1589–1665: Reconstruction and Defeat* (Cambridge: Cambridge University Press, 1997), 79.

AGS: Archivo General de Simancas
GyM: Guerra y Marina
CCA: Cámara de Castilla; CED: Libro de registro de cédulas
AHU: Archivo Histórico Nacional
AMN: Archivo del Museo Naval de Madrid
Guerra y Marina: GyM
Cámara de Castilla: CCA, Libro de registro de cédulas (CED)
BUS: Biblioteca de la Universidad de Sevilla

Bibliography

Alberdi Lonbide, Xavier. (2012), "Conflictos de intereses en la economía marítima guipuzcoana. Siglos XVI–XVIII." PhD diss., Universidad del País Vasco, 2012.

Álvaro Aragón Ruano, Álvaro. *El bosque guipuzcoano en la Edad Moderna: aprovechamiento, ordenamiento legal y conflictividad.* San Sebastián: Sociedad de Ciencias Aranzadi, 2001.

Ayerbe Iríbar, María Rosa. "Universidad de *Sancti Spiritus* de Oñate. Fuentes y líneas de investigación." In *Universidades Hispánicas. Modelos territoriales en la Edad Moderna (II): Valencia, Valladolid, Oñate, Oviedo y Granada*, ed. by Luis Enrique Rodríguez San Pedro Bezares and Juan Luis Polo Rodríguez, 97–163. Salamanca: Ediciones Universidad de Salamanca, 2007.

Ballester Rodríguez, Mateo. *La identidad española en la Edad Moderna (1556–1665). Discursos, símbolos y mitos.* Madrid: Tecnos, 2010.

Bunes Ibarra, Miguel Ángel de. "La defensa de la cristiandad; las Armadas en el Mediterráneo durante la Edad Moderna." *Cuadernos de Historia Moderna. Anejos,* 5 (2006): 77–99.

Campmany de Montpalau, Antonio. *Memorias históricas sobre la Marina, Comercio y Artes de la antigua Ciudad de Barcelona.* Madrid: Imprenta de D. Antonio de Sancha, 1779.

Cano Martínez de Velasco, José Ignacio. *Los usos jurídicos. Derecho y Sociología.* Madrid: Dilex, 2008.

Carande, Ramón. "La navegación y el comercio en el Mediterráneo durante el siglo XVI." In *Estudios de Historia nº1, Temas de historia de España.* Barcelona: Crítica, 1989, 321–65.

Carrera i Pujal, Jaime. *Historia política y económica de Cataluña. Siglos XVI al XVIII.* Tomo 2, Barcelona: Bosch, 1947.

Casado Soto, José Luis. "Barcos para la guerra. Soporte de la Monarquía Hispánica." *Cuadernos de Historia Moderna. Anejos* 5 (2006): 15–53.

Casado Soto, José Luis. "Los barcos de la expansión oceánica española", in *Estudios del Mar* (*XVIII Semana de Estudios del Mar*), ed. by Asociación de Estudios del Mar, 199–222. Pontevedra: Fundación ASESMAR, 2000.

Cerezo Martínez, Ricardo. *Las Armadas de Felipe II*. Madrid: Editorial San Martín, 1988.

Clavero, Bartolomé, Pedro Ruiz de Torres and Francisco J. Hernández Montalbán. *Estudios sobre la Revolución Burguesa en España*. Madrid: Siglo XXI de España Editores, 1979.

Cortes de Cádiz, *Actas de las sesiones secretas de las Cortes Generales extraordinarias de la nación española, que se instalaron en la Isla de León el día 24 de setiembre de 1810 y cerraron sus sesiones en Cádiz el 14 de igual mes de 1813*. Madrid: Imprenta de J. Antonio García, 1874.

Cortes de Cádiz. *Colección de Decretos y Órdenes de las Cortes de Cádiz*. Vol. 1. Madrid: Publicaciones de las Cortes Generales, 1987.

Cortes de Cádiz, *Diario de Sesiones de las Cortes Generales y Extraordinarias*. Nº 368, Cádiz: Imprenta Real, 1811.

Crespo Solana, Ana. "Las reformas del comercio gaditano a comienzos del siglo XVIII en el contexto europeo de las políticas navales y comerciales." In *La Casa de la Contratación y la navegación entre España y las Indias*, coord. by Enriqueta Vila Vilar, Antonio Acosta Rodríguez and Adolfo Luis González Rodríguez, 183–203. Sevilla: Universidad de Sevilla-Consejo Superior de Investigaciones Científicas, 2004.

de Diego García, Emilio. "Estructuras y Organización Naval: Departamentos y Arsenales Peninsulares." In *XXVI Jornadas de Historia Marítima "Arsenales y Construcción naval en el siglo de la Ilustración"*, ed. by Instituto de Historia y Cultura Naval, 41 (2002), 14–39.

de la Cruz Aguilar, Emilio. *La destrucción de los montes* (*claves histórico-jurídicas*). Madrid: Servicio de Publicaciones de la Facultad de Derecho de la Universidad Complutense, 1994.

Díaz Ordóñez, Manuel. *Amarrados al negocio. Reformismo borbónico y suministro de jarcia para la Armada Real* (*1675–1751*). Madrid: Ministerio de Defensa – Secretaría General Técnica, 2009.

Domínguez Nafría, Juan Carlos. *El Real y Supremo Consejo de Guerra* (*siglos XVI–XVIII*). Madrid: Centro de Estudios Políticos y Constitucionales, 2001.

Elliot, John. H. *La rebelión de los catalanes* (*1598–1640*). Madrid: Siglo XXI de España Editores, 1977.

Evelyn, John. *Sylva, or discourse of Forest-Trees and the propagation of Timber in his Majesties Dominions*. London: Jo. Martyn & Ja. Allestry, Printers to the Royal Society, 1664). Available at, http://www.archive.org/details/sylvaordiscourse00eveluoft.

Fernández Duro, Cesáreo. *A la mar madera. Libro Quinto de las Disquisiciones Náuticas*. Madrid: Imprenta de Aribau, 1880.

Gómez Rivero, Ricardo. "Aproximación a la Superintendencia en Guipúzcoa (s. XVII)." In *Centralismo y Autonomismo en los siglos XVI–XVII*, coord. by Aquilino Iglesia Ferreirós and Sixto Sánchez-Lauro, 161–71. Barcelona: Servicio de Publicaciones de la Universidad de Barcelona, 1989.

Gómez Rivero, Ricardo. "La Superintendencia de construcción naval y fomento forestal en Guipúzcoa (1598–1611)." *Anuario de Historia del Derecho Español* 56 (1986): 591–636.

Goodman, David. *El Poderío naval español. Historia de la Armada del siglo XVII*. Barcelona: Ediciones Península, 2001.

Goodman, David. *Poder y penuria. Gobierno, tecnología y ciencia en la España de Felipe II*. Madrid: Alianza Universidad, 1990.

Juan Vidal, Josep and Martínez Ruiz, Enrique. *Política interior y exterior de los Borbones*. Madrid: Ediciones Istmo, 2001.

Lalinde Abadía, Jesús. *La Institución Virreinal en Cataluña (1471–1716)*. Barcelona: Instituto Español de Estudios Mediterráneos, 1964.

Lynch, John. *Los Austrias (1516–1700)*. Barcelona: Crítica, 2000.

Martínez González, Alfredo José. "Bosques guipuzcoanos para la construcción de navíos y recopilación normativa por el Real Consejo de Castilla (1569)." *Tiempos Modernos. Revista electrónica de Historia Moderna* 39 no. 2 (2019): 321–42

Martínez González, Alfredo José. "La Ordenanza de Montes de Marina, de 31 de enero de 1748, y sus fuentes jurídicas durante los Austrias." In *Estudios jurídicos e históricos sobre navegación en la Edad Moderna: visiones cruzadas*, dir. by Alfredo José Martínez González, 101–32. Granada, Comares, 2021.

Martínez González, Alfredo José. *Francisco Mier y Terán: vida y perspectiva de un jurista de la Marina y de la Real Audiencia de Canarias en la transición del Antiguo al Nuevo Régimen*. Granada: Comares, 2021.

Mira Caballos, Esteban. *Las Armadas imperiales. La guerra en el mar en tiempos de Carlos V y Felipe II*. Madrid: La Esfera de los Libros, 2005.

Nieto Soria, José Manuel. *Medievo constitucional. Historia y mito político en los orígenes de la España contemporánea (ca. 1750–1814)*. Madrid: Akal, 2007.

Nisbet, John. *Sylva or a discourse of Forest Trees: by John Evelyn F.R.S. with an essay on the life and works of the author by John Nisbet*. London: Published by Arthur Doubleday & Company Limited at York Building Adelphi, 1908, vol. 1. Available at, http://www.archive.org/details/sylva00eveluoft.

Odriozola Oyarbide, Lourdes. "Agustín de Ojeda". In *Diccionario Biográfico de la Real Academia de la Historia*, ed. by Real Academia de la Historia, Madrid. Available at: https://dbe.rah.es/biografias/72037/agustin-de-ojeda.

Onaindía, Mario. *La construcción de la nación española. Republicanismo y nacionalismo en la Ilustración*. Barcelona: Ediciones B, 2002.

Ortega y Medina, Juan A. *El conflicto anglo-español por el dominio oceánico (siglos XVI y XVII)*. Málaga: Editorial Algazara, 1992.

Parker, Geoffrey. *La revolución militar. Innovación militar y apogeo de occidente (1500–1800)*. Madrid: Alianza, 2002.

Pérez-Mallaína Bueno, Pablo Emilio. *Andalucía y el dominio de los espacios oceánicos. La organización de la Carrera de Indias en el siglo XVI*. Sevilla: Fundación Corporación Tecnológica de Andalucía, 2010.

Pi Corrales, Magdalena de Pazzis. *Felipe II y la lucha por el dominio del mar*. Madrid: Editorial San Martín, 1989).

Reglà, Joan. *Els Virreis de Catalunya*. Barcelona: Vicens-Vives, 1980.

Sánchez-Arcilla Bernal, José. *Historia de las instituciones político-administrativas contemporáneas (1808–1875)*. Madrid: Dykinson, 1994.

Serra i Puig, Eva. "La vida parlamentària a la Corona d'Aragó: segles XVI i XVII. *Una aproximació comparativa*." In *Actes del 53è Congrés de la Comissió Internacional per a l'Estudi de la Història de les Institucions Representatives i Parlamentàries*, coord. by Jaume Sobrequés i Callicó et al., vol. 1, 501–36. Barcelona: Parlament de Catalunya / Museo d'Història de Catalunya, 2005.

Sobrequés Callicó, Jaume. "La práctica política del pactismo en Cataluña." In *El pactismo en la Historia de España*, ed. by VV. AA., 49–74. Madrid: Instituto de España – Cátedra "Francisco de Vitoria", 1980.

Soria Sesé, Lourdes. *Derecho municipal guipuzcoano. Categorías normativas y comportamientos sociales*. Bilbao: Administración de la Comunidad Autónoma de Euskadi, 1992.

Thompson, Irving A.A. "Las galeras en la política militar española en el Mediterráneo durante el siglo XVI." *Manuscrits: Revista d'història Moderna* 24 (2006): 95–124.

Thompson, Irving A.A. *Guerra y decadencia. Gobierno y administración en la España de los Austrias, 1560–1620*. Barcelona: Crítica, 1981.

Urteaga González, José Luis. *La tierra esquilmada. Ideas sobre la conservación de la naturaleza en la cultura española del siglo XVIII*. Barcelona: Ediciones del Serval S.A. – Madrid: C.S.I.C., 1987.

Vallet de Goytisolo, Juan. "Valor jurídico de las Leyes Paccionadas." In *El pactismo en la Historia de España*, ed. by VV. AA, 75–110. Madrid: Instituto de España – Cátedra "Francisco de Vitoria", 1980.

Vigón Sánchez, Ana María. *Guía del Archivo Museo "D. Álvaro de Bazán*. Viso del Marqués: Instituto de Historia y Cultura Naval, 1985.

Zudaire Huarte, Eulogio. *El Conde-Duque y Cataluña*. Madrid: Escuela de Historia Moderna- C.S.I.C., 1964.

CHAPTER 4

The King's Woodlands: Managing the Forestry Resources of Military Orders in Southern Castile during the Sixteenth Century

Francisco Fernández Izquierdo and Francisco J. Moreno Díaz del Campo

1 Introduction

The development of a modern state in the kingdom of Castile from the early sixteenth century onwards was enabled by the implementation of new management methods and systems for all assets and resources controlled by the Crown. To fund such a colossal governmental and administrative enterprise, the Habsburgs had a copious stream of economic resources at their disposal. Taxation, trade promotion, and the management of the American treasure were some of the components of a many-sided fundraising policy.

Woodlands were among the economic and jurisdictional spaces governed by the monarchs. In a recent work, Koldo Trapaga-Monchet has established a classification of the areas that were exploited for timber and other woodland products by the Hispanic Monarchy in the Iberian Peninsula.[1] Not content with compiling a mere listing of territories, the author builds a hierarchical framework linked to the financial, political, and institutional structures that harnessed these resources, based on the binomial property/land use. This procedure has yielded three distinct spheres which are inevitably closely related and interdependent.

The first of these domains are the Royal Woodlands (*bosques del rey*), which Trapaga Monchet equates with the *sitios reales* – the residential and leisure areas created for the monarchs' enjoyment when the court settled in Madrid. The renowned string of palaces that ringed the capital of the kingdom were seasonal residences for the royal family and their entourage. These estates

1 Koldo Trapaga-Monchet, "Las políticas forestales en los reinos de Castilla y Portugal (siglos XV–XVIII)," in *Il bosco. Biodiversità, diritti e cultura del medioevo al notro tiempo*, ed. Alessandra Dattero (Rome: Viella, 2022), 85–103.

DOI:10.1163/9789004745698_006

were managed by an institution created for the purpose under the name of *Junta de Obras y Bosques*.[2]

The second type were the Woodlands of the Realm (*bosques del reino*), which did not belong to the monarchs but to their attendant administrative and political institutions – chiefly local authorities, but also private landowners such as the nobility or the Church. Many of these woodlands were protected from agricultural land use by ordinances and by the creation of game reserves.

The Portuguese and Castilian monarchs' concern about these spaces began in the midfifteenth century, and even earlier in the case of Portugal. Castile's legislation was the more advanced in its management of forestry resources,[3] which included the demarcation of communal *montes* and *dehesas* belonging to municipalities, along with uncultivated lands used for livestock grazing.[4] These were key spaces for the supply of timber, firewood and charcoal, vital resources for economic and everyday life in the 1500s, a century marked by population growth and urban development in the Castilian interior. Woodlands became increasingly important to the Crown as a source of timber for shipbuilding from the last third of the fifteenth century. Until that time, timber had been obtained from the Royal Woodlands, which had been used for a variety of purposes besides entertaining the court. Indeed, this function, like other activities such as hunting and farming, was soon relegated as timber

2 There is an immense body of literature on this topic. For a recent general approach, see Concepción Camarero Bullón and Félix Labrador Arroyo, dirs., *La extensión de la Corte: los Sitios Reales* (Madrid: UAM Ediciones, 2017).

3 Trapaga Monchet, "Las políticas forestales," 102–103.

4 The word "*monte*" denotes a complex concept relating to both landscape and legal status. In principle, *monte* denotes open woodland, be it mountainous or flat, with shrubs, grasses, and scattered trees growing naturally. The municipal authorities had communal *montes* within their jurisdictions where local people could gather firewood, timber, etc. and graze their animals under the watchful eye of foresters. A *monte* is not as dense as a *bosque* or forest, where trees predominate over shrubs or pastures. In the period studied in this paper, *bosques* were associated with the presence of big game and were easily differentiated from *montes*. The term "*dehesa*" – etymologically "*defesa*" from the Latin "*defendere*" – defines a public or private land enclosure used essentially for animal grazing. The derivative "*monte o bosque adehesado*" is the result of human action on natural woodlands populated by trees of the *Quercus* family – common, holm and cork oak – and other species such as ash or chestnut, creating a landscape of scattered trees which are regularly pruned to maximise summer shade. The open areas between trees are covered with pasture or even dryland crops in some cases. See Francisco Fernández Izquierdo, "Paisajes rurales del siglo XVI en tierras de la Alcarria," in Paisajes de tierra y agua. Gentes y ecosistemas naturales en Castilla y La Mancha (siglos XVI–XVIII), coords. Francisco Javier Moreno Díaz del Campo and Miguel Fernando Gómez Vozmediano (Toledo: Biblioteca Añil, 2022), 82.

extraction became the principal form of land use in these domains.[5] This shift was partly due to the monarchy's new defence requirements, and partly to the implementation of a policy which, although changeable and faltering, never failed to demand preferential use of certain wooded areas for the benefit of the Crown, to the detriment of its subjects' interests.[6]

In this paper we explore a different type of open space directly administered by the Crown where forestry resources were exploited for profit, but are not included in the above classification. We are talking about the vast properties accumulated by the military orders in the wake of the Christian conquest of the southern Iberia, which were incorporated into the Crown's estate when the Catholic Monarchs obtained permission to personally manage the enormous assets that had hitherto been in the hands of the orders' masters and commanders.[7]

A large proportion of the territories formerly belonging to dignitaries of the military orders – particularly the parcels of land known as *mesas maestrales*, which were the exclusive domain of the masters – became extensive *dehesas* that were usually leased to the owners of transhumant flocks of sheep for winter grazing. The *dehesas* and their facilities, such as watermills, enjoyed a special status halfway between the Royal Woodlands and the Woodlands of the Realm. They were directly managed by royal officials and monitored

5 A clear example is the evolution of land use in the Soto de Roma, which has been studied by Félix Labrador Arroyo and Koldo Trapaga-Monchet in "La configuración del espacio y la explotación forestal de un enclave singular: el Real Sitio del Soto de Roma durante la dinastía Habsburgo," *Studia Histórica. Historia Moderna* 39 (2017): 293–327.

6 Koldo Trapaga-Monchet and António Santos, "Forestry and timber supply in the royal forests of the Iberian Peninsula through the 16th century," *Skyllis. Journal for underwater Archaeology* 15, no. 1 (2016): 66. For a discussion of these issues in the case of Portugal, see Koldo Trapaga-Monchet, "No es madera para vasallos, sino del rey: las políticas forestales de los Habsburgo en Portugal," *Obradoiro de Historia Moderna*, 28 (2019): 105–34. For an overview, see John T. Wing, "Keeping Spain afloat: State Forestry and Imperial Defense in the Sixteenth Century," *Enviromental History* 17, no. 1 (2012); 116–45; and John T. Wing, *Roots of Empire: State Formation and the Politics of Timber Access in Early Modern Spain, 1556–1759* (Leiden: Brill, 2015).

7 On the process of incorporating the *mesas maestrales* or *maestrazgos* to the Crown, see Carlos Ayala Martínez, "La corona de Castilla y la incorporación de los maestrazgos," *Militarum Ordinum Analecta* 1 (1997): 259–90; Luis Suárez Fernández, "Cuando los maestrazgos se incorporan a la Corona," *Revista de Historia Militar* extra-1 (2000): 223–30; Elena Postigo Castellanos, "Reyes y maestres. Incorporación y encomienda perpetuas de los maestrazgos de las órdenes monástico-militares ibéricas (1523–1587)," *Palacios, plazas y patíbulos: la sociedad española moderna entre el cambio y las resistencias*, coord. James Amelang et al., (Valencia: Tirant Lo Blanch, 2018), 503–14. On the Order of Calatrava in particular, see Francisco Fernández Izquierdo, *La orden militar de Calatrava en el siglo XVI. Infraestructura institucional, sociología y prosopografía de sus caballeros* (Madrid: Editorial CSIC, 1992), 48–56.

by administrators or foresters, who were in charge of enforcing regulations specifically designed to protect, maintain, and increase the natural resources that directly or indirectly benefited the Crown. The particular features of these wooded areas and the demarcation of open spaces where trees of the oak family – *Quercus robur, ilex,* and *suber*, which yielded acorns, firewood and timber – grew side by side with grasses and dryland crops were a key element of that policy. These were not the large, dense forests that grew in mountain regions or in the areas favoured by the moisture and rain of the Atlantic, but in drier, coarser soils less suitable for forestry which were home to a combination of scattered trees, grasses, and Mediterranean scrub. Located a long way from the court, in the heart of the Castilian plateau, these areas were rich in game, attracting the attention of Philip II (1556–98), who visited them for hunting or leisure in the course of his travels across the kingdom.

This paper examines three cases of woodland exploitation in *dehesas*. First, the commanderies[8] of Otos and Aceca, on the banks of the river Tagus, near the towns of Borox, Añover de Tajo and Villaseca de la Sagra, in the province of Toledo, which were part of the Real Bosque de Aranjuez. Second, the *dehesa* of Zacatena, Daimiel, in the province of Ciudad Real, a former property of the Order of Calatrava. The third case study relates to the *dehesa* of Alcobaza, Jerez de los Caballeros, in the province of Badajoz, formerly a property of the Order of Santiago, where trees were extracted as a one-off gift from the Crown to some important figures in the court of Prince Philip – the future Philip II – during his time in England as king consort to Mary Tudor (Figure 4.1).

These three territories were managed from the early sixteenth century onwards by the Council for the Military Orders (*Consejo de las Órdenes Militares*) as assets of the Hispanic Monarchy. Although this paper focuses primarily on woodland resources, it should be borne in mind that these areas were chiefly used as a source of pasture, over and above other products as firewood and timber. Our focus lies in this particular combination of features because, despite their differences, these three examples allow us to examine the context, methods, and strategies deployed by the Hispanic Monarchy to

8 Commanderies (*encomiendas*) were lands and rights ruled by a commander, who was knight of a military order. They normally consisted of a territory with one or more human settlements, including medieval castles where the commander had his residence. He was in charge of dispensing justice, collecting on seigneurial rights from the population, and avail himself of the proceeds of various resources in his territory (fields, mills, kilns, etc.). In exchange, he contributed soldiers and weapons to the Order's troops and paid certain taxes to the Crown as a member of the clergy. In practice, commanders were like feudal lords, but were subject to the king in the period under study, who controlled them through the Council for the Military Orders.

FIGURE 4.1 Location of case-studies
SOURCE: AUTHORS

exploit their natural resources with the unquestionable aim of obtaining material profit, while supposedly striving to preserve the environment.

2 On the Banks of the Tagus River: Otos and Aceca

The orders of Calatrava (founded in 1158) and Santiago (founded in 1175) settled on the banks of the upper and middle Tagus river in the second half of the twelfth century, thanks to several commanderies and fortresses granted by King Alfonso VIII (r. 1158–1214). The aim of these concessions was to consolidate the hitherto unstable frontier between Castile and al-Andalus. Both orders subsequently extended their dominions westward from the villages of Auñón and Berninches – in the present-day province of Guadalajara – and along the Tagus basin to Villaseca de la Sagra, some twenty kilometres from the city of Toledo, where both orders had churches and properties. Like his grandparents and predecessors Ferdinand and Isabella, the Catholic Monarchs (r. 1474–1516), Emperor Charles V (r. 1520–58) had a penchant for hunting in pleasant riverside landscapes. Charles decided to set up a residence for his own leisure and rest by extending the Crown's properties to incorporate the Santiago commanderies of Alpajes (1536) and Oreja (1539), and the Calatrava's Otos and Aceca (1535). These were the core territories that formed the estate known as Real Bosque de Aranjuez, which eventually became a self-contained district with the addition of other parcels of land.[9]

The cost of building palaces, landscaping gardens, digging irrigation canals, maintaining infrastructures, and paying household staff on the estate demanded ongoing investments which were financed from the proceeds of the exploitation of agricultural resources in the annexed commanderies and *dehesas*. The Otos commandery, in the municipal district of Borox, included orchards and houses in the village itself, saltflats and plots of arable land in neighbouring Ciruelos, plus the *dehesas* of La Higuera and Alhóndiga on the north bank of the Tagus, the latter measuring 1,259 hectares[10] and including a bridge-side inn, a mill, and several islets on the river. The main source of funds was the Otos *dehesa*, on the southern bank, measuring 3,793 hectares[11] divided into ten *millares*.[12] An additional, albeit modest, income was obtained from hunting and fishing rights, firewood, and sundry seigneurial levies. Yearly revenues totalled 1.06–1.65 million *maravedíes* between 1536 and 1543.

9 Magdalena Merlos Romero, *Aranjuez y Felipe II: idea y forma de un real sitio* (Madrid: Dirección General de Patrimonio Cultural, 1998); Pedro Andrés Porras Arboledas, *La Orden de Santiago en el siglo XV: la provincia de Castilla* (Madrid: Dykinson, 1997); Enrique Rodríguez Picavea, *La formación del feudalismo en la meseta meridional castellana: los señoríos de la Orden de Calatrava en los siglos XII–XIII* (Madrid-Mexico: Editorial Siglo XXI, 1994); Francisco Fernández Izquierdo, "A poniente del Real Bosque de Aranjuez: Otos, Aceca y Alhóndiga en los primeros tiempos de su gestión (1531–1570)", *Pragmatismo e ilusión: el agua y la gestión del espacio y territorio en Aranjuez y otros sitios cortesanos (siglos XVI–XVIII)*, dirs. Félix Labrador Arroyo and Magdalena Merlos Romer (Madrid: Sílex, 2023), 55–105.

10 The 1,259 hectares are calculated from a 1753 measurement of Alhóndiga at 3,350 *fanegas* of 400 *estadales*, a *fanega* being equivalent to 3,758 m^2. Concepción Camarero Bullón and Laura García Juan, "Geografía histórica de los espacios reales: Alóndiga, Aceca y Barciles, despoblados del rey en la vega del Tajo," *Estudios Geográficos* 79, no. 284 (2018): 220.

11 The Otos *dehesa* was measured in 1707 at 10,092 *fanegas*. Juan Álvarez de Quindós y Baena, *Descripción histórica del Real Bosque y Casa de Aranjuez: dedicada al rey nuestro señor* (Madrid: Imprenta Real, 1804), 81. Quindós y Baena applied the *fanega* of Toledo, equivalent to 400 *estadales*, or 3,758 m^2, resulting in a total of 3,793 hectares, as per the table included in Dirección General del Instituto Geográfico y Estadístico, *Equivalencias entre las pesas y medidas usadas antiguamente en las diversas provincias de España y las legales del sistema métrico-decimal* (Madrid: Imprenta de la Dirección General del Instituto Geográfico y Estadístico, 1886), 51. Rental income from this *dehesa* was between 800,000 and 1,000,000 *maravedíes* per year in round figures.

12 A *millar* was a sufficient extension of pasture to feed one thousand head of sheep. The two *dehesas* were incorporated into the Royal Woodland and are today within the municipal district of Aranjuez, in the Madrid region.

Income from woodland resources amounted to barely 1% of total revenues, save for exceptional years like 1536, when the sale of timber and firewood yielded 61,930 *maravedíes*.[13] The wood was extracted from the species that grew along the banks of the Tagus: white poplar (*Populus alba*) and ash (*Fraxinus excelsior*) for timber; tamarisk (*Tamarix gallica, Tamarix canariensis*), orache (*Atriplex halimus*), and other shrubs for firewood. Areas were rotated for pruning every three or four years to allow the vegetation to recover naturally. Livestock keepers also used to clear the undergrowth to maximise pasture for their flocks. Tamarisk was low quality firewood, the price of a cartload fluctuating between half a real and 20–24 *maravedíes* depending on the year.[14] White poplars from the riverside grove Soto del Peral were used to make barrel staves, but they were too few to generate a regular income. It took twenty years for a tree to regrow and many failed to recover, such as the willows that stood below the bridge at Alhóndiga, which were lost for good.[15]

Aceca was located west of Otos along the north bank of the Tagus, near Villaseca de la Sagra. It included a dilapidated fortress which had been the headquarters of its commander and was refurbished by Philip II for his own accommodation to break the journey between Aranjuez and Toledo.[16] The king visited the estate on several occasions, signing and dating deeds and royal orders from the house. It was set among *dehesas* – Castillo, Aceca, and El Soto – meadows, and farmland, including several islets in the river and an inn by the name of Bel. Like Otos, the Aceca commandery incorporated certain seigneurial rights. The water mills, fishing rights, and the ferry across the Tagus yielded some 100,000 *maravedíes* out of a total yearly revenue of 820,000–830,000 between 1531 and 1534.[17] Woodland exploitation was limited to firewood and

13 Archivo Histórico Nacional (AHN), Órdenes Militares (OM), Archivo de Toledo (AT), exp. 88036.

14 Money was usually counted in *maravedíes* in mid-sixteeenth century Castile, but for substantial transactions the real (34 *maravedíes*) and the ducat (375 *maravedíes*) were used. A labourer's wage varied between half a real per day (15–17 *maravedíes*) and three reals depending on the skills required. On salaries in this period see Juan Aranda Doncel, "Los salarios de los trabajadores en Córdoba a mediados del siglo XVI," *Boletín de la Real Academia de Córdoba de Ciencias, Bellas Letras y Nobles Artes* 56, no. 108 (1985): 33–48.

15 AHN, OM, AT, exp. 45.142. The valuation was prepared for the transfer of the commandery from the Order of Calatrava to the Real Bosque de Aranjuez. Logging for firewood in riverside groves was temporarily forbidden when they were incorporated into the Royal Woodlands in 1535.

16 For details of this transformation, see Antonio José Díaz Fernández, "Aceca, de castillo a palacio," *Anales Toledanos*, 27 (1990): 81–96. The buildings were designed by the royal architect Gaspar de Vega under the king's direct supervision in 1556. Building works seem to have been ongoing in the following years, including a new workshop and stables in 1561 under the direction of Juan Bautista Muñoz.

17 AHN, OM, AT, exp. 45140.

barely amounted to 1% of the total income, or 9,000 *maravedíes*. Added to these revenue sources, a modest toll was charged at mill-side weirs in Alhóndiga and Aceca for the pine logs that were driven downstream to Toledo from the mountain forests of Molina (Guadalajara) and Cuenca. Both locations offered log drying facilities and sawmills for the local timber trade.

The combined revenues experienced a marked increase once the Otos-Alhóndiga and Aceca commanderies were incorporated into the Real Bosque de Aranjuez, from 2.2–2.7 million *maravedíes* in 1540–1543 to 3.3 million in 1561, as recorded in the accounts submitted by the steward, Melchor de Torres,[18] who was also a *jurado* in Toledo.[19] The income from woodland resources, which was almost negligible in previous accounts, had risen to 8% of the total by 1561.

It can be deduced from the prices paid that each tamarisk cartload contained 10.2 loads, and each orache cartload 8.5 loads. These equivalences are applied to rationalise the available figures as cartloads.

Timber sales were limited to 36 poplars from San Remondo, sold at auction to an Alonso Pérez for 67,000 *maravedíes*. The largest sum came from logging in Soto de la Madre Vieja in Otos, which yielded 189,770 *maravedíes* under a direct management system whereby woodcutters were paid a wage and the firewood sold out of carts in the neighbouring villages, prior announcement by a crier.[20]

Thanks to a report dated 1561 we are able to examine the administration of these former commanderies. Woodlands were managed by overseers (*vaderos* or *veedores*), who monitored firewood extraction and handled sales. The post holders at the time were two Borox men; Miguel Rodríguez was in charge of the northern side of the *dehesa*, and Pedro Galán of the southern, known as the Castillejo or Yepes *dehesa*. They earned a daily wage of three reals, although Galán's was cut to 2.5 reals in the last few days of the season because of the reduced workload. This tells us that the Crown was always happy to rely on the services of people who were knowledgeable about the local environment.

The report lists the species that grew in the riverside groves – mainly tamarisk and white poplar, plus a few cartloads of orache which were collected in

18 Archivo General de Palacio (AGP), Administración General (AG), caja 165, exp. 4, Aranjuez.

19 *Jurados* were representatives of the local population in some Castilian municipal councils. Their role was to oversee the council's performance, including economic matters such as procurement, which is the reason why *jurados* from Toledo were chosen to manage these former commanderies. Francisco José Aranda Pérez, *Poder municipal y cabildo de jurados en Toledo en la edad moderna: siglos XV–XVIII* (Toledo: Concejalía del Área de Cultura, 1992); Francisco José Aranda Pérez, *Poder y poderes en la ciudad de Toledo: gobierno, sociedad y oligarquías urbanas en la Edad Moderna* (Cuenca: Universidad de Castilla-La Mancha, 1999), 103–31.

20 The total figure appears as 189,861 *maravedíes* in the ledgers due to minor calculation errors.

TABLE 4.1 Firewood obtained from Madre Vieja in Otos in 1561 (in *maravedíes*)

Weeks	Tamarisk		Orache	Income	Expenditure				Net profit	% costs/sales
	Borox	Yepes	B+Y	Sales	Wood-cutters	Over-seers	Other costs	Total		
21 Jul–27 Jul	105.7	60.5	5.1	17,124	2,876	1,428	369	4,673	12,452	27.3
28 Jul–3 Aug	105.7	60.5	5.1	17,124	2,876	1,428	369	4,673	12,452	27.3
4–10 Aug	192.2	84.5		28,218	4,903	1,428	272	6,603	21,615	23.4
11–17 Aug	275.1	69.2		35,118	5,939	1,428		7,367	27,751	21.0
18–24 Aug	275.1	69.2		35,118	5,939	1,428		7,367	27,751	21.0
25–31 Aug	197.5	48.9		25,124	6,170	1,428		7,598	17,527	30.2
1–7 Sep	197.5	48.9		25,124	6,170	1,428		7,598	17,527	30.2
8–17 Sep	124.6	32.0		15,980	3,936	1,428		5,364	10,617	33.6
15–21 Sep	124.6	32.0		15,980	3,936	1,428		5,364	10,617	33.6
22–28 Sep	56.1	17.9		7,546	1,856	1,428		3,284	4,262	43.5
28 Sep–4 Oct	27.2	52.8		8,158	1,957	1,428	68	3,453	4,705	42.3
5–11 Oct	27.2	52.8		8,158	1,957	1,428		3,385	4,773	41.5
12–18 Oct		52.8		5,385	1,293	714		2,007	3,378	37.3

TABLE 4.1 Firewood obtained from Madre Vieja in Otos in 1561 (in *maravedíes*) (*cont.*)

Weeks	Tamarisk		Orache	Income	Expenditure				Net profit	% costs/sales
	Borox	Yepes	B+Y	Sales	Wood-cutters	Over-seers	Other costs	Total		
19–25 Oct		48.2		4,916	1,152	714		1,866	3,050	38.0
26 Oct–1 Nov		48.2		4,916	1,152	714		1,866	3,050	38.0
2–8 Nov		48.2		4,916	1,152	714		1,866	3,050	38.0
9–15 Nov		48.2		4,916	1,152	714		1,866	3,050	38.0
16–22 Nov		48.2		4,916	1,152	714		1,866	3,050	38.0
23–29 Nov		7.1		720	144	893		1,037	–317	–144.0
30 Nov–7 Dec		7.1		720	144	893	272	1,309	–589	–181.7
	1,708.5	936.9	10.2	270,178	55,854	23,205	1,349	80,408	189,770	29.8
Total cartloads:		2,655.6	460 kg each at 40 *maravedíes*: 1,221,570.6 kg. Net profit 5,581 reals.							

SOURCE: AGP, AG, CAJA 165, EXP. 4

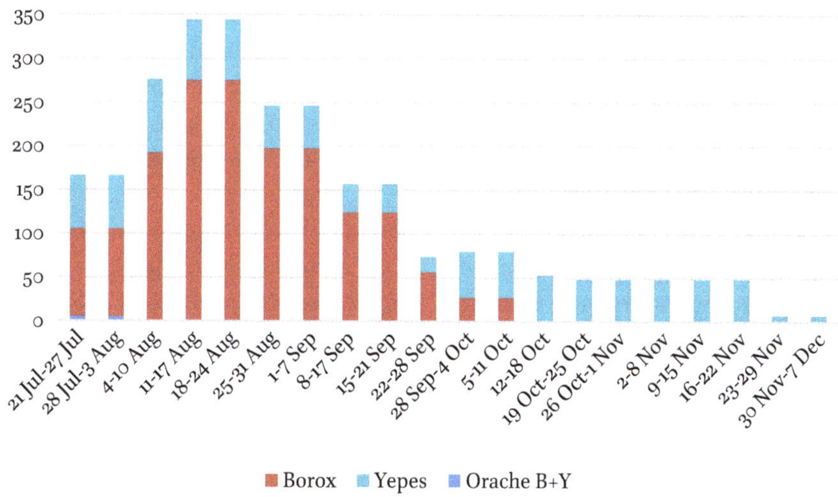

CHART 4.1 Firewood obtained from Madre Vieja in Otos in 1561 (cartloads)
SOURCE: AGP, AG, CAJA 165, EXP. 4

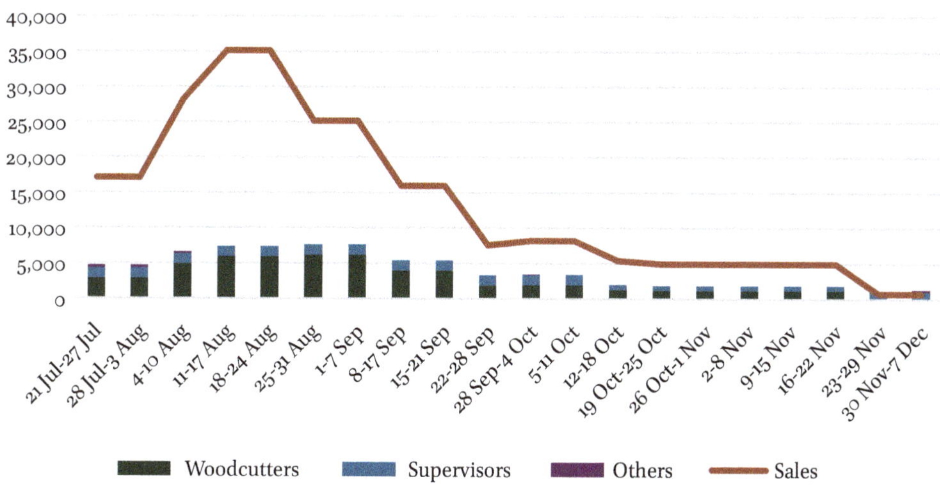

CHART 4.2 Firewood sold at made Vieja in Otos in 1561 (in *maravedies*)
SOURCE: AGP, AG, CAJA 165, EXP. 4

the first few weeks of the season. The logging campaign began on 20th July and ended in December, at a rate of 24 cartloads per day in July and early August, rising to 49 later in the month, 22 by mid-September, 11 in October, and 7 in the following weeks until the end of November. The work was at its most intense between July and September in Borox, and in October and November in the southern *dehesas*. Because of the Tagus water cycle, logging operations had to be performed in the summer and early autumn, when the water retreated from the banks after the spring floods. Since the river at this point meanders back and forth across a shallow valley, its banks usually flooded on a yearly basis. Riverside vegetation thrived thanks to this natural input, which is now artificially controlled by the large dams built in the twentieth century in the upper reaches of the Tagus.

Specific data contained in the 1561 report allows us to appraise the financial significance of these operations. A cartload of tamarisk was priced at three reals, showing a sixfold rise since 1535. Large loads were paid at half a real each and small ones at ten *maravedíes*.[21] A cartload of orache cost one real, and a load – as much as a man can carry on his back – four *maravedíes*. Woodcutters were paid half a real per cartload of tamarisk, and two *maravedíes* per load.[22] Carts in the region carried an average of 460 kg (10 *quintales* or 40 *arrobas*).[23] If one *arroba* is equivalent to 11.5 kg, then the 2,655.6 cartloads recorded in the accounts would have totalled 1,221,570.6 kg. The sale of this firewood, after a deduction of 29.8% for expenses, yielded a profit of 5,581 reals (see Table 4.1 and Charts 4.1 and 4.2). Revenues for woodland exploitation increased by a third in the years 1562 and 1563, followed by a fallow break in the remaining years for which financial data are available (Table 4.2).

TABLE 4.2 Yearly income from the sale of firewood and timber from Otos, Alhóndiga and Aceca (*maravedíes*)

1561	1562	1563	1568	1570	Average
256,681	409,817	404,513	0	0	214,238

SOURCE: AGP, AG, CAJAS 165, EXP. 4; 165, 6; 166, 2; 170, 5 AND 11

21 A "load" is understood to be a bundle of firewood that can be carried by a person or a small pack animal, which is far too vague to compare with precise measurement units. The difference between small and large loads may refer to a human vs. an animal carrier, but this is mere guesswork.

22 These 1561 prices are very similar to those at the *dehesa* of Zacatena on the banks of the rivers Guadiana and Cigüela, near Daimiel, in the middle years of the sixteenth century. Firewood from the holm oaks in the *dehesa* was sold at 3–4 reals a cartload, while smaller loads fetched between half a real (17 *maravedíes*) and 19 *maravedíes*. Francisco Fernández Izquierdo, "La extracción de leña y madera de la dehesa de Zacatena (Daimiel, Ciudad Real) en el siglo XVI y su situación a mediados del siglo XVIII," *Memoria y Civilización* 25 (2022): 70.

23 The carts that supplied stone for the repair of the weir serving the Aceca mills between 1545 and 1549 had a capacity of 40 *arrobas*. Archivo General de Simancas (AGS), Contaduría Mayor de Cuentas (CMC), 3ª época, leg. 2002, nº 7.

Bearing in mind the scheduled logging campaigns and fallow breaks to allow the trees to recover, the figures show the Crown's wish to protect its woodlands, both to ensure sustainable exploitation and for hunting purposes. After all, big-game hunting was the ultimate royal sport and practised in the king's estates, which required a suitable environment for the most highly valued animals and birds to thrive. On the other hand, rabbit hunting in Otos was leased out and delivered higher revenues than firewood – 405,000 *maravedíes* in 1561 and 244,000 in 1562. Fishing rights were also leased by auction, but the profits were far smaller.[24] Although riverside copses were exploited for timber, particularly white poplars, these plantations were not encouraged because the constant flow of mountain pines downstream was a better option both in terms of quality and profit.

3 Among Pastures and Water: Zacatena

As in the Tagus, the lands surrounding the upper Guadiana were annexed to Christian territories by donation to the military orders, in this case the Order of Calatrava, which undertook the task of defending and repopulating the area. The Order had been founded in the mid-twelfth century in the fortress of Qal'at Rabah, on the bank of the Guadiana, near the present-day town of Carrión de Calatrava. The knights extended their dominions from that base until they owned a large portion of the south-western half of today's province of Ciudad Real, later obtaining smaller concessions in the provinces of Jaén, Guadalajara, and Teruel.[25]

The *dehesa* of Zacatena lay a short distance from the fortress-town. In its early days, the Order of Calatrava managed its pasturelands directly, both in Zacatena and elsewhere, but in the fourteenth century, according to Almagro Vidal, the knights opted for a lease arrangement.[26] The main beneficiary of this move was the *Honrado Concejo de la Mesta* – the association of livestock breeders – whose extremely advantageous legal status led many of its "brethren" to

24 AGP, AG, cajas 165, 4; 165, 6, and 166, 2.

25 On the establishment of the Order of Calatrava in the region and its impact on the landscape, see Clara Almagro Vidal, *Paisajes medievales en el Campo de Calatrava* (Madrid: La Ergástula, 2016).

26 Clara Almagro Vidal, "La dehesa de Zacatena en la organización territorial de la Orden de Calatrava (siglos XIII–XV)," in *Montes, pastos y caza a la vera del Guadiana en las Tablas de Daimiel: la Real Dehesa de Zacatena en la Edad Moderna*, coords. Francisco Fernández Izquierdo and Francisco Javier Moreno Díaz del Campo (Granada: Comares, 2022), 69–80.

choose the *dehesas* belonging to the military orders as wintering grounds for their flocks.

Despite its considerable size of approximately ten thousand hectares, Zacatena was not a particularly important *dehesa* in the Campo de Calatrava region. About two thirds of its surface area was pastureland, divided into 24 or 25 *quintos*, with a total capacity to support 12,000–12,500 sheep.[27] While it was one of the most extensive estates in La Mancha, its location on the edge of the region meant colder – and therefore less desirable – pastures than, for instance, the large open fields in the valley of Alcudia. This had a negative effect on its profitability and, although prominent nonlocal *posesioneros* in Zacatena are certainly documented,[28] its main users were local farmers.[29]

On the other hand, Zacatena's range of natural resources made it very appealing not only to the local population but to the Crown. Due to its location at the confluence of two rivers, water was the main feature of the area, not only from a visual point of view but particularly in economic terms. The two watercourses, Guadiana and Cigüela, spilt their waters over the flatlands and joined the emerging groundwater, creating a singular ecosystem of scattered ponds. This peculiar landscape gave rise to two trades that were specific to this *dehesa*. The first was fishing, which was regulated by leasing arrangements until the last third of the sixteenth century, when hunting and fishing were banned altogether on Philip II's initiative, even beyond the limits of the estate.[30] Even more important was the trade generated by the watermills on the banks of the Guadiana which, although not the exclusive property of the

27 A *quinto* was a measure of pastureland where five hundred sheep could graze. Like *millares* (see note 12), as these plots were defined in terms of productive capacity, they had no clear-cut boundaries. Zacatena acquired a settled shape around the late sixteenth century, but it was not until the seventeenth that the toponyms begin to appear in a fixed form in the sources. However, an alternative measurement unit – the *cuarto* (quarter) – is also found in early documentation from the Early Modern period, which shows that the particulars of livestock rearing were far from established. Fernández Izquierdo, "La extracción de leña," 51.

28 Jerónimo López-Salazar Pérez, *Mesta, pastos y conflictos en el Campo de Calatrava* (Madrid: CSIC, 1987), 108; Jerónimo López-Salazar Pérez and Aitana García Navarro, "Los señores de ganado de la dehesa de Zacatena," in *Montes, pastos y caza a la vera del Guadiana en las Tablas de Daimiel: la Real Dehesa de Zacatena en la Edad Moderna*, coords. Francisco Fernández Izquierdo and Francisco Javier Moreno Díaz del Campo (Granada: Comares, 2022), 275.

29 In the Mesta's terminology, a *posesionero* was a livestock keeper who leased pastures.

30 AHN, OM, AT, exp. 39191.

Order nor subject to specific seigneurial rights, were a generous source of income for the Crown.[31]

Zacatena also accommodated other infrastructures, such as shelters or pens for livestock (*majadas*), lime kilns, fishermen's lodges, and a very few beehives.[32] But the most important facility in the *dehesa* was the so-called "guards' lodge", which is first mentioned in the 1560s, when Philip II himself is recorded as having stayed overnight, probably to enjoy the rich hunting opportunities it offered.[33] It is known that punitive measures were implemented at various times to prevent poaching and to protect the various species of game.[34]

Finally, the archives provide plenty of information on the exploitation of woodland resources. It was supported on two pillars: regulation and institutions. A curious body of ordinances governed every aspect of economic activity in the *dehesa* from the mid-fifteenth century onwards, with the exception of livestock, which was regulated by lease contracts. The first ordinances were dictated in 1450 by Master Pedro Girón and updated in 1567. Their normative scope included the management of fishing, hunting, firewood, and timber resources, regulating the circumstances and conditions under which these activities could be performed. Not surprisingly, they lay down a system of penalties to be issued by the head forester (*guarda mayor*),[35] although only the local judiciary, represented by the governor of Almagro or the mayor of Daimiel depending on the period, had the power to enforce such fines.

31 Over a dozen mills used to operate in the Zacatena area in the sixteenth century. Nine were located on the edges of the *dehesa* and five were totally or partially owned by three of the Order's institutions: the Daimiel commandery, the Sacro Convento, and the *mesa maestral*. Full listing in AHN, OM, AT, exps. 38285 (1433); 39448 and 40967 (1550–1552), and 38285 (1567).

32 Francisco Javier Moreno Díaz del Campo, Francisco Fernández Izquierdo, and Francisco Miguel Gómez Vozmediano, "Notas sobre los molinos de harina de la dehesa de Zacatena en la Edad Moderna,", in *I Simposio anual de Patrimonio Natural y Cultural ICOMOS España*, eds. Alfonso López-Menchero Bendicho and Victor Manuel Maldonado Zamora (Valencia: Editorial Universitat Politècnica de Valencia, 2020), 21.

33 Francisco Fernández Izquierdo, "La administración y vigilancia de la dehesa de Zacatena: su amojonamiento, las ordenanzas, el guarda mayor y la supervisión desde el Consejo de Órdenes Militares en el siglo XVI," in *Montes, pastos y caza a la vera del Guadiana en las Tablas de Daimiel: la Real Dehesa de Zacatena en la Edad Moderna*, coords. Francisco Fernández Izquierdo and Francisco Javier Moreno Díaz del Campo (Granada: Comares, 2022), 124.

34 A special order was issued in 1562 to protect the game birds on the banks of the Guadiana as a complement to the Zacatena ordinances. AHN, OM, libro 336, fol. 53 r.

35 Francisco Fernández Izquierdo, "La dehesa de Zacatena en el siglo XVI: aproximación a sus ordenanzas y guardería mayor", in *V Jornadas de Historia de Daimiel*, coord. Clemente Espinosa (Daimiel: Ayuntamiento de Daimiel, 2020), 173–215.

Traditionally, with some punctual exceptions, the position of head forester was linked to the office of commander of Daimiel but was sold off in the early seventeenth century, mainly due to the Crown's financial difficulties. Thereafter it was occupied by members of the aristocracy who usually resided at the court and used the title for purely decorative purposes, delegating the executive role to deputy foresters (*tenientes de guarda*). These local men effectively bought the management of the *dehesa*, paying considerable sums to the title-holder in order to gain access to the post.[36]

The issue of the management of the *dehesa* is very significant because several factors interacted to build an extremely complex scenario. First, the duality between proposed fines and imposed penalties, along with the conflicts of interest between foresters and judges. Second, the fossilisation of the penalties themselves. Since they were not updated with inflation, by the seventeenth and eighteenth centuries it had become increasingly cheap to break the laws governing conservation in the *dehesa*. The third factor was the deputy foresters' pressing need to recover as quickly as possible the sums disbursed to obtain the post. This led to a paradoxical situation where the foresters applied themselves vigorously to surveillance, with the subsequent issuing of fines, while indulging in profit-making activities in the *dehesa* which were not always compatible with the ordinances. The fourth and final factor was the corrective role of the Council for the Military Orders – and of the Crown by extension – which kept a fairly effective control over the exploitation of woodland resources in the *dehesa* at the behest of the local people and the local judiciary.

Besides the elms and white poplars that grew on the banks of the Guadiana and Cigüela, the tree population in Zacatena consisted chiefly of holm oak (*Quercus ilex*), with a scattering of other species of the oak family. Shrubs such as broom, kermes oak (*Quercus coccifera*), and cistus also thrived. As in Otos, other species such as tamarisk were present thanks to its regular flooding pattern, but they were considered of little value and are seldom mentioned in the sources. In contrast, holm oaks are frequently documented, particularly in relation to the acorns used as pig fodder.

Timber and firewood were the main two resources exploited Zacatena's woodlands, as reflected in the ordinances. Timber extraction was very limited and used almost exclusively for repairs in the local flour mills belonging to the Order. The trees targeted for this purpose were usually mature, robust specimens which, given the slow growth of the oak family, were likely to be

36 Francisco Javier Moreno Díaz del Campo, "La gestión de los recursos forestales en la dehesa de Zacatena durante el siglo XVII y el cruce de intereses entre la Corte y el campo," *Manuscrits. Revista d'Història Moderna* 42 (2020): 111–32.

of considerable age. The felling of large trees was allowed prior authorisation from the Council for the Military Orders, but disagreements did arise occasionally on account of the price and quality of the timber, as the millers had to pay for it themselves, or when it was suspected that the wood would be used for other purposes than those originally stated.[37]

Firewood collection which also created problems. It was managed through a licensing system whose beneficiaries could be individuals or collectives. The latter included convents and charities in the region, as well as certain commanderies and institutions belonging to the Order of Calatrava, most of which had prior written authorisation from the Council for the Military Orders. Individual licences were normally issued by the head forester to locals, including people who lived in the *dehesa* such as shepherds, millers, livestock owners, and even the head forester himself and his assistants.

The first documented concessions of firewood from Zacatena as royal gifts were granted by the Catholic Monarchs in the fifteenth century.[38] The reports are extremely vague and barely mention any coercitive measures, which leads us to assume that firewood was sufficiently abundant at the time not to be an issue, as restrictions were introduced in the late 1510s.[39] Nevertheless, and despite limitations on the free extraction of firewood and timber, the sources suggest that, in those early years, both the Crown, as the landowner, and the livestock keepers as lesees, were aware that its potential wealth was greater than current use of resources implied, and that its prospective growth might hinge on keeping the tree mass under control. Was Zacatena, as it were, a *dehesa* in development? Although livestock had grazed in Zacatena for centuries, as medieval sources attest, the fact is that there are numerous reports of the authorities' particular interest in these woodlands around the mid 1400s.[40] At all events, Zacatena was first and foremost a *dehesa*, and this meant that, regardless of its forestry resources and its hunting and fishing appeal, the

37 AHN, OM, AT, exp. 46566. This case is studied in detail in Francisco Javier Moreno Díaz del Campo, "Obras y reparos en el molino. La Orden de Calatrava y la gestión de sus infraestructuras industriales en la Edad Moderna," in *Pagesia, indústria i món rural. VIIIè Congrés sobre Sistemes agraris, organització social i poder local*, coord. Enric Vicedo Rius (Lleida: Institut d'Estudis Ilerdecens, 2016: 177–200.

38 AHN, OM, AT, exp. 39448. Further details in Fernández Izquierdo, "La extracción de leña," 52.

39 The right to unlimited firewood enjoyed by the treasurers who leased land in *mesas maestrales* was curbed in 1508 and regulated more strictly in the lease contract for the years 1511–15. AHN, OM, AT, exp. 35234 y AHN, OM, libro 323, fol. 28 r-v, Real Cédula, December 10 1511, Burgos.

40 For the conceptual difference between *monte* and *bosque*, see note 4.

primary source of revenues for the Crown was the lease of pastures, whose profitability might be curbed by an excessively dense tree layer and unchecked undergrowth. As the river flowed through the *dehesa*, water was also a crucial factor, but as far as is known, no attempt was made to drain the banks of the Guadiana and Cigüela until much later,[41] whereas woodland clearing operations began as early as the sixteenth century.[42] Proof of this are the tree fellings undertaken in 1542, when the head forester obtained permission from the Council for the Military Orders to "thin out" the *dehesa* by systematically removing trees. Attempts to encourage pasture expansion continued in 1551, also at the head forester's express request and, despite local opposition, areas that had previously been virtually inaccessible were opened to livestock, particularly in the northern sector of the *dehesa*. The case ended up in court and, as part of the proceedings, the Council instructed the Order of Calatrava's attorney, Frey Francisco Ortiz, to carry out a visual inspection of the *dehesa* and take witness statements. His findings showed that not only the tree layer had regenerated considerably in a few short years, but the amount of pasture had also increased.

The sources appear to introduce at this time a terminological change which may have been deliberate and, if confirmed, may well be linked to the new state of affairs. References to plots allocated to pasture had been referred to as *cuartos* (quarters, as they were four in number) until at least 1527.[43] By the 1550s, around the time of the clearing and weeding operations, they began to be known as *quintos* (resulting in 23.5–25 portions at different times). Was the timing coincidental? Were the terms misleading? At any rate, this is a working hypothesis that should be assessed against the background of the authorities' new-found appreciation of the territory's potential for livestock farming. To confirm the extent of this new attitude it would be useful to have access to the valuations of pastures in Zacatena attached to the lease contracts which were

---

41 The first projects to drain the flood plains of the *dehesa* were drawn up as late as the eighteenth century. It may have been technically impossible or simply unwanted, given the king's fondness of hunting and the scarce lobbying power of the local peasants. Alberto Celis Pozuelo, "Las Tablas de Daimiel entre 1751 y 1887. Las raíces históricas de su desecación," in *II Jornadas de Historia de Daimiel*, ed. Diego Clemente Espinosa (Daimiel: Ayuntamiento de Daimiel, 2013), 277–91; Francisco Javier Moreno Díaz del Campo, "El Guadiana y la Real Dehesa de Zacatena a finales del Antiguo Régimen. Proyectos de desecación, cambios ambientales y conflictividad institucional," *Estudios Geográficos* 8, no. 291 (2021): 1–17.

42 Fernández Izquierdo, "La extracción de leña," 114–16.

43 López-Salazar Pérez, *Mesta, pastos y conflictos*, 17; Fernández Izquierdo, "La extracción de leña," 175.

signed before the Consejo de Hacienda – the Crown's revenue office. However, the dehesas belonging to the *mesa maestral*, including Zacatena, were treated as a unit until 1593,[44] which unfortunately deprives us of a piece of information that could confirm whether the "thinning out" of the *dehesa* increased its potential as a resource for livestock rearing.

At any rate, there is no doubt that these logging and clearance operations marked a turning point in the management of the *dehesa*, since no further actions of a similar nature are documented after that. On the contrary, everything seems to indicate that surveillance was stepped up, chiefly as a result of updates to the ordinances, and that these updated rules were diligently applied, to say the least, in the final years of the sixteenth century. This was in marked contrast to the problems that arose in the wake of the institutional changes introduced in the seventeenth century outlined above.

Good examples of the new state of affairs can be found in the firewood extraction registers kept by Gonzalo de Oviedo Basallán during his time as head forester between 1578 and 1595. His detailed annual reports to the Council for the Military Orders are an invaluable tool that helps us evaluate the extent of firewood and timber exploitation in Zacatena. They tell us when and how firewood was removed from the *dehesa*, following a very different cycle than that at Otos and Aceca, with its lowest point in the summer.[45] The reason for this was probably that the potential market for this firewood was local and therefore did not require major transport operations or early stockpiling for winter. The sources do not mention the existence of charcoal storage or production, but only firewood in general, although some of it may have been used for that purpose. In terms of quantity, between 1579 and 1590 the *dehesa* supplied the local population with an average of 1,160 cartloads of firewood per year, or some 667,000 kg (approximately 111 kg per hectare). These figures show that firewood collection was even more modest here than in the *dehesas* on the banks of the Tagus.[46]

---

44 Jerónimo López-Salazar Pérez, "Un importante conflicto entre la Corona y los ganaderos mesteños: la medición del Valle de Alcudia de 1590", *Estudios geográficos* 44, no. 172 (1983): 395–434.

45 Francisco Fernández Izquierdo and Francisco Javier Moreno Díaz del Campo, "Leña y madera de Zacatena, ¿una explotación sostenible en la época moderna?: los guardas mayores en los siglos XVII y XVIII," in *Montes, pastos y caza a la vera del Guadiana en las Tablas de Daimiel: la Real Dehesa de Zacatena en la Edad Moderna*, coords. Francisco Fernández Izquierdo and Francisco Javier Moreno Díaz del Campo (Granada: Comares, 2022): 193–94. See Chart 4.1 for a comparison.

46 Detailed figures in Fernández Izquierdo, "La administración y vigilancia," 138–43; Fernández Izquierdo and Moreno Díaz del Campo, "Leña y madera de Zacatena," 195; Fernández

Looking at these facts as a whole, it can be seen that vigilance became intense and strict in Zacatena as a result of the new ordinances for the management of the *dehesa* implemented in the last quarter of the sixteenth century.[47] But after the demise of the model head forester Gonzalo de Oviedo Basallán, he was succeeded by Antonio de Toledo, head huntsman of the recently crowned Philip III (r. 1598–1621), who delegated his duties to a deputy. His appointment was accompanied by permission to cut 20,000 cartloads of firewood, causing the livestock farmers to protest. In the event, the firewood yielded only 9,891 reals[48] which, at the prices known for that time, would amount to the sale of 3,200 cartloads. The dates for these fellings coincide with the clearing activities in the *dehesas* of Extremadura which are explored in the next section.

4 Industrial Uses of Woodland Resources: Alcobaza

The Order of Santiago owned a vast territory named "*Provincia de León*", in the Extremadura region, where sixty *dehesas* have been documented as belonging to the *mesa maestral* and various commanderies.[49] The *dehesa* of Alcobaza, located in Jerez de los Caballeros, in the province of Badajoz, was part of this *mesa maestral*. Its 2,972 hectares had been leased to the Mesta's sheep farmers since the Middle Ages. Both the Order's capitular laws and the local ordinances established penalties aimed to protect the woodlands from soap makers, charcoal burners, and tanners who cut down cork oaks (*Quercus suber*), the most abundant species in the *dehesa*.[50] The case study we are about to present is at the heart of the struggle between land use and conservation, with special emphasis on the monetary value of a specific product of the cork oak: phloem (*casca*). The phloem is a layer of bark that covers the woody parts of the trunk and is protected by the outer bark which is harvested for cork.[51] This

Izquierdo, "La extracción de leña y madera," 73. For load and firewood yield calculations, see Fernández Izquierdo, "La extracción de leña y madera," 70–1. For a comparison with Otos and Aceca, see Charts 4.1 and 4.2.

47 Fernández Izquierdo, "La extracción de leña," 141–42.

48 AHN, OM, libro 202, fols. 188v–189v, marginal note dated 1603 recording the profit, which was shared among the officers of the queen's stables.

49 Daniel Rodríguez Blanco, "Ganados y señores en la Extremadura medieval," in *Trashumancia y cultura pastoril en Extremadura* (Mérida: Asamblea de Extremadura 1993), 78 ff.

50 Daniel Rodríguez Blanco, *La Orden de Santiago en Extremadura en la Baja Edad Media (siglos XIV y XV)* (Badajoz: Diputación Provincial de Badajoz, 1985), 249–52.

51 The exploitation of phloem from the cork oak is confirmed by María Pilar Jiménez Sancho in *Las regiones de procedencia de "Quercus suber" L. en España* (Madrid: ICONA, 1995), 21.

intermediate layer carries the sap to feed the boughs and nourish the tree and is very rich in tannins, which were employed to cure leather.[52] Its extraction is documented since the days when Extremadura and the Alentejo were part of al-Andalus, from the eighth to the thirteenth centuries.[53] It can be 2–3 centimetres thick in very old trees which have never had their bark removed. It was clearly a valuable product, since its removal results in the death of the tree from the loss of the channel that distributes the sap.[54]

Phloem was far more valuable than the parts of the tree that were cut for timber, firewood, charcoal, or ash – the latter being used to make bleach and soap. The people who harvested it had no interest in any other uses of the trees and left them standing, at risk of catching fire or being torn down by the wind. This was a well-known fact and offenders were prosecuted by the authorities, whose measures to curtail this harmful practice appear frequently in the sources.[55]

In Alcobaza, as in Zacatena, successive monarchs granted favours in the form of firewood from the *dehesas* reserved to *mesas maestrales*, once the lands belonging to the military orders were incorporated into the royal estates. In the case of Alcobaza, the beneficiaries of these gifts were men in the service of Emperor Charles v (r. 1520–56), as documented on at least two occasions

52 By immersing hides in water with powdered phloem, a natural, very resistant and flexible tanning was achieved. Lina Falcão and Maria Eduarda M. Araújo, "Vegetable Tannins Used in the Manufacture of Historic Leathers," *Molecules* 23, no. 5 (2018): 1081.

53 J. Esteban Hernández Bermejo et al., "Landscapes and Forest Flora of Al-Andalus: A Reconstruction from Textual Historical Documentation", *Mediterranean Botany* 40, no. 1 (2019): 73.

54 José Jordana y Morera, *Notas sobre los alcornocales y la industria corchera de la Argelia* (Madrid: Colegio Nacional de Sordomudos y Ciegos, 1884), 37–8, 87.

55 For example, a by-law (*Real Provisión*) was issued in Granada on 18 October 1501 whereby Cáceres city council was to require cobblers and tanners, who had permission to extract phloem, to leave half of it lengthwise on the tree to ensure it was not left completely bare and perished. AGS, RGS, leg. 1501.10, 207 (accessible online through PARES). The Mérida ordinances of 1537 also included a recommendation to respect a part of the trunk so as to preserve the trees. Rodríguez Blanco, *La Orden de Santiago*, 254. The 1489 ordinances for Valencia de Alcántara demanded that a third of the trunk be left untouched to ensure the survival of cork oaks, subject to a penalty of 600 *maravedíes*. The Galisteo ordinances of 1531 quote an identical fine for unlawful phloem extraction, a strictly regulated activity due to the damage it caused. Domingo Bohórquez Jiménez, *Ordenanzas del Concejo de Valencia de Alcántara* edición y estudio preliminar por Domingo Bohórquez Jiménez. Cáceres: Institución cultural El Brocense de la Diputación de Cáceres, 1982, 130–131. Julián Clemente Ramos, *Ordenanzas de Galisteo: 1531*. Cáceres: Universidad de Extremadura, Servicio de Publicaciones, 2001, 87. An extended edition of this work was published in 2016. Last two references supplied by courtesy of Professor Luis Vicente Clemente Quijada, Universidad de Chile.

during the sixteenth century. The first of these donations dates from 1545, when the emperor's physician, Doctor Moreno, informed him that many of the cork oaks in Alcobaza were barkless and dead, so that not only did they produce no acorns, but they were detrimental to the healthy trees and the pasture. He therefore asked for the gift of one thousand of those dead trees that stood in the *dehesa*. The administrators of the Council for the Military Orders proceeded with caution, opening an investigation on the viability of the proposal.[56] The petitioner's witnesses agreed on the potential benefits of this move, as Nuño Martín from Jerez declared:

> los dichos alcornoques secos y acernados es en daño y perjuicio del pasto y de los otros árboles questán sanos en la dicha dehesa, porque si un alcornoque seco se quema de los fuegos que dan en el término desta çibdad, aquel fuego haze daño a los questán cabe él, e quitándolos quedan holgados y desabahagados los otros árboles verdes.
>
> The said dead and bare cork oaks are damaging and harmful to the pasture and to the other trees in the said dehesa which are healthy, because if a dead cork tree burns whe there is a fire in the area around this city, that fire damages those that are near it and by removing them the other green trees are les crowded.

Diego Rodríguez Parrado, a tanner who knew the *dehesas* well, concurred, because if they were to be removed, the land and the woodland would be less crowded and the other trees would give more fruit and the sun would reach the soil for the grass to grow.[57] This assessment was confirmed by other witnesses, such as the butcher and tanner Diego Rodríguez de la Pandura, who added that not only were there many dead trees which had been stripped to the heartwood, but also some of the boles had had the phloem extracted or had fallen:

56 AHN, OM, AT, exp. 7965. The investigation was set in motion by Real Provisión 1.9.1545, Valladolid, addressed to the magistrate or resident judge in Jerez, near Badajoz, who was to look into the matter and hear witness statements, which were recorded by Licentiate Palomares. A copy of the instruction is in the archives of the Order of Santiago, AHN, OM, AT, exp. 78345. Reference supplied by courtesy of Pedro A. Porras Arboledas, Universidad Complutense de Madrid. This case is mentioned in Rodríguez Blanco, *La Orden de Santiago*, 255.

57 'porque quitándose se desavaharía la tierra y el monte, y dan más fruto los otros árboles y alcança el sol a dar en la tierra para creçer la yerva'.

> aunque se saquen de la dicha dehesa quatro ni çinco mill alcornoques, que no le haría daño, antes le sería muy mucho provecho para el pasto de la dicha dehesa e para el fruto de los otros árboles, pues darían fruto e ansí no lo dan unos con otros.
>
> Even if four or five thousand cork oaks are removed from the said *dehesa*, it will come to no harm but it would be very beneficial for the pastures in the said *dehesa* and for the fruit from the other trees because they would give fruit and as they are do not because of all the others.

The city of Jerez responded by putting forward its own witnesses, who held the opposite view to those who had testified before, accusing them of having vested interests in the operation, since Alcobaza was one of the best *dehesas* in the district and every effort should be made to ensure 'that no cork oaks whatsoever are cut in the said *dehesa* because, along with the dry and barkless ones, they will also cut and remove many of the others, which give much fruit and profit'.[58]

García de Arjona the Younger stated that he had known the *dehesa* for forty years and that the witnesses who had testified were phloem collectors, ashmen, cobblers, and tanners, who removed the phloem for themselves. He believed that many good cork oaks in Alcobaza would be damaged,

> como lo hazen en otras dehesas de particulares, questando sus dueños enzima de los caxqueros le[s] hurtan de noche otros árboles fuera de los que le tienen vendidos y los descaxcan, que lo mismo y mucho más se haría en la dicha dehesa de Alcobaça por tener muchos árboles muy buenos.
>
> as they are in other *dehesas* belonging to individuals where, since the owners keep a sharp eye on the phloem collectors, they steal other trees at night that haven't been sold to them and strip the phloem, and they would do the same and more in the said *dehesa* of Alcobaça because it has many and very good trees.

Other witnesses argued that the loss of healthy trees would result in less revenue for the *mesa maestral*, which was valued at 300,000 *maravedíes* a year.

---

58 'que no se corten alcornoques algunos en la dicha dehesa, porque a buelta de los secos y açernadados sacarían y cortarían muchos de los otros que dan fruto e renta'.

The local population was also against the royal physician's concession because it contravened their traditional right to cut down stripped cork oaks after two years, a practice that was beneficial both to themselves and to the land. If the people of Jerez were deprived of these dead trees, they would help themselves to live ones, which would be detrimental to the *dehesa* and to the king.

Doctor Moreno replied in defence of his own interests, and was followed by the head accountant of the Order of Santiago, who was asked how much the rent for Alcobaza was, since it would be necessary to compensate the *posesioneros* Diego López de Ribera, Gabriel Vázquez, and Mateo de Arévalo, who had instructed the Mesta's attorney Gómez de Carvajal in 1546 to stop all logging activities on the grounds that felling cork oaks would result in a reduced yield of acorns in a *dehesa* that specialised in pig and cow farming.[59]

The result of this lawsuit is unfortunately unknown because the archives are incomplete, but the episode obviously left an impression in the collective memory. Years later, in 1553, an application was made for a notarised copy of the documentation of the proceedings regarding the gift of trees from Alcobaza to Doctor Moreno and to the emperor's servant Bartolomé Costilla.[60] The request was prompted by the Council for the Military Orders' instruction to Francisco de Luzón, governor of the León estate, to select any trees that could be cleared from the *dehesa*, as they were to be a gift from the then prince Philip – the future king Philip II (r. 1556–98) – to Juan Vázquez de Molina, royal secretary and a member of the State Council, and Ruy Gómez de Silva, squire of the body and a close friend of the prince's. By the terms of the concession, the beneficiaries would acquire 19,669 trees, to be divided equally between them. In this case, however, the gift consisted of live, uncut trees instead of dead ones to avoid any potential conflict with the rights of the local population. The governor appointed Hernando Guillén and Francisco Blanco to select the trees to be culled and, having been sworn in, they and six others made their way to the *dehesa* on 4 April. They marked the expendable trees – both cork and other species – scattered across the nineteen sections (*trazos*) that made up the territory. Each section was identified by a number and its boundaries (Table 4.3).

Vázquez de Molina and Gómez de Silva delegated the task of felling and selling the trees to Domingo Idiáquez, a servant of the former. An auction was

59 The answer to this question is not given by the source, but the *dehesa* is known to have been leased for 480,000 *maravedíes* a few years later. In 1600–03 the rent was 814,200 *maravedíes* per year, accommodating 878 cows and 3,000 pigs. Manuela Caballer Navarro and Julio Fernández Nieva, "Las Órdenes Militares en la Extremadura moderna," *Revista de estudios extremeños* 38, no. 1 (1982): 26.

60 AHN, OM, AT, exp. 53009.

TABLE 4.3 Cork oak trees selected for felling in Alcobaza in 1553

Section	Cork oaks	Section	Cork oaks
1	1,065	11	456
2	1,041	12	1,488
3	1,025	13	1,522
4	928	14	1,061
5	357	15	930
6	1,073	16	1,275
7	1,642	17	525
8	799	18	1,146
9	1,403	19	1,075
10	858	Total	19,669

SOURCE: AHN, OM, AT, EXP. 53,009

held on 10 May 1553 and won by a group of citizens from Jerez,[61] who signed a contract on the 26th for the sum of 13,400 *ducados*, payable in four instalments due on St Michael's day (29 September) every year from 1553 to 1556. Should an unmarked specimen be cut down, they would be liable to pay 6,000 *maravedíes* which would be shared equally between the estate overseer, Hernando Guillén, who had taken part in the selection, and his Majesty's Chamber. They were allowed to extract phloem, timber, cork, and ash, as long as the latter was produced in kilns where it would not cause damage.

At that price, each tree was valued at 255.5 *maravedíes* or 7.5 reals, double the 3–4 reals paid per 50 *arroba* (575 kg) cartload of firewood in Otos-Aceca and Zacatena at the time. The girth of these cork oaks was such that each may have filled two cartloads, which explains the high price. By contrast, while the timber – very hard and difficult to work with – and certain products (cork, phloem, and ash) were considered valuable, other parts of the tree were

61 AHN, OM, AT, exp. 53009. The successful bidders were Francisco Moriano and his wife Isabel Vázquez; Lorenzo de Senabria, son of Lope de Senabria; Juan de Senabria and his wife María de Selis; Arias Salaver Senabria and his wife Juana de la Costa; Juan de Vargas Sotomayor and his wife Ana de Silva; Rodrigo Sirgado and his wife Francisca de Céspedes; Mayor Mejía, mother of Francisco Moriano; Juan de Losada and his wife Costanza de Losada; Francisco Moriano on behalf of his brother Gonzalo Mejía, resident in Nogales, Condado de Feria.

discarded, such as the leaves and twigs which could have been used for animal feed and firewood.

Jerez city council reasserted its objection to the felling of cork oaks as soon as the contract had been signed. In a complaint to the Council for the Military Orders, they alleged that the successful bidders were overstepping the mark and cutting four trees for each one included in the contract. At that rate, the 480,000 *maravedíes* rent for Alcobaza would dwindle to 100,000.[62] There was also a territorial issue at stake. As the city's population had no access to common woodlands, they resorted to gathering firewood in private land. The council's grievance was that since time immemorial it had been in its power to issue licences for cutting cork oaks within its jurisdiction and to impose serious penalties for unauthorised felling.

Once the culling had begun, it was reported in August that the phloem had been stripped off a singularly mighty cork tree which stood downhill from the Tapias fountain, down the stream, in a glade on the right-hand side next to a knoll, which was very thick and large according to witness Pedro Monago:

> que tiene este testigo por cierto que otro tal como él no ay en toda Alcobaça, e muy sano, sin ninguna toquera, ni carcoma, ni cosa que le hiciese daño, e que el pie tiene de grosura en tocón cuatro brazas medidas por Bartolomé García de Gata, e que oyó decir al dicho Hernando Guillén y a Francisco Blanco que aquel árbol no se había señalado para lo cortar, cuando los otros los señalaron, e que el dicho Hernando Guillén le dijo que había requerido a Lorenzo de Sanabria y a sus compañeros que no lo cortasen aquel árbol porque no era de los señalados para se cortar, e que el dicho alcornoque está descajcado, e cortado los ramos todos e el pie, salvo un ramo.

> that this witness is certain that there is no tree like it in all Alcobaza and [was] very healthy, without cavities, or woodworm, or any harmful thing, and its girth at the base was four brazas[63] as measured by Bartolomé García de Gata, and that he heard the said Hernando Guillén and Francisco Blanco say that the tree had not been marked for cutting when the others were marked, and that the said Hernando Guillén told him that he had asked Lorenzo de Sanabria and his mates not to cut that tree because it was not among those marked for cutting and that the said cork

62 AHN, OM, AT, exp. 53009, years 1553–1556.

63 1 *braza* = 2 Castilian yards or 1.6718 metres, equivalent to an English fathom. A girth of 6 *brazas* would be a 6.69 metre circumference.

> tree has been stripped bare of phloem [including] the bole and all the branches but one.

The complaint succeeded and the contractors, Rodrigo Sirgado and Lorenzo de Senabria, were jailed by Deputy Magistrate Buendía for what today would be classified as an environmental offence. The accused unsuccessfully argued that the tree had been marked for felling by Governor Luzón, and subsequently appealed to the Council for the Military Orders, where the court case continued in July 1554 with new witness statements.[64] The judge delivered his ruling on 18 August, sentencing the men to pay the 6,000 *maravedíes* stipulated in the contract for such offences, plus the money obtained from the sale of the phloem. Sirgado and Senabria appealed this decision as well.

Parallel to these proceedings, the Concejo de la Mesta filed a suit in August 1553 on behalf of the farmers who had the lease on the *dehesa*. On being interrogated, the witnesses specified the procedures followed in the selection of trees.[65] Vázquez and Silva replied that the marking of trees for felling was correct, and that the leaseholders' contract only included grass and acorns. Besides, the chosen trees were "the useless ones because they are dead and very old and very harmful to the said *dehesa*", and did not produce the fruit the opposing party claimed would be lost if they were cut down, whereas clearing the woodland would encourage pasture and increase profit. The Council for the Military Orders heard the farmers' protest and halted the logging while information was received from all parties. The first local hearing ratified the prince's donation on the grounds that the trees had been correctly marked.[66] As matters stood, the mayor of Jerez, Cristóbal de la Cueva, was to execute the culling, leaving no more than ten paces between trees and respecting unmarked specimens at all times. Juan Vázquez and Rui Gómez were authorised to appoint a person to supervise the operation, as were the farmers. Additionally, if the farmers should sustain any damage from the loss of acorns as a result of the cull, Vázquez and Silva would be liable for any compensation due.

The farmers took their appeal one step further, with new statements recorded in December 1554. The witnesses who testified for Vázquez and Silva insisted that there should be a distance of more than ten feet between the cork oaks left standing or, in other words, that no harm would result if more trees were removed. They claimed that this had been done by the municipality of

---

64 AHN, OM, AT, exp. 9237.

65 AHN, OM, AT, exp. 14325.

66 The sentence did not include costs and was signed by Drs Goñi and Ovando, and Licentiate Argüello in Valladolid on 18 September 1554.

Cabeza la Vaca in one of its *dehesas* and that the same policy had been applied in the town of Jerez itself and across the whole region,[67] leaving around fifteen or sixteen paces between trees. If trees were cleared at ten paces in Alcobaza, as the gentlemen of the Council for the Military Orders had determined, it would still be very dense, "because it is a very thick woodland and, should it be cleared to a distance of at least twelve paces, it would prosper and revenues would increase henceforth". The witnesses explained the benefits of removing some of the trees, as stated by Pedro Periáñez, a servant of Juan de Silva, who had cleared his master's *dehesa* in Valcabado y Ciervas, leaving a distance of fifteen paces between trees to obtain more pasture and acorns, resulting in increased profits. Similar views were expressed by Pedro Puertocarrero and María de León, lady of La Pulgosa.

Many different opinions were aired in the process. Some witnesses considered the ten paces established in the sentence more appropriate, while others thought thirteen or fourteen would be better. Gutierre de Acosta, who had owned an adjacent *dehesa* for thirty years, had seen that the trees marked for removal in Alcobaza were less than five paces from one another, because the remaining ones, being further apart, would produce more fruit. Alonso Moreno added that the measurement used for thinning out the *dehesa* La Cabra, in the Villanueva jurisdiction, was "soga toledana".[68] Juan de Logroño, owner of Los Carros estate, explained that the size of each tree should be considered in the selection, giving larger ones more clearance, perhaps up to twenty paces. The witness statements were completed in January 1555.

Assuming that the size of Alcobaza at the time was roughly the same as its known measurements – 2,972 hectares – it may be possible to estimate the density of the tree layer on the basis of the distance in paces between trees suggested in the above comments. Although the source does not specify whether the measurements refer to military or geometrical paces, the former are more likely. In view of the results shown in Table 4.4, the royal gift appears to have amounted to a sensible cull whichever distance was chosen, since a

67 The witnesses mentioned other *dehesas* in the vicinity by the names of their owners or leaseholders: La Ramira and La Cabra, in Villanueva del Fresno. In Jerez, El Campillo, Valcabado and Ciervas; El Enzinal; Torre de Pedro Serrano; La Pulgosa; Heredad de la Margarita; Campo Cebadilla; Las Ciervas; El Moriano or Marianes; Buharda; Heredad de los Carros; Cuartillo de Chanca; Heredad de los Buenos; La Granjera; La Capilla; Corcovado; Pingallos.

68 *Soga toledana:* a measurement equal to 33 handbreadths, or 6.9 metres. Ignacio González Tascón (com.), *Felipe II. Los ingenios y las máquinas: ingeniería y obras públicas en la época de Felipe II. Real Jardín Botánico, CSIC, Pabellón Villanueva, 10 septiembre–10 noviembre 1998* (Madrid: Librería Solón, 1999), 401.

TABLE 4.4 Density of tree layer in the Alcobaza dehesa under square planting system, as per witness proposals for distance between trees

Distance between trees	Military pace[a]				Geometrical pace[b]			
	m	Trees per 100 m	Trees per hectare	Trees in 2,972.16 hectares	m	Trees per 100 m	Trees per hectare	Trees in 2,972.16 hectares
8 paces	5.92	16.9	285	848,064	11.14	9.0	81	239,326
Soga toledana	6.9	14.5	210	624,272				
10 paces	7.4	13.5	183	542,761	13.93	7.2	52	153,169
13 paces	9.62	10.4	108	321,160	18.11	5.5	30	90,632
14 paces	10.36	9.7	93	276,919	19.50	5.1	26	78,147
15 paces	11.1	9.0	81	241,227	20.90	4.8	23	68,075

a 1 military pace = 0.74 m or 2.5 feet, RAE, *Diccionario de autoridades*, Tomo V (1737). In Valencia and Murcia, paces and *varas* (yards) could be equivalent (1 Castilian yard = 3 feet = 0.8359 m). Felipa Sánchez Salazar, "Medidas de superficie tradicionales y su equivalencia con el sistema métrico decimal," *Agricultura y sociedad* nº 49 (1988), 467–81. p. 476.

b One geometrical pace = 1,393 metres. *Diccionario de autoridades, ibidem*, and Lorenzo Blanco Nieto, María del Carmen Cruz Cancho, Ricardo Luengo González, and Vicente Mellado Jiménez, «Estudio de pesas y medidas tradicionales en Extremadura», *Campo abierto: Revista de educación*, nº 2 (1983): 29–52, p. 33.

SOURCE FOR THE DIFFERENT DISTANCES BETWEEN CORK OAKS IN PACES: AHN, OM, AT, EXP. 14.325

considerable number of cork oaks would have remained standing at the usual density in *dehesas*. The felling of 19,669 trees leaving a ten pace clearance around remaining trees would imply the removal of 4% (military pace) or 13% (geometrical pace) of the whole tree cover, with a final density of 183 (military pace) or 52 (geometrical pace) trees per hectare.[69] It should be noted that this elementary calculation of tree density obviates various factors relating to physiography (altitude, orientation, gradient), climate, and soil, not forgetting the girth and species of the trees, etc. A lower density would probably result if these variables were considered.[70]

The judges finally issued their next ruling on 16 July 1555, ordering a commissioner from the Council for the Military Orders to deal with the matter on site because the litigation was becoming a source of concern, given the proximity of Vázquez and Silva to Philip, king consort of England at the time. A minister from the Council for the Military Orders, Doctor Ovando, travelled to Jerez in March 1556 to restart the fellings but, due to the loss of markings in forest fires since the original selection, the trees had to be identified again.[71] Ovando ranged over Alcobaza between 11 and 20 March selecting specimens to be removed, leaving a distance of ten paces between remaining trees. He also preserved three shelters for cattle and one for pigs. Ovando was then informed that the phloem could only be extracted during the four summer months, when the sap was active. At that rate, it would take six seasons to remove the selected cork oaks, whereas the sentence stipulated that it was to be enforced within one year. On being informed of this circumstance, the Consejo authorised the removal of trees to continue until 1558 and even beyond.[72] Being unable to extend his stay further, Doctor Ovando delegated the supervision of the operation to Francisco Sánchez Cecioso before his departure from Alcobaza.

69 This figure is compatible with the usual tree layer density of 30–60 trees per hectare, as per Olea, L., and Alfonso San Miguel Ayanz. "The Spanish *Dehesa*. A Traditional Mediterranean Silvopastoral System Linking Production and Nature Conservation," *Sustainable Grassland Productivity: Proceedings of the 21st General Meeting of the European Grassland Federation, Badajoz, Spain, 3–6 April 2006*, ed. Jaume Lloveras i Vilamanya et al., 3–13. Madrid: Organizing Committee of the 21st General Meeting of the EGF, 2006. Costa, J. C., A. Martín, R. Fernández, and M. Estirado. *Dehesas de Andalucía: caracterización ambiental* (Sevilla: Consejería de Medio Ambiente. Junta de Andalucía, 2006), 32.

70 Montoya, J.M. *El árbol y el pasto. Hojas divulgadoras*, 2104 (Madrid: Ministerio de Agricultura, 1993). Junta de Extremadura. *Plan Forestal de Extremadura* (PFEx), Anejos 2008, 51–52. http://extremambiente.juntaex.es/files/2008/Ordenacion%20y%20Gestion%20Forestal/PlanForestal/PFE-Anejos.pdf.

71 AHN, OM, AT, exp. 14325.

72 *Real cédula* issued on 5 March 1558 extending the deadline for the removal of trees from Alcobaza. AHN, OM, libro 50, fols. 215v–216r.

No further controversies of this type are recorded in the following years, which suggests that the king stopped using the thinning out of the *dehesas* in the Order of Santiago's *mesa maestral* as currency to grant favours, given the controversy it aroused, but this does not imply that the woodlands were not cleared as and when required. Indeed, an unrelated incident occurred in February 1552, not long before the events discussed above, when the Order of Santiago's prosecutor accused the livestock farmers who leased the Order's *dehesas*, including Alcobaza, of breaking their lease contracts by ploughing fields without authorisation between 1541 and 1546. He demanded a sum between 600,000 and 750,000 *maravedíes* from their profits. The *posesioneros* – Diego López, Mateo de Arévalo, and Gabriel Vázquez – were fined 150,000 *maravedíes*, although this was reduced to the payment of costs on appeal.[73]

Finally, the removal of trees from Alcobaza for the recipients of Philip II's favour was disrupted by fire on at least two further occasions. The first, in August 1560, affected the districts of Sierra del Águila, Las Tapias, Navalguijo, Pizarrillo, and Cuarto de la Casa. The second broke out in the García Álvarez *dehesa*, adjacent to Alcobaza, in late September. Juan Zambrano, Juan Rodríguez de Alburquerque, and their associates were burning boughs to make ash, when fire got out of control, entering Alcobaza at the Tapias fountain. Zambrano was arrested but his companions escaped. In the ensuing trial, the prosecutor was Tirso de Granda, appointed by Alonso González de la Rúa, prosecutor for the Order of Santiago, on behalf of the Council for the Military Orders. The court sentenced the perpetrators to a fine of 20,000 *maravedíes*, one hundred lashes astride a donkey for public scorn, and six years' banishment. Pedro Puertocarrero, a resident of Jerez and owner of the *dehesa* where the blaze started, was fined 3,000 *ducados* and sentenced to six months' military service in Oran, Tunisia, at his own expense, including weapons, a horse and two servants, but he appealed. In the investigation of the damage, Martín Gómez and his associates Rodrigo Chacón and Diego Pena, both residing in Fregenal, claimed losses of 4,000 *ducados* in the fire, since they had bought the phloem and the cork oaks that were yet to be removed as per the licence granted by the king. This turned against them when, in the course of two inspections to ascertain the extent of the fire damage to the *dehesa*, a number of trees which were not among those selected by Doctor Ovando were seen to have been marked for felling, according to the statements of witnesses Alonso Díaz Verjano and Diego López Zaynos. Martín Gómez was jailed and sentenced to finance the marking of the trees to be preserved by a trustworthy person designated by the judge, Doctor

73 AHN, Diversos, Mesta, 28, n. 9. *Ejecutoria* dated 26 December 1556.

Mora. The appointed person was to declare the number of specimens chosen and the mark used to identify them so that it would be clear that they must be spared. Should any of them be cut, Martín Gómez and his associates would be fined 10,000 *maravedíes*, and should Gómez fail to comply, he would be fined 100 *ducados* payable to his Majesty's Chamber. Furthermore, the inspections found that two trees which had not been marked by Doctor Ovando had already been cut down and stripped of phloem, whereupon Gómez and his associates were fined 10 *ducados*. The judge was unable to ascertain whether all the trees felled in Alcobaza were among the authorised set, so he ordered his loyal appointee to be present at every future tree cutting. For the avoidance of doubt, he transferred the case to the judges of the Council for the Military Orders for review, but at the same time sentenced Martín Gómez and his associates to pay 100 *ducados* to his Majesty's Chamber, plus the costs incurred by the prosecutor Tirso de Granda, including his court expenses and salary; by the overseers who performed the two inspections of the *dehesa*; and by the judge himself. Further enquiries were conducted in May 1561 and the proceedings were still at the appeal stage in the Council for the Military Orders by 1565.[74]

On the basis of these documents, it seems clear that a general woodland clearing operation took place in the region's *dehesas* in the mid-sixteenth century. Its aim was to extend the pastures and to clear fields for ploughing and growing crops, as well as to obtain profit from the cork oaks. While it was necessary to clear the woodlands, at the same time they needed to be protected from potential excesses that might put their ongoing sustainable exploitation at risk. The legal proceedings launched by the Council for the Military Orders prove that the conservation of woodland assets was a matter of interest which led to the prosecution of people who caused forest fires or unlawfully cut trees.

In the following decades, be it as a result of protective measures or negligence, the woodlands became badly overgrown. King Philip II was consulted in April 1597 about the *dehesas* located in the municipality of Jerez, Alcobaza, la Bóveda, el Potroso, el Prado de Broza, el Rincón and dehesa del Rey. Their income had decreased greatly because the vegetation had become so thick that it interfered with the pasture and prevented the cork oaks and holm oaks from producing enough acorns for the sow cattle. The technician Juan Domingo Quirico and the governor of Jerez, accompanied by experts, recommended clearing the forest, which would make grass grow for sheep and acorns for pigs. 80,000 *ducados* could be obtained from the sale of timber, phloem and firewood, and the income from the pastures would increase by 6,000 *ducados* per

74 AHN, OM, AT, exp. 20,485.

year. Although the king died in September 1598 without resolving this matter, in 1602 the Council for Military Orders calculated the specimens that had to be culled from the *dehesas* of the Order of Santiago for greater usefulness and use. A total of 66,278 cork oaks were selected for removal, distributed among various *dehesas* as follows: Rey (36,740), Alcobaza (27,736), La Bóveda (1,471), and El Potroso (331).[75] On April 10, 1604, the Treasury Council reconsidered his father's query to Philip III. The king responded with his own hand that what the council proposed be done, and that he be kept informed by sending a trusted person. The increase in income should be used to manufacture ships and navies.[76] These operations were closely monitored to prevent theft and malpractice, particularly in relation to the highly valuable phloem. Constant vigilance and prosecution of unauthorised cases of phloem extraction, with severe penalties for offenders, are mentioned frequently in the sources.[77]

5 Conclusions – Sustainability at the Ancien Règime Way: Common Uses, Profitability and Vigilance

The exploitation of firewood and timber in the *dehesas* belonging to the military orders in sixteenth century Spain was managed using different formulas aimed at fulfilling a variety of requirements, such as fundraising for building works and other operations in the royal estates, supplying the local population with firewood, and producing materials for industrial use. However, the management of woodland resources in areas associated with livestock farming went a step further, becoming one of the components of the complex dynamics involved in the goal of making those open spaces more profitable. And when all is said and done, profitability in a livestock setting is invariably linked to the number of animals that can be fed subject to the size and quality of the pastures.

Despite their differences, the *dehesas* of Alcobaza, Zacatena and Otos-Aceca are a perfect showcase of these circumstances because their management practices had very similar aims and followed similar procedures. Being under

75 AHN, OM, AT, exp. 53,781. This information is included in a 1611 dossier regarding a gift of 800 Alcobaza cork oaks to be shared between the convents of Nuestra Señora de Aguas Santas, the barefoot Franciscan friars in Jerez to build a roof over their church, and the nuns of Portaceli de Zarzosa, in the Salamanca diocese.

76 AGS, Consejo y Juntas de Hacienda, 441,16.

77 AHN, OM, AT, exp. 6,675. Juan Abril and Pedro Hernández Bermejo were fined and banished for stripping the phloem off a cork tree in Alcobaza without a licence in 1576. The density of other Santiago dehesas in the late sixteenth century is confirmed by Rodríguez Blanco in *La Orden de Santiago*, 255.

the Crown's control, their appeal resided in their ability to put their resources to work for the king. The actions taken by the Council for the Military Orders – the institution responsible for managing the potential of these spaces – and the Treasury Council clearly demonstrate the Hispanic Monarchy's interest in preserving their natural wealth. Proof of the Crown's appreciation of good management and rational exploitation of these resources is the dispatch of high-level magistrates from the capital to resolve legal disputes in Zacatena and Alcobaza.

Centuries of experience in the development of woodlands into *dehesas* and the management of riverside vegetation led to respectful exploitation practices in a fragile natural environment with slow growth rates. Nevertheless, it would be anachronistic to judge this rational behaviour by current conservation parameters. The ongoing existence of those resources was based on experience, and experience dictated that only the right exploitation and protection practices would ensure healthy profits. It was, so to speak, sustainability Ancien-Règime style.

Abbreviations

AGP: Archivo General de Palacio
  AG: Administración General
AGS: Archivo General de Simancas
  CJH: Consejo y Juntas de Hacienda
  CMC: Contaduría Mayor de Cuentas
AHN: Archivo Histórico Nacional
  OM: Órdenes Militares
  AT: Archivo de Toledo

Acknowledgements

This work is part of the research project *Representaciones y prácticas políticas en la Monarquía Hispánica desde la plena Edad Media a la Edad Moderna*, itself part of the consolidated research group *DeReHis* (UCLM) (Ref.: 2022-GRIN-34065) financed by FEDER funds.

Bibliography

Almagro Vidal, Clara. *Paisajes medievales en el Campo de Calatrava.* Madrid: La Ergástula, 2016.

Almagro Vidal, Clara. "La dehesa de Zacatena en la organización territorial de la Orden de Calatrava (siglos XIII–XV)". *Montes, pastos y caza a la vera del Guadiana en las Tablas de Daimiel: la Real Dehesa de Zacatena en la Edad Moderna*, coordinated by Francisco Fernández Izquierdo and Francisco Javier Moreno Díaz del Campo, 69–80. Granada: Comares, 2022.

Álvarez de Quindós y Baena, Juan Antonio. *Descripción histórica del Real Bosque y Casa de Aranjuez: dedicada al rey nuestro señor*. Madrid: Imprenta Real, 1804.

Aranda Doncel, Juan. "Los salarios de los trabajadores en Córdoba a mediados del siglo XVI." *Boletín de la Real Academia de Córdoba de Ciencias, Bellas Letras y Nobles Artes* 56, no. 108 (1985): 33–48.

Aranda Pérez, Francisco José. *Poder municipal y cabildo de jurados en Toledo en la edad moderna: siglos XV–XVIII*. Toledo: Concejalía del Área de Cultura, 1992.

Aranda Pérez, Francisco José. *Poder y poderes en la ciudad de Toledo: gobierno, sociedad y oligarquías urbanas en la Edad Moderna*. Cuenca: Ediciones de la Universidad de Castilla-La Mancha, 1999.

Ayala Martínez, Carlos. "La corona de Castilla y la incorporación de los maestrazgos." *Militarum Ordinum Analecta* 1 (1997): 259–290.

Caballer Navarro, Manuela and Julio Fernández Nieva. "Las Órdenes Militares en la Extremadura moderna." *Revista de estudios extremeños* 38, no. 1 (1982): 5–36.

Camarero Bullón, Concepción and Félix Labrador Arroyo (dirs.). *La extensión de la Corte: los Sitios Reales*. Madrid: Ediciones UAM, 2017.

Camarero Bullón, Concepción and Laura García Juan. "Geografía histórica de los espacios reales: Alóndiga, Aceca y Barciles, despoblados del rey en la vega del Tajo." *Estudios Geográficos* 79, no. 284 (2018): 209–35.

Celis Pozuelo, Alberto. "Las Tablas de Daimiel entre 1751 y 1887. Las raíces históricas de su desecación." In *II Jornadas de Historia de Daimiel*, edited by Diego Clemente Espinosa, 277–291. Daimiel: Ayuntamiento de Daimiel, 2013.

Martín Vicente, Ángel, Rocío Fernández Alés and María Estirado Oliet. *Dehesas de Andalucía: caracterización ambiental*. Sevilla: Junta de Andalucía, 2006.

Díaz Fernández, Antonio José. "Aceca, de castillo a palacio." *Anales Toledanos* 27 (1990): 81–96.

Dirección General del Instituto Geográfico y Estadístico, *Equivalencias entre las pesas y medidas usadas antiguamente en las diversas provincias de España y las legales del sistema métrico-decimal*. Madrid: Imprenta de la Dirección General del Instituto Geográfico y Estadístico, 1886.

Falcão, Luisa and Maria Eduarda M. Araújo. "Vegetable Tannins Used in the Manufacture of Historic Leathers." *Molecules* 23, no. 5 (2018): 1081. DOI: 10.3390/molecules23051081.

Fernández Izquierdo, Francisco. *La orden militar de Calatrava en el siglo XVI. Infraestructura institucional, sociología y prosopografía de sus caballeros*. Madrid: CSIC, 1992.

Fernández Izquierdo, Francisco. "Gestión privada y pública en las encomiendas de las órdenes militares. Aceca (1531–1543), Bolaños (1593) y Montanchuelos (1582–83)." *Palacios, plazas y patíbulos: la sociedad española moderna entre el cambio y las resistencias*, coordinated by James Amelang et al., 515–530. Valencia: Tirant Lo Blanch, 2018.

Fernández Izquierdo, Francisco. "La navegación de madera en el río Tajo con destino a los Reales Sitios en el siglo XVI." *Tiempos Modernos: Revista Electrónica de Historia Moderna* 9, no. 39 (2019): 283–320.

Fernández Izquierdo, Francisco "La dehesa de Zacatena en el siglo XVI: aproximación a sus ordenanzas y guardería mayor." In *V Jornadas de Historia de Daimiel*, edited by Diego Clemente Espinosa 173–215 (Daimiel: Ayuntamiento de Daimiel, 2020).

Fernández Izquierdo, Francisco. "Paisajes rurales del siglo XVI en tierras de la Alcarria." In *Paisajes de tierra y agua. Gentes y ecosistemas naturales en Castilla y La Mancha (siglos XVI–XVIII)*, coordinated by Francisco Javier Moreno Díaz del Campo and Miguel Fernando Gómez Vozmediano, 67–106. Toledo: Almud, 2022.

Fernández Izquierdo, Francisco. "La administración y vigilancia de la dehesa de Zacatena: su amojonamiento, las ordenanzas, el guarda mayor y la supervisión desde el Consejo de Órdenes en el siglo XVI." In *Montes, pastos y caza a la vera del Guadiana en las Tablas de Daimiel: la Real Dehesa de Zacatena en la Edad Moderna*, coordinated by Francisco Fernández Izquierdo and Francisco Javier Moreno Díaz del Campo, 107–148. Granada: Comares, 2022.

Fernández Izquierdo, Francisco. "La extracción de leña y madera de la dehesa de Zacatena (Daimiel, Ciudad Real) en el siglo XVI y su situación a mediados del siglo XVIII." *Memoria y Civilización* 25 (2022): 47–88. DOI: https://doi.org/10.15581/001.25.042.

Fernández Izquierdo, Francisco and Francisco Javier Moreno Díaz del Campo. "Leña y madera de Zacatena, ¿una explotación sostenible en la época moderna?: los guardas mayores en los siglos XVII y XVIII". In *Montes, pastos y caza a la vera del Guadiana en las Tablas de Daimiel: la Real Dehesa de Zacatena en la Edad Moderna*, coordinated by Francisco Fernández Izquierdo and Francisco Javier Moreno Díaz del Campo, 183–222. Granada: Comares, 2022.

González Tascón, Ignacio (com.). *Felipe II. Los ingenios y las máquinas: ingeniería y obras públicas en la época de Felipe II. Real Jardín Botánico, CSIC, Pabellón Villanueva, 10 septiembre-10 noviembre 1998*. Madrid: Librería Solón, 1999.

Hernández Bermejo, J. Esteban et al. "Landscapes and Forest Flora of Al-Andalus: A Reconstruction from Textual Historical Documentation." *Mediterranean Botany*, 40, no. 1 (2019): 71–80. DOI: 10.5209/MBOT.63045.

Jiménez Sancho, María Pilar. *Las regiones de procedencia de «Quercus suber L.» en España*. Madrid: ICONA, 1995.

Jordana y Morera, José *Notas sobre los alcornocales y la industria corchera de la Argelia*. Madrid: Colegio Nacional de Sordomudos y Ciegos, 1884.

Labrador Arroyo, Félix and Koldo Trápaga-Monchet. "La configuración del espacio y la explotación forestal de un enclave singular: el Real Sitio del Soto de Roma durante la dinastía Habsburgo." *Studia Histórica. Historia Moderna* 39 (2017): 293–327. DOI: https://doi.org/10.14201/shhmo2017392293327.

López-Salazar Pérez, Jerónimo. "Un importante conflicto entre la Corona y los ganaderos mesteños: la medición del Valle de Alcudia de 1590." *Estudios geográficos* 44, no. 172 (1983): 395–434.

López-Salazar Pérez, Jerónimo. *Mesta, pastos y conflictos en el Campo de Calatrava.* Madrid: CSIC, 1987.

López-Salazar Pérez, Jerónimo and Aitana García Navarro. "Los señores de ganado de la dehesa de Zacatena." In *Montes, pastos y caza a la vera del Guadiana en las Tablas de Daimiel: la Real Dehesa de Zacatena en la Edad Moderna*, coordinated by Francisco Fernández Izquierdo and Francisco Javier Moreno Díaz del Campo, 253–332. Granada: Comares, 2022.

Merlos Romero, Magdalena María. *Aranjuez y Felipe II: idea y forma de un real sitio.* Madrid: Dirección General de Patrimonio Cultural, 1998.

Moreno Díaz del Campo, Francisco Javier. "Obras y reparos en el molino. La Orden de Calatrava y la gestión de sus infraestructuras industriales en la Edad Moderna." In *Pagesia, indústria i món rural. VIIIè Congrés sobre Sistemes agraris, organització social i poder local*, coordinated by Enric Vicedo Rius, 177–200. Lleida: Institut Estudis Ilerdencs, 2016.

Moreno Díaz del Campo, Francisco Javier. "La gestión de los recursos forestales en la dehesa de Zacatena durante el siglo XVII y el cruce de intereses entre la Corte y el campo." *Manuscrits. Revista d'Història Moderna*, 42 (2020): 111–132. DOI: 10.5565/rev/manuscrits.298.

Moreno Díaz del Campo, Francisco Javier. "El Guadiana y la Real Dehesa de Zacatena a finales del Antiguo Régimen. Proyectos de desecación, cambios ambientales y conflictividad institucional." *Estudios Geográficos*, 8, no. 291 (2021) 1–17. DOI: https://doi.org/10.3989/estgeogr.202183.083.

Moreno Díaz del Campo, Francisco Javier, Francisco Fernández Izquierdo and Miguel Fernando Gómez Vozmediano, "Notas sobre los molinos de harina de la dehesa de Zacatena en la Edad Moderna." In *I Simposio anual de Patrimonio Natural y Cultural ICOMOS España*, edited by José Luis Lerma García, Victor Manuel López- Menchero Bendicho and Alfonso Maldonado Zamora, 19–26. Valencia: Editorial Universitat Politècnica de Valéncia, 2020.

Olea-Márquez de Prado, Leopoldo and Alfonso San Miguel-Ayanz. "The Spanish *Dehesa.* A Traditional Mediterranean Silvopastoral System Linking Production and Nature Conservation." In *Sustainable Grassland Productivity: Proceedings of the 21st General Meeting of the European Grassland Federation, Badajoz, Spain, 3–6 April 2006*, edited by Jaume Lloveras i Vilamanya, 3–13. Madrid: Organizing Committee of the 21st General Meeting of the EGF 2006.

Porras Arboledas, Pedro Andrés. *La Orden de Santiago en el siglo XV: la provincia de Castilla*. Madrid: Dykinson, 1997.

Postigo Castellanos, Elena. "Reyes y maestres. Incorporación y encomienda perpetuas de los maestrazgos de las órdenes monástico-militares ibéricas (1523–1587)." In *Palacios, plazas y patíbulos: la sociedad española moderna entre el cambio y las resistencias*, edited by James Amelang et al., 503–514. Valencia: Tirant Lo Blanch, 2018.

Rodríguez Blanco, Daniel. *La Orden de Santiago en Extremadura en la Baja Edad Media (siglos XIV y XV)*. Badajoz: Diputación Provincial de Badajoz, 1985.

Rodríguez Blanco, Daniel. "Ganados y señores en la Extremadura medieval." In *Trashumancia y cultura pastoril en Extremadura*, 69–88. Mérida: Asamblea de Extremadura, 1993.

Trapaga Monchet, Koldo and António Santos. "Forestry and timber supply in the royal forests of the Iberian Peninsula through the 16th century." *Skyllis. Journal for underwater Archaeology* 15, no. 1 (2016), 62–68.

Trapaga Monchet, Koldo. "No es madera para vasallos, sino del rey: las políticas forestales de los Habsburgo en Portugal." *Obradoiro de Historia Moderna* 28 (2019), 105–34.

Trapaga Monchet, Koldo. "Las políticas forestales en los reinos de Castilla y Portugal (siglos XV–XVIII)." In *Il bosco. Biodiversità, diritti e cultura del medioevo al notro tiempo*, edited by Alessandra Dattero, 85–103. Rome: Viella, 2022.

Rodríguez Picavea, E., *La formación del feudalismo en la meseta meridional castellana: los señoríos de la Orden de Calatrava en los siglos XII–XIII* (Madrid-México: 1994).

Suárez Fernández, Luis. "Cuando los maestrazgos se incorporan a la Corona." *Revista de Historia Militar*, extra 1 (2000), 223–230.

Wing, John. "Keeping Spain afloat: State Forestry and Imperial Defense in the Sixteenth Century." *Enviromental History* 17, no. 1 (2012), 116–145.

Wing, John T. *Roots of empire: State formation and the politics of timber Access in early modern Spain, 1556–1759*. Leiden: Brill, 2015.

CHAPTER 5

Fire Hazards, Forest Protection and Production in Portugal: the Pinewood of Leiria during the Habsburg Dynasty (1580–1640)

Koldo Trapaga-Monchet

1 Introduction

The pinewood of Leiria is by far the most emblematic Portuguese woodland. It is not only the pride of Portuguese foresters, it has also been regarded as part of Portuguese national identity.[1] The National Woodland of Leiria (*Mata Nacional de Leiria*) is in the municipality of Marinha Grande, and is currently recovering from a huge fire that destroyed 86% of its forest cover in October 2017. It encompasses over 11,027 hectares, stretching along the coast of the Atlantic Ocean, and *pinheiro-bravo* (*Pinus pinaster*) is the predominant tree species.[2]

Although some research has been conducted on the history of this pinewood,[3] Portuguese nationalism has led historians to produce biased accounts or overlook the subject of forestry practices in Leiria during the Habsburg dynasty because up to only a few decades ago this period had been considered a time of Spanish oppression.[4] In the past few decades, some scholars have provided new information on the history of the pinewood of Leiria during the late medieval and Early Modern ages.[5] Special mention should be made of

1 António Arala Pinto, *O Pinhal do Rei – Subsídios* (Alcobaça: Oficina de Oliveira Junior, 1938), vol. 1, 13; Bernhard E. Fernow, *History of forestry in Europe, the United States and other countries* (Toronto: Toronto University Press, 1911), 360–63.

2 Observatorio Técnico Independiente, *Recuperação da Mata Nacional de Leiria após os incêndios de outubro de 2017* (Lisbon: Assambleia da República, 2020).

3 António Arala Pinto's book is still the most thorough research on the history of the pinewoods of Leiria, Pinto, *O Pinhal*, 2 vols. For a brief state of the art see Koldo Trapaga-Monchet & Raúl Romero-Calcerrada, "Forest policies, administration, and management of the Leiria pinewood in Portugal (13th–18th centuries)," *Management & Organizational History* 17, no. 3–4 (2022): 138–40.

4 This narrative is very clear in Pinto's *O Pinhal*, 2 vols.

5 Nicole Devy-Vareta, "Para uma geografia histórica da Floresta portuguesa. As matas medievais e a «Coutada velha» do Rei," *Revista da Faculdade de Letras – Geografia* 1 (1985): 54, 60 and 68; Nicole Devy-Vareta, "Para uma geografia histórica da floresta portuguesa. Do Declínio das matas medievais à política florestal do Renascimento (séc. XV e XVI)," *Revista*

© KOLDO TRAPAGA-MONCHET, 2026 | DOI:10.1163/9789004745698_007

the most recent contributions of Cristina Joanaz de Melo on forestry practices in the pinewood in the second half of the eighteenth century,[6] and an essay by Trapaga-Monchet and Romero-Calcerrada giving a historical overview of the administration and management of the pinewood of Leiria from the thirteenth to the early nineteenth century.[7] Francisco Rego, in turn, has provided data on the pinewood from its origins, as well on the plantings and maps of the pinewood from the late nineteenth to late twentieth century.[8]

However, what is missing is a contribution that addresses in depth the history of the pinewood of Leiria during the Habsburg dynasty (1580–1640). This essay aims to demonstrate that this historical period was of great significance to the history of the pinewood. A natural hazard, the fire of 1613, not only gave rise to major inquiries from the Crown, but also had a long-lasting impact on the management and administration of the pinewood. The inquiries conducted by the monarchy left an archival corpus that has hitherto remained unknown and is the main documentary source used to unravel the dynamics of the royal pinewood during these decades.[9]

The first part of this contribution provides an overview of the royal pinewood of Leiria prior to 1580, paying special attention to the first ordinances issued by the Crown in 1524 to regulate the office of the main guard (*guarda-mor*) of the pinewood of Leiria, as well as the creation and evolution of the legally established royal forest in the physical pinewood (woodland) and the administrative staff recruited by the Portuguese Crown to enforce the legislation on the ground. The second part constitutes the backbone of this essay, as it addresses three different aspects of the history of the pinewood of Leiria from 1580 to 1640: 1) the conservation, forestry policies and material uses to which the pinewood was put from 1580 to 1612; 2) the fire of 1613, examined in a brief but separate section; and 3) the inquiries conducted after the huge fire and the alternatives sought by the Portuguese crown to supply the necessary

da Faculdade de Letras – Geografia 1 (1986): 20, 24–25, 27 and 33; Carlos Leite, "Os trabalhos de Reinaldo Oudinot em Leiria nos finais do século XVIII: um plano global de ordenamento hidráulico, agrícola e florestal" (Masters Thesis: University of Coimbra, 2016).

6 Cristina Joanaz de Melo, "Menos coutadas melhores pinhais: imperio, inundações, fisiocracia, guerra e especialização das matas reais em Portugal (1777–1824)," *Tiempos Modernos. Revista Electrónica de Historia Moderna* 9, no. 39 (2019): 456–87; Cristina Joanaz de Melo, "Guerra, impérios e Corte Joanina nas Coutadas de Caça: Alavancas de Regeneração Florestal em Portugal, em meados do século XVIII," *Manuscrits. Revista d'Història Moderna* 42 (2020): 199–218.

7 Trapaga-Monchet & Romero-Calcerrada, "Forest policies," 138–61.

8 Francisco Rego, *Florestas públicas* (Lisbon: Direcção Geral das Florestas, 2000), 38–42.

9 Arquivo Nacional de Torre Tombo (ANTT), Manuscritos da Livraria, 1113, doc. 16, fols. 198r–206r.

timber components of *Pinheiro-bravo* due to the inability of the pinewood of Leiria to carry on meeting these requirements.

2 The Pinewood of Leiria Prior to 1580

Scholars' studies have largely revolved around the origins of the pinewood of Leiria. Earlier nationalistic literature linked the creation of the pinewood of Leiria to King Dinis I (1279–1325) and his wife Queen Isabella of Portugal (1282–1325), who is thought to have introduced *Pinus pinea* seeds from the south of France. The planting of the pinewood was mainly conducted to stop sands from advancing inland.[10] That the origins of the pinewood are truly related to this romantic narrative is doubtful; it is more likely that the roots of the so-called 'King's Pinewood' (*Pinhal do rei*) lie in the administrative establishment of the pinewood of Leiria. In other words, during the twelfth and thirteenth centuries the Crown acted upon an existing ecosystem.[11]

At this point, it is important to stress the difference between the 'woodland' (pinewood) and the notion of 'royal forest' derived from the concept of forest. It is nothing new to state that during the Early Modern Age the term 'forest' referred to a legal concept: an area where the king and noblemen exercised their hunting rights.[12] A forest could encompass a much larger array of ecosystems than woodlands: from agricultural lands to wood-pasture, or mixed woodlands, and, to lesser extent, monoculture woodlands with agricultural and/or pasture lands. It could also include different types of land tenure, for whose administration the Crown could either rely on the local inhabitants, overseen by a royal minister, or directly manage those territories through royal officers who, at least for the pinewood of Leiria, were recruited from among the local inhabitants.[13]

In the case of Portugal, the term 'coutada' referred to a legal 'forest' and this category included several ecological or physical realities besides the woodlands.[14] As with England, it is not known when the Portuguese monarchy established the legal forest ('royal pinewood' as 'coutada') in the physical pinewood of Leiria. The account provided in this section is reconstructed from the

10 Pinto, *O Pinhal*, vol. 1, 57–69, and vol. 2, 414–15.

11 For a short state of the art on the administrative foundation and evolution of the pinewood of Leiria see Trapaga-Monchet and Romero-Calcerrada, "Forest policies," 146–48.

12 Paul Warde, "Fear of Wood Shortage and the Reality of the Woodland in Europe, *c.* 1450–1850," *History Workshop Journal* 62 (2006): 34.

13 Oliver Rakcham, *Woodlands* (United Kingdom: Harper Collins, 2006), 141–42.

14 Devy-Vareta, "Para uma geografia," 1985, 60.

historical sources retrieved from volumes 1 to 4 of Carlos Baeta Neves's *História florestal*,[15] the regulations on the *guarda-mor* of the pinewood of Leiria issued in 1524,[16] the historical references supplied by António Arala Pinto in his essay on the pinewood of Leiria, and other authors who are quoted from where necessary.

The Portuguese Crown had been harvesting timber for shipbuilding since at least the fourteenth century, if not earlier.[17] Information about a permanent administrative staff for the protection (rather than management) of the pinewood dates back at least to early fifteenth century.[18] From the very beginning, the Crown relied on local inhabitants to transport timber from the pinewood. In this connection, King Fernando I (1367–1383) granted privileges to the villagers of Pederneira in exchange for transporting timber from the pinewood.[19] Although we should not find this information surprising, it is important to emphasize the long-standing and strong connections between the Crown and the local inhabitants when it came not only to conserving and managing the pinewood, but also to harvesting and transporting its commodities. Sometimes this led to grievances from other authorities, such as the representatives of the city of Leiria at the Portuguese *Cortes* of 1455, who complained that the inhabitants of the municipality of Leiria were neglecting agriculture because it was more profitable to work on harvesting, sawing and transporting timber from the pinewood of Leiria to Lisbon, Santarém, Montemor-o-Velho and other parts of Portugal.[20]

According to Devy-Vareta, the pinewoods located in Leiria, Pederneira and Ribatejo were diminished by overcutting and fires.[21] The regulations governing the post of *guarda-mor* of the pinewood of Leiria, adopted in 1524, provide new insights on the history of the pinewood, as this document was hitherto unknown.[22] The Crown stated in these regulations that the pinewood had been managed according to the tradition or custom upheld and handed down by the junior wardens (*guardas pequenos*), as well as through a number of royal orders, as there was no legislative act regulating the duties and privileges of the office. This legislation did not deal with the management of the

15 Carlos Baeta Neves, *História Florestal, aquícola e cinegética. Colectânea de documentos existentes no Arquivo Nacional da Torre do Tombo* (Lisbon: Ministerio de Agricultura e Pesca, 1980–1982), vols. 1 to 4.

16 Biblioteca da Faculdade de Direito da Universidade de Lisboa (BFDUL), Ms. 2–12–6. I am grateful to prof. Isabel Graes for providing me with this document.

17 Pinto, *O Pinhal*, vol. 1, 129–36, and vol. 2, 439.

18 Pinto, *O Pinhal*, vol. 1, 115–16, 211–12; Neves, *História florestal*, vol. 2, 45–6.

19 Pinto, *O Pinhal*, vol. 1, 113.

20 Devy-Vareta, "Para uma geografia," 1986, 20.

21 Devy-Vareta, "Para uma geografia," 1986, 24.

22 BFDUL, Ms. 2–12–6.

pinewood; rather, it established the duties and privileges given to the administrative staff for the purpose of preserving the pinewood for the future. In this connection, Karl Appuhn has coined the term 'negative legislation' to explain the scope and contents of the Early Modern forestry legislation of the Republic of Venice. This legislation, which follows a Europe-wide pattern, was mainly grounded in an array of prohibitions on the uses of and access to forest resources. It did not include positive measures to encourage human-induced planting and conservation of the woodlands. The author concluded that negative forestry did not consider the needs of the local population.[23] However, it is very likely the reality on the ground differed greatly from the contents of the royal legislation.[24]

Likewise, the Portuguese Crown regarded forest regulations as one of the key strategies for ensuring the enforcement of its interests on the ground, namely the protection of the woodlands for material purposes. It is worth noting that the regulations of 1524 did not make any mention of the destruction of the pinewood. The administrative staff specified in the regulations were, therefore, viewed as another means of preserving the pinewood by enforcing the regulations issued by the Crown. This might explain why the scope of these regulations was restricted to the main guard of the pinewood and not extended to managing the pines and the pinewood themselves. The main guard of the pinewood was the head of the administrative staff, who were the king's eyes and hands on the ground. Although the regulations do not provide any insights on this matter, other historical sources suggest that the administrative chief would have hailed from a well-known and highly regarded family:[25] someone respected and even feared by the lower echelons of society and of sufficient social status not to be ignored by the upper echelons.[26] By 1524, the main guard of the pinewood was Jorge da Costa,[27] who would have belonged to the nobility as he was referred to by King João III (r. 1521–57) as 'my beloved and dear

23 Karl Appuhn, "Forests, Forestry, and State Power in Renaissance Venice," *The Journal of Modern History* 72, nº 4 (2000): 871–73.

24 On this matter, see Paolo Grossi, *Mitología jurídica de la Modernidad* (Madrid: Trotta, 2003); Paolo Grossi, *El orden jurídico medieval* (Madrid: Marcial Pons, 1992); António Manuel Hespanha, *As vésperas do Leviathan. Instituições e poder político. Portugal (séc. XVII)* (Lisbon: Pedro Ferreira, 1987), vol. 1.

25 In 1617, the Portuguese Crown founded the legal forest in the pinewood of Cabeção (Moura, Portugal). To ensure its protection the Crown proposed as main guard a local inhabitant that everybody in the surroundings of the pinewood feared. Arquivo Histórico Ultramarino (AHU), Conselho Ultramarino (CU), Reino, box 2, folder 16.

26 AHU, CU, Reino, box 2, folder 16.

27 He took up the position in August 1523 after replacing Gonçalo Rodriguez, his father-in-law, Neves, *História florestal*, vol. 5/1, 69–70.

[esteemed] cousin.'[28] This did not mean to say that he was a blood relative of the king; rather, as Manuel Rivero pointed out, it was a strategy used by Early Modern European monarchies to create a symbolic union or kinship between the king and the nobility.[29] That Jorge da Costa Silva was a knight of the 'Casa do Mestre' is further evidence of his noble status.

The first clause of the 1524 regulations clearly shows their intentions.[30] Nobody was permitted either to cut or to collect '*pão*' within the boundaries established by the outer firebreaks. '*Pão*,' meaning 'stick' or 'tree,' refers in this case to pines[31] that were tall and straight enough to be used to construct important ship components. It is therefore very likely they were left to grow for decades to ensure they met the future needs for imperial shipbuilding timber. *Pinheiro-bravo* is a fast-growing species,[32] taking around 70 to 80 years to produce sturdy timber for shipbuilding, after which point it begins to rot.[33] The second clause prohibited hunting game within the limits of the pinewood. This measure did not result from the perception of a conflict between hunting and shipbuilding. Hunters were blamed for starting fires that could trigger a wildfire with unpredictable consequences for the pinewood. This was especially true in the months from June to September, when rainfall decreases and the temperature rises, making it easier for fire to spread.[34] Moreover, the biology of the pinewood was an additional factor that contributed to the rapid spread of fire: *Pinus pinaster* and its management result in highly dense pinewoods, as can be seen in Figure 5.1.

---

28 'meu muito amado e presado primo', BFDUL, Ms. 2–12–6.

29 Manuel Rivero Rodríguez, *La edad de oro de los virreyes. El virreinato en la Monarquía Hispánica durante los siglos XVI y XVII* (Madrid: Akal, 2011), 135–38.

30 BFDUL, Ms. 2–12–6. The following lines are based on *ibid.*

31 The *pinheiro-bravo* (*Pinus pinaster*) was, and still, it is, the predominant tree species in Leiria pinewood.

32 Alenxadre Vaz Correia, Ângelo Carvalho Oliveira and António Fabião, "Biologia e ecologia do pinheiro-bravo," in *Pinhais e eucaliptais. A floresta cultivada*, ed. Joaquim Sande Silva (Lisbon: Fundação Luso-American para o Desenvolvimento, 2007), 18.

33 Pinto, *O Pinhal*, vol. 2, 259–61 and 267–69. Still in 1843, Francisco Maria Pereira da Silva and Caetano Batalha considered that the pines were reserved for civil and naval constructions of the highest quality: Francisco Maria Silva and Caetano Maria Batalha, *Memoria sobre o Pinhal Nacional de Leiria. Suas madeiras e productos rezinosos* (Lisbon: Imprensa Nacional, 1859), 16–8. The first edition came out in 1843.

34 Friderico Varnhagen identified the fire as one of the main enemies of the pinewood of Leiria and considered it important to control the underwood ('mato'), especially after the warmest months of the year, Friderico Varnhagen, *Manual de instrucções praticas sobre a Sementeira, cultura e corte dos pinheiros* (Lisbon: Typografia da Academia, 1836), 50–2. António Arala Pinto also devoted several pages to the management and improvement of the pinewood to avoid fires, Pinto, *O Pinhal*, vol. 2, 272–309.

FIGURE 5.1 Estrada da Vieira: Pinhal das Areias (Pinhal de Leiria)
SOURCE: ANTT, ANTERO DE SEABRA, BOX 21, N. 1389. DOCUMENT PROVIDED BY THE ANTT

The third clause refers to the harm caused to the pinewood as a result of mismanagement by the forestry administration and other officials with some responsibilities over its governance. The licences granted by the Crown for the harvesting of timbers either for private use or to supply timber to the royal fleets had not previously gone through the main guard of the pinewood.

However, from now onwards he would be the only official able to channel the permits or licences for cutting trees to obtain timber and wood. A combination of both factors had resulted in overcutting. The forestry policy henceforth required all cutters to show their licences to the main guard, who would indicate the area where trees could be cut causing the least harm to the pinewood. The junior wardens would be responsible on the ground for checking that the main guard had signed the licences.

The regulations confirmed a long-standing tradition on making pitch (*pez*) from the pinewoods (clause 4). The Crown had granted an unknown marquis economic rights over the production of pitch. The bailiff (*almoxarife*) of the pinewood was the official entrusted with issuing the documents specifying the quantities of pitch that the grantee would be allowed to make. He was to forward these documents to the main guard of the pinewood to sign them. For the sake of the conservation of the pinewood the junior wardens of the pinewood could neither make pitch themselves nor hire contractors for this purpose.

Although we have not located the original contract between the Crown and the marquis, the regulations stated (clause 5) that its purpose was to conserve the pinewood. The contract specified the people who were permitted to cut trees, so neither the bailiff (*almoxarife*) nor the marquis were allowed to issue permits to cut wood. This shows that the Crown perceived the pinewood as a place that provided wood and timber resources: its economic or material uses were therefore the key reason for its conservation. The main strategy for its conservation relied on severely restricting the use of and access to the pinewood. This recalls the negative prohibitions established by Venetian forestry legislation, which did not introduce positive practices (such as afforestation) until 1531.[35]

The regulations of 1524[36] addressed the technique of making firebreaks for the conservation of the pinewood. Every year after Easter the main guard would give orders for the outer firebreaks (*aceiros*) of the pinewood to be broken up and cleared, as had been traditional practice. The main guard was allowed to change the location of the outer firebreak as long as he requested permission to do so from the Crown (clause 6). The outer firebreak was intended to prevent fires from spreading inside the pinewood. Some archival references confirm that the administrative staff of the pinewood had been putting firebreaks in place since the mid-1400s at least.[37] It is very likely that this practice can be traced back to the beginning of the century (and perhaps even earlier) and

35 Appuhn, "Forests," 139–40.

36 BFDUL, Ms. 2–12–6.

37 Neves, *História florestal*, vol. 2, pp. 45–6.

the Crown simply codified in a law a tradition or custom practiced by local inhabitants. This documentary corpus provides evidence of how fires were regarded as the main challenge to the preservation of the pinewood.

It might come as a surprise to learn that this forest code had not addressed the planting of *pinheiro-bravo* in the pinewood of Leiria. Such shortcomings might be explained by the biological characteristics of *Pinus pinaster*. This tree species has a significant capacity for natural regeneration,[38] and it is very likely that the Portuguese officers were aware of this fact. In addition, the Crown regarded restrictive legislation and the administrative staff for its enforcement as the two key strategies for the conservation of the pinewood.

In this connection, the main guard was empowered to appoint up to 16 junior wardens (*guardas*). They would not receive a salary for their job; instead, they would be granted certain privileges and tax exemptions and 100 *reis* for each day's work when cutting wood for private individuals (chapter or clause 8). The main guard was to be paid in kind, as he would receive all the ends (*pontas*) of the trees cut both for private individuals and for the Crown, as well as dry and fallen wood (*madeira seca e derribada*) and the sum of 20 *reis* for each licence (*alvará*) to produce pitch (*pez*) (clause 9).

Consequently, the regulations were quite short, as they only contained information about the duties and privileges of the administrative staff (especially the main guard) entrusted with the conservation of the pinewood. There is no doubt, however, that the Crown's true motive for conserving the pinewood was to ensure it continued to be a source of material commodities. The industrial uses of the pinewood of Leiria (shipbuilding and, and to lesser extent, the production of pitch) had been the underlying main reason for its conservation throughout the historical period addressed here. Although such a concern is far removed from ecological awareness, it is possible to argue that the Crown achieved its main purpose of conserving the pinewood to produce material commodities, especially timber components for the royal fleets. Negative legislation and some positive measures (mainly the creation and clearing of firebreaks) were regarded as two of the main strategies for ensuring the present and future existence of the pinewood for industrial uses.

Thus, in the ensuing year (1525) King João III adopted a highly contradictory measure when he gave orders at the *Cortes* of 1525 (held in Torres Novas) for the royal pinewood (*coutada*) to be dismantled in response to the petition made by the representatives of Leiria. In the eyes of the administrative staff of the pinewood, this measure was a step backwards for the conservation of the

38 Correia, Oliveira and Fabião, "Biologia," 18.

pinewood. Among other things, it would permit the local inhabitants to hunt in the surroundings and inside the pinewood, with the risk of starting new fires.[39]

In 1530, Jorge da Costa, the main guard of the pinewood of Leiria, claimed that the number of 16 junior wardens (*couteiros*) was insufficient to ensure the effective conservation of the pinewood, as it stretched for 24 kilometres (4 Portuguese leagues or *léguas*) along the coast and 6 kilometres (in Portuguese *léguas*) inland.[40] This made it impossible to keep the firebreaks clear. The Crown increased the number of junior wardens to 20 due to the material uses to which the pinewood was constantly put. The new guards would be exempted from paying some taxes on bread and wine (*'oitavo e jugada de pão e vinho'*[41]), as well as from involvement in transporting the timber harvested from the pinewood. The main guard would appoint the new guards from among the local inhabitants, excluding farmers from the choice as King Don Manuel I had ordered.[42] The underlying reasons for this restriction are unclear, as it greatly reduced the Crown's options.

During the 1530s, the Crown reinforced the interconnections between the administration of the pinewood and the shipbuilding industry. In 1534, it confirmed the privileges granted to the 20 junior wardens of the pinewood of Leiria due to the ever substantial need for timber for the royal fleets and the much work that needed to be done on policing tasks as well as on maintaining and clearing the firebreaks.[43] In the same year, the Crown reinstated the royal forest (royal pinewood) in Leiria, after years of complaints from the main guard of the pinewood. The penalties specified in the regulations of 1524 were also brought back.[44] Probably as a result of restoring the royal pinewood, in 1535 the Crown confirmed the privilege granted to the local inhabitants of Pataias (located southeast of the pinewood) to hunt rabbits all year round, as the inhabitants of Pederneira had to deal with the damage they caused to their crops.[45] It is very likely this was due to an overabundance of rabbits.[46]

39 Neves, *História florestal*, vol. 5/2, 120–21.

40 Leonor Freire Costa, *Naus e galeões na ribeira da Lisboa. A construção naval no século XVI para a Rota do Cabo* (Cascais: Patrimònia, 1997), 320–35; Pinto, *O Pinhal*, vol. 1, 138–45.

41 The 'oitavo' was a proportional tribute that entailed paying an eighth of something, and 'jugada' refers to a royal tribute of payment in kind (in this case it might have been wheat).

42 Neves, *História florestal*, vol. 5/2, 74–5.

43 Neves, *História florestal*, vol. 5/2, 117–19.

44 Neves, *História florestal*, vol. 5/2, 120–21.

45 Neves, *História Florestal*, vol. 5/2, 126.

46 I am grateful to Cristina Joanaz de Melo for pointing out this idea.

If attention is paid to the ecology and geographical location of the pinewood, we find a delicate triple contradiction in the contents of these regulations passed by the Crown. Firstly, the pinewood of Leiria is in a sandy area that is highly unsuitable for agriculture. This greatly limited the settlement of population in the areas around the pinewood, and in turn entailed a problem for the Crown when it came to hiring local wardens as there is archival evidence that it preferred to appoint them from among the population living near the forest.[47] António Pinto asserted that the origins of the villages of Marinha Grande and Engenho are a consequence of the pinewood of Leiria. In 1527, Marinha Grande had only 80 inhabitants, a number that rose steadily throughout the eighteenth century (from 550 inhabitants in 1712, to 1,100 in 1758) due to the resin, ceramics, glass and sawing industries that stemmed from the local natural resources extracted from the pinewood.[48] It is very likely that the local economy was primary-sector based with a high importance of agriculture and cattle (or hunting), which clashed with the Crown's prohibitions on bringing cattle into the pinewood.

Secondly, this contradiction had even more serious consequences for the Crown's harvesting of timber commodities for shipbuilding. As the Crown did not have a permanent administrative staff for harvesting timber, it needed to rely on the local inhabitants for felling, sawing and transporting the timber from the pinewood to the ports from where it was taken to Lisbon and other shipyards. Thirdly, as the Crown barred local farmers from acting as junior wardens, these positions would have been filled by the local elites who had no interest in them except for the tax exemptions they entailed. As it is very likely that they consumed more wood than the poor farmers, they may have neglected their policing duties. In addition, they might have been the only inhabitants of the surroundings with the economic and logistical means to ensure the felling and transportation of timber for trade. It should be borne in mind that the guards (and their property, chiefly carts and oxen) were exempted from taking part in the harvesting and transportation of timber.

During the ensuing decades, the Crown continued appointing officers for the preservation of the pinewood. To mention but a few, in 1567 João Rodrigues Barba was chosen as *guarda-mor* of the pinewood after wedding Helena da Costa, the daughter of the former *guarda-mor* Jorge da Costa.[49] In 1577, the office passed to Helena's nephew Jorge da Silva.[50] Jorge da Silva Costa held the

47 ANTT, Manuscritos da Livraria (Ms. Livraria), 1113, doc. 16.
48 Pinto, *O Pinhal*, vol. 1, 222.
49 Neves, *História florestal*, vol. 6, 55–6.
50 Neves, *História florestal*, vol. 6, 101–2.

position in the early seventeenth century,[51] and it is likely that he was the one who had been appointed in 1577. This is not surprising considering that Philip II (r. 1581–98) decided to maintain all the Portuguese officers in their posts after Portugal came under Habsburg rule.[52]

3 The Pinewood of Leiria during the Habsburg Dynasty (1580–1640)

In terms of forestry, scholars have regarded this period as one of administrative innovation. Some of the laws passed from the 1580s to 1600 formed the backbone of Portuguese forestry legislation until the second half of the 1700s, including the 1597 and 1598 regulations concerning the pinewood of Leiria and the 1605 regulations on the *Monteiro-mor*, in which the pinewood of Leiria was described.[53] This section will address the history of the pinewood of Leiria by paying special attention to a report written in 1614 by Doctor Gonçalo Sousa after the fires of 1613. This hitherto unknown document provides new insights into the protection, management and material uses of the pinewood, as well as the relations between the local inhabitants and the Crown. As this contribution argues that the fire of 1613 marked a turning point in the history of the pinewood, it is arranged into three chronological sections: 1) conservation, forestry policies and material uses to which the pinewood was put from 1580 to 1612; 2) the fire of 1613, in a brief but separate section; and 3) the inquiries conducted after the huge fire, and the alternative sources sought by the Portuguese Crown for supplying the required timber components of *Pinheiro-bravo* due to the impossibility of continuing to rely solely on the pinewood of Leiria.

51 Archivo General de Simancas (AGS), Secretarías Provinciales (SSP), lib. 1472, fols. 246r–247r).

52 Félix Labrador Arroyo, *La Casa Real en Portugal* (Madrid: Polifemo, 2009), 42–55, 237–38.

53 Nicole Devy-Vareta and António Monteiro Alves, "Os avanços e os recuos da floresta em Portugal – da Idade Média ao Liberalismo," in *Floresta e sociedade. Uma história em comum*, coord. Joaquím Sandes Silva (Lisbon: Fundação Luso-Americana para o Desenvolvimento, 2007), 64–5. For an updated study on the forestry policies of this period see Koldo Trapaga-Monchet, "A destruction that preserves": Maritime warfare, empirical forestry and sustainability in Portugal (13th–17th centuries)," in *Roots of Sustainability in the Iberian Empires: Shipbuilding and Forestry, 14th–19th Centuries*, eds. Koldo Trapaga-Monchet, Álvaro Aragón-Ruano and Cristina Joanaz de Melo (United Kingdom: Routledge, 2023), 187–203.

3.1 *Conservation, Material Uses, and Administration of the Pinewood (1580–1612)*

In 1596, don Pedro de Castilho, then bishop of Leiria, proposed that King Philip II conduct plantings of pines near the church of Nossa Senhora da Nazaré in Pederneira because of the great shortages of timber for the royal fleets.[54] Similarly, the governors of Portugal wrote to the monarch pointing out the appropriateness of performing plantings in a stretch of barren land ('*charneca*') situated between the south border of the pinewood of Leiria and the territories of the monastery of Alcobaça (see Figure 5.2).[55] Scholars who have studied the monastery of Alcobaça have pointed out that the religious house had a very well thought-out policy for strengthening its boundaries that was in place as early as the fourteenth century.[56] In other words, the limits of the monastery had been established and marked for at least a century, and extended well into the north, encompassing the 'lake of Pataias' (lake of Pataias).[57]

From the perspective of the Crown, new plantings made it necessary to appoint new guards not only to protect the seedlings, but also to oversee the harvesting of timber for shipbuilding. On 20 September 1597, King Philip II accordingly increased the number of junior or lesser wardens (*couteiros*) of the pinewood of Leiria by eight because he had recently given orders for pines to be planted from the beginning of the old pinewood to the church of Nossa Senhora de Nazaré. These guards were granted the same privileges – in the form of tax and service exemptions – that those of the pinewood had enjoyed, and they replaced eight carters (*carreteiros*) of the works of the monastery of Batalha.[58]

In the same year, Philip II decreed not only the limits of the pinewood of Leiria but also its enlargement to ensure the future abundance of suitable timber for shipbuilding.[59] Although the ministers had warned the king of the importance of dealing with the monastery of Alcobaça,[60] the Crown did not pay heed, and this triggered major issues with the monastery. In 1598 and during the first years of Philip III's reign, large-scale plantings of pines were attempted at least twice, but failed because the wrong pine seeds had

54 ANTT, Colecção de Cartas, Núcleo Antigo 878, doc. 10.

55 ANTT, Núcleo Antigo, 877, no. 292.

56 Iria Gonçalves, *O Património do Mosteiro de Alcobaça nos séculos XIV e XV* (Lisbon: Universidade Nova de Lisboa, 1989), 19–23, 46 and 351–56.

57 Pedro Barbosa, *Povoamento e estructura agrícola na Estremadura Central: Séc. XII a 1325* (Lisbon: Instituto Nacional de Investigação Científica, 1992), 113–17.

58 AHU, CU, Reino, box 6, folder 34.

59 Pinto, *O Pinhal*, vol. 1, 159–62.

60 ANTT, Núcleo Antigo, 877, nº 292, AHU, CU, Reino, box 1, folder 62.

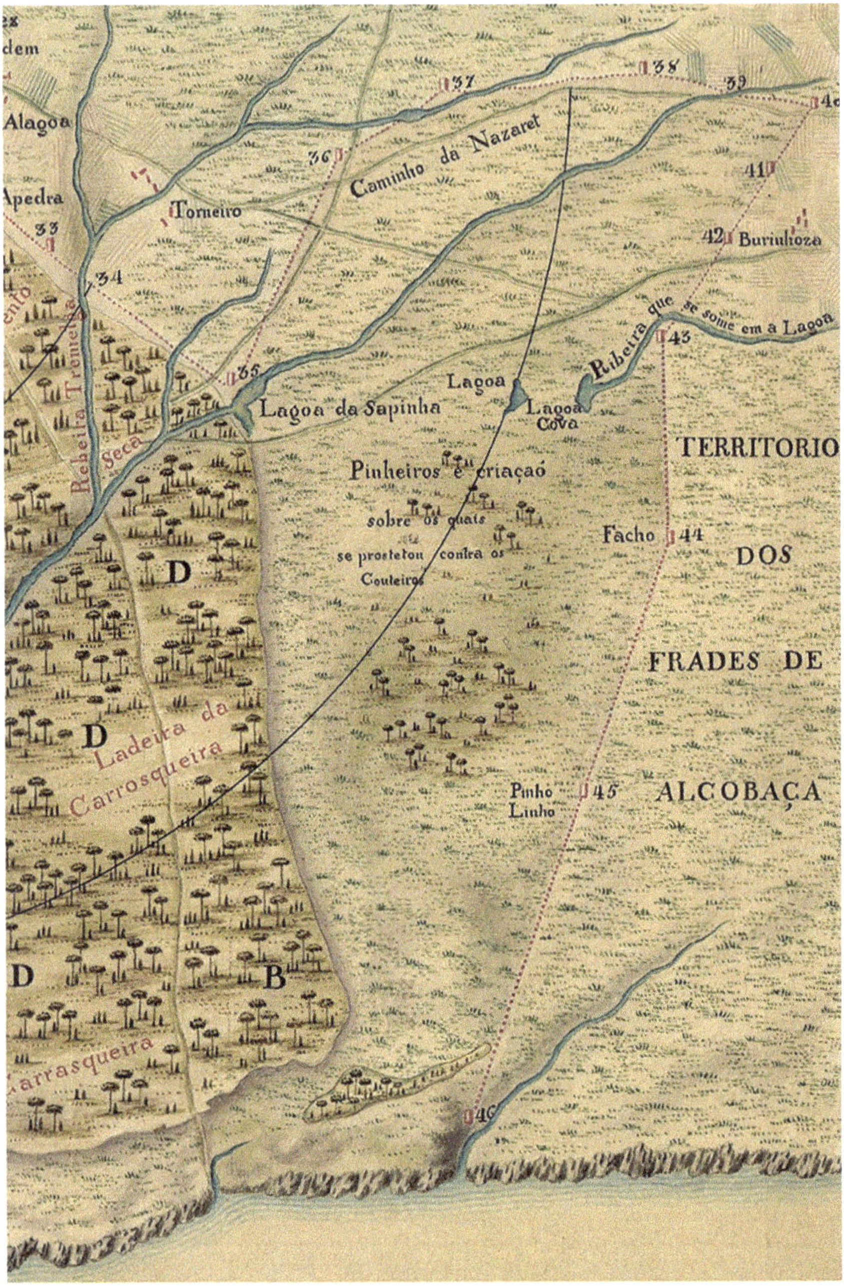

FIGURE 5.2 The limits of the royal pinewood in the south, bordering on the territories of the monastery of Alcobaça
SOURCE: DGT, IGP, CA-112, THE ROYAL PINEWOOD OF LEIRIA IN 1769

been chosen (*Pinheiro-manso* instead of *Pinheiro-bravo*).[61] The 1769 map of the pinewood (see Figure 5.2) shows not only the area where the plantings were conducted but also how Philip II had extended the royal pinewood (*coutada*) with the purpose of establishing the royal presence in the territory as a strategy for permitting the growth and conservation of the pines.

Consequently, the dispute that had arisen between the Crown and the monastery during the reign of Philip III still lingered on in 1769. The *comendatário* of Alcobaça reported having visited the abbey's territories around 1608, where a significant number of inhabitants of Pederneira and Pataias complained about the enlargement of the forest (*coutada*), and also about the behaviour of the main guard of the pinewood of Leiria with respect to the conservation of the recently planted pines. The local farmers, who were responsible for transporting timber for His Majesty, had not only lost their hives and vineyards but, worse still, they had been deprived of the lands where they used to graze their cattle. Moreover, due to the *guarda-mor*'s overzealous performance of his duties, the underwood (*mato*) had grown, leading to an increase in the population of wolves.[62] This was also a risk if a fire broke out, as it facilitated its spread.

This document shows three highly interesting and interconnected realities. Firstly, the Crown perceived the presence of the local inhabitants as detrimental to the growth of young pinewoods. It is true that some cattle could have entered the woods and endangered the growth of the very young pines during the first years, but they would not have posed a threat to older pines. Secondly, the effectiveness of the regulations depended on how zealously they were enforced by the administrative staff. According to the *comendatário* of the abbey of Alcobaça, the *guarda-mor* of the pinewood had implemented them very strictly. However, there are three objections to this argument. The first is that it is highly doubtful whether the *guarda-mor* had a constant presence in this area as he was entrusted with the conservation of the whole woodland, which spanned more than 20 kilometres along the coast. Secondly, the *guarda-mor* was probably based in the city of Leiria, more than 20 kilometres away from Pataias – a considerable distance to cover without an overnight stay. Lastly, it is very likely that timber harvesting was carried out in the north of the pinewood, meaning that the *guarda-mor* would have been far away from the new royal pinewood.

61 José Justino Silva, *Colecção chronologica da legislação portuguesa compilada e anotada (1603–1612)* (Lisbon: Imprensa de J. J. A. Silva, 1854), 120–21; AHU, CU, Reino, box 1, folder 62.

62 AHU, CU, Reino, box 1, folder 62. The following lines are based on this document.

The third interconnected factor is that the Treasury Council (*Conselho da Fazenda*) was opposed to enlarging the royal forest in this area because it would hamper the transportation of timber harvested for the royal fleets. If this was true, it meant that not only had timber had been transported through Nazaré to Lisbon (and the other final destinations), but also that the Crown permitted (and possibly even encouraged) the presence of local settlers near the pinewood for the purpose of ensuring the transportation of timbers. These local inhabitants were allowed to earn a livelihood there, which included the granting of permits for their cattle to enter for grazing.

There are fewer doubts about the Crown's interest in conserving the pinewood for the sake of the royal navy. From the perspective of the Crown, the material uses to which the pinewood was put were the driving force behind its conservation. And this is the key point to be addressed: the necessary work involved not only in conserving the pinewood, but also in cutting, sawing, and carting the timber to the seaports from where it was transported to Lisbon. Although the historical sources do not go into much detail[63] about the timber harvested in the pinewood of Leiria for the royal fleets from 1580 to 1612, they do provide some insights that are worth highlighting.

In March 1607, the Board of the Portuguese Treasury[64] (*Junta da Fazenda de Portugal*) met to establish the needs of the Portuguese *Carreira da Índia* of 1608. King Philip III had ordered the construction of three ocean-going ships of large dimensions (*naus*) and the repair of another two. The cutting and sawing of the timber had been underway since at least early January.[65] On 23 March 1607, a board made up of Vasco Fernándes Cesar, then the purveyor of the king's warehouse (*Provedor-mor dos armazéns*), and another unidentified minister (see the right-hand signature in Figure 5.3) issued in Lisbon a report on the *Quercus suber*, *Pinus pinea* and *Pinus pinaster* timber that private individuals had been enlisted to supply for the royal fleet, as well the contract price and the money the Crown owed the contractors.[66]

63 From 1580 to 1640 there is not a single reference to harvesting timber in the pinewood of Leiria in the archive of the Monteiro-mor, Biblioteca e Arquivo Histórico do Ministerio de Obras Púbricas (BAHMOP), Montaria-mor do Reino (MMR), núcleo 9.

64 For this Junta, see Santiago Luxán Meléndez, "El control de la hacienda portuguesa desde el poder central: la Junta de Hacienda de Portugal (1602–1608)," in *Política y Hacienda en el Antiguo Régimen*, eds. José Ignacio Fortea Pérez and Carmen Mª Cremades Griñán (Murcia: Universidad de Murcia, 1993), vol. 1, 377–88; Diego J. Martínez Gutiérrez, *La Junta de Hacienda de Portugal* (Pamplona: Newbook Ediciones, 1996).

65 AGS, SSP, lib. 1466, fols. 149r–151v.

66 AGS, SSP, lib. 1466, fols. 152r–154v.

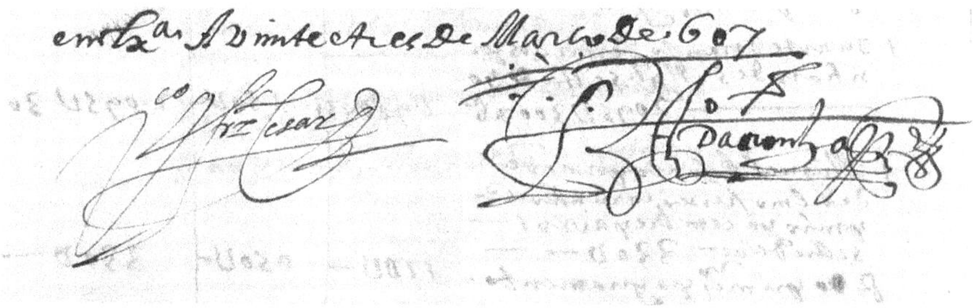

FIGURE 5.3 The signature of the ministers who wrote the report on timber supplied in 1607
SOURCE: AGS, SSP, LIB. 1466, FOL. 154V

TABLE 5.1 Pine timbers of from Pederneira supplied by Diogo Lobo and Pero Fernandes Lobo

Type of timber	Price (in *reis*)	Total price
200 dozen 'taboado de costado' (lateral planks)	9,900	1,980,000
100 dozen 'dal Caixa' (hull planking between the wales)	5,500	550,000
300 dozen 'forro' (ceiling planks)	2,420	726,000
15 'paos de lemes' (trees or sticks to make rudders)	16,666	250,000
		3,506,000

SOURCE: COMPILED BY THE AUTHOR FROM AGS, SSP, LIB. 1466, FOL. 152V

The report stated that the price of the timber to be supplied amounted to 15,344,500 reis, which was the lowest sum that had been paid. The Crown entered into a contract with Diogo Lobo and Pero Fernandes Lobo to supply pine timber components from Pederneira (see Table 5.1). Although the tree species were not specified, comparison with other years' timber supplies indicates that it is very likely that the timbers of lateral planks (*tabuado de costado*) and '*tabuado dal Caixa*' (hull planking between the wales[67]) were made of *pinho-manso* (stone pine), and that those of ceiling planks (*tabuado de forro*) were made of *pinho-bravo* (maritime pine).

The Crown hired Duarte de Araujo, then *feitor* of the pinewood of Leiria, to supply timber components most likely produced by the pinewood of Leiria (see Table 5.2).

67 I am thankful to Prof. Dr. Filipe Castro for his help with translating this terminology into English. All the mistakes are mine.

TABLE 5.2 Timbers supplied by Duarte de Araujo in 1607

Type of timber	Price per unit (in *reis*)	Total price
1,900 'latas' (deck beams)	330	627,000
600 dozen 'taboado de cuberta' (deck planking)	2,310	61,306,000[a]
24 'mesas de guarnição' (channels)	1,320	31,680
16 'apostizas'	1,320	21,120
10 'bombas' (pumps)	715	7,150
8 'paos bravos' (sticks or trees) for 'lemes' (rudders)	2,750	22,000
		2,095,300

a The document merely mentions 600 dozen 'taboado de cuberta' (deck planking) and the overall price of 61,306,000. Therefore, the calculation of the price per unit (10,217) is mine. It is more likely the total price would have been 6,130,600

SOURCE: COMPILED BY THE AUTHOR FROM AGS, SSP, LIB. 1466, F. 152V

TABLE 5.3 Timbers of *pinho-manso* of Pederneira supplied by António de Freitas Lobo in 1607

Type of timber	Price per unit (in *reis*)	Total price
21 dozen 'taboado de costado' (lateral planks)	9,900	207,900
35 dozen 'taboado dal Caixa' (hull planking between the wales)	5,500	192,500
		400,400

SOURCE: AGS, SSP, LIB. 1466, FOL. 153R

The report went on to list the different tree species supplied for the repairs on the ships for the *Carreira da Índia*. In addition, António de Freitas Lobo also undertook to supply the Royal Treasury with *pinho-manso* (*Pinus pinea*) for 1,000 cruzados (see Table 5.3). He committed to deliver the pines at the river and beach (*ribeira e praia*) of Pederneira.

The area that supplied pine timber was extended to the south of Lisbon, especially the pinewoods located in the municipality of Alcácer do Sal for which the Crown signed different agreements with private individuals.[68]

68 AGS, SSP, lib. 1466, fols. 153v–154r.

TABLE 5.4 Proposal of Martin Luis to supply timber components in 1608

Type of timber	Final price in 1607 (in *reis*)	Price of Martin Luis (in *reis*)	Reduction of the price
'Madeira de costado' (lateral plank, or hull plank) from stone pine (*pinho-manso*)	9,000 per dozen	7,000 per dozen	22.2%
'Madeira dal caixa' (hull planking between the wales) from stone pine (*pinho-manso*)	5,000 per dozen	4,000 per dozen	20%
'Madeira de forro' (ceiling plank) to cut in the pinewoods of Batalha from maritime pine (*pinho-bravo*)	2,200	2,000	10%

SOURCE: COMPILED BY THE AUTHOR FROM AGS, SSP, LIB. 1472, FOLS. 248R–249R

The following year the Crown enlisted contractors to ensure a supply of timber for shipbuilding and ships' repairs. In January 1608, Martin Luis had submitted a bid to supply timber for the *naus* sailing from Lisbon to India. His prices were significantly lower than those paid by the monarchy in 1607 (Table 5.4). Indeed, the final 1608 prices show a reduction compared to those offered in 1607 (Tables 5.1 to 5.3), as can be seen in Table 5.4. Martin Luis undertook to deliver the timbers to the port of Pederneira, from where the Crown would be responsible for transporting them to the shipyards of Lisbon.

In January 1608, Jorge da Silva submitted a bid for supplying timbers for lateral planks (*madeira de costado*), hull planking between the wales (*madeira dal caixa*), and ceiling planks (*madeira de forro*). His prices, summarized in Table 5.5, are significantly lower than those of Martin Luis. Later on, Jorge da Silva, *guarda-mor* of the pinewood of Leiria, undertook to deliver deck beams (*latas*) and deck planking (*taboado de coberta*) at the prices detailed in the table. The proposal entailed a significant reduction in the prices paid by the Crown, ranging from 12.5% to 25%.

In January 1609, Jorge da Silva reported that he had been forced to supply the *Pinheiro-manso* and *Pinheiro-bravo* timbers with the same conditions and prices that Pedro Fernandes Lobo and his partners had agreed for the construction of three new ships. Jorge da Silva agreed to deliver the timbers to the port of Pederneira, from where the Crown would be responsible for transporting them to Lisbon's shipyards. The Treasury Council (*Conselho da Fazenda*) voted to accept the contract on condition that Jorge da Silva committed to

TABLE 5.5 Timbers supplied by Jorge da Silva da Costa in 1608

Type of timber	Prices proposed by Martin Luis in 1608 (in *reis*)	Prices proposed by Jorge da Silva Costa in 1608 (in *reis*)
'Madeira de costado' (lateral plank, or hull plank?)	7,000 per dozen	5,500 per dozen
'Madeira dal caxa'	4,000	3,500 per dozen
'Madeira forro' (ceiling plank)	2,000	1,500 per dozen
'Latas' (deck beams)		260 each
'Taboado de cuberta' (deck planking)		1,600 per dozen

SOURCE: COMPILED BY THE AUTHOR FROM AGS, SSP, LIB. 1472, FOLS. 246R–247R

TABLE 5.6 Timbers supplied by Jorge da Silva Costa in early 1609

Ship's component	Price (in *reis*)
'Madeira de costado' (lateral planking) from stone pine (*Pinheiro-manso*)	5,500 per dozen
'Madeira dal Caixa' very likely from stone pine (*Pinheiro-manso*)	3,500 per dozen
'Madeira de forro' (ceiling plank) from maritime pine (*Pinho-bravo*)	1,500 per dozen
'Madeira de cuberta' (deck planking) from maritime pine (*Pinho-bravo*)	1,900 per dozen
'Latas' (deck beams)	260 each

SOURCE: COMPILED BY THE AUTHOR FROM AGS, SSP, LIB. 1.472, FOL. 243R–V

lowering the overall price to 600,000 *reis* less than that paid for the previous year's timber. The final prices of the timber he was hired to supply are summarized in Table 5.6.[69]

Although the economic history of the supply of timber will be addressed elsewhere, it is important to stress the significant reduction in the prices agreed for the supply of pine timber components from 1607 to 1608, and again

69 AGS, SSP, lib. 1472, fols. 243r–244r.

from 1608 to 1609. Moreover, it is worth noting that the Crown eventually purchased at the lowest price from Jorge da Silva, who in theory possessed sound first-hand knowledge of the availability of timber from the pinewood of Leiria because he held the office of *guarda-mor* there. It is very unlikely that such a price reduction was merely due to lower costs, and it probably stemmed to a large extent from an increased availability of timber components from the pinewood of Leiria and the surroundings of Pederneira. In February 1612, the king informed the viceroy that he had signed an *asiento* (contract) with Jorge da Silva, *guarda-mor* of the pinewood of Leiria, to provide the timbers for the construction of the ships that would sail to India in 1613. The prices could not exceed those established in the previous *asiento*.[70]

However, the fires of 1613 not only compromised the Crown's capacity to ensure the supply of the timber from the pinewood of Leiria, but also led it to carry out significant inquiries, which had a lasting impact on the administration of the pinewood of Leiria.

3.2 *The Fires of 1613 and the First Inquiries*

In September 1613, two fires largely destroyed the pinewood of Leiria. The following lines are based on a report produced by Manuel Veloso Cabral, who held the position of judge of the city of Leiria when the fire broke out.[71] According to his testimony, he was in Matoeira with his notary (*tabalião*) when a junior or lesser warden (*couteiro*, although he called him *mateiro*) of the pinewood of Leiria came to report on a fire. The judge ordered the *couteiro* to go and see the *corregedor* of Leiria and have him gather as many people as possible to set to work on extinguishing the fire that was then ravaging the north of the pinewood, the area from which most of the shipbuilding timbers were harvested. The junior warden was also entrusted with informing the judge of *vintenas* – settlements of at least 20 houses – in an attempt to enlist as many people as possible to attend to the pinewood.[72]

Meanwhile, Veloso and his notary departed with the local inhabitants of Matoeira. The judge stated that the pair of them worked very hard throughout the night to create a firebreak (*aceiro*) above 'Lago Limpa' (see Figure 5.4) in a place called 'Carreira dos Sellos' (perhaps he was referring to 'Caminho das Sellas') (see Figure 5.4), which stretched for more than one league and half

70 Biblioteca de Ajuda (BA), Ms. 51-VI-28, f. 45v.

71 AHU, CU, Reino, box 1a, folder 19, report of 1 October 1613.

72 Rafael Bluteau, *Diccionario da Lingua Portugueza* (Lisbon: Officina de Simão Thaddeo Ferreira, 1789), vol. 2, 527–8.

FIGURE 5.4 The places 'Lago Limpa' and 'Caminho das Sellas' where a 9-km-long firebreak reaching as far as the sea was constructed

SOURCE: DGT, IGP, CA-112. THE PINEWOOD OF LEIRIA IN 1769

(9 kilometres[73]) as far as the sea. He aimed not only to set an example for the rest, but also to separate the pinewood that was burning from the pinewood from which the Crown harvested shipbuilding timber. Therefore, creating a firebreak was again the main management strategy used by the ministers to protect the pinewood from fires.

Once judge Veloso finished making the firebreak, the bishop of Leiria and the *corregedor* informed him that another large fire had broken out in São Pedro de Moel and was destroying most of the pinewood. Judge Veloso marched with his entourage to São Pedro de Moel, where he found that people were not doing anything because they were afraid of the fire. He worked very hard at building another firebreak with the purpose of encouraging the others. When the firebreak was completed, he went to the north of the pinewood to ensure that the first fire would not cause any further damage.[74] A report of 1634 noted that the fire (very likely referring to the combination of both) had spread from the mouth of the Lis River to São Pedro de Muel.[75]

By 30 September 1613 the fires were under control, and judge Veloso went to the epicentre of the fire that had broken out inside the royal pinewood, in Coucinheira. He found evidence of arson, and began the first inquiries soon afterwards. However, the Crown took this task very seriously and headed the inquiries, which had a long-lasting impact on the administration of the pinewood.

What happened after the fires of 1613 is known because of the protests made by the administrative staff of the pinewood of Leiria, who during the 1630s bitterly complained that the Crown had not fulfilled the promises it had made in 1615. The *desembargador* (judge) doctor Gonçalo de Sousa was appointed to conduct the inquiries into the fires. The junior wardens or guards (*couteiros*) of the pinewood of Leiria were detained on charges of not having firebreaks in place. However, Sousa left and by 1615 had been replaced by doctor Jerónimo de Souto, who also held the position of *desembargador* of *Casa de Suplicação*. The Crown appointed him to make firebreaks stretching from the mouth of the Lis River up to the Ermida de São Pedro all around the pinewood, in a circle encompassing around 24 kilometres. The firebreaks were eventually built in two parts: 1) from the mouth of the Lis to a place called Lagoa Limpa; and 2) from the Ermida of São Pedro to 'Lagoa Sa Pinha.' Figures 5.5 and 5.6 show the approximate areas where these two firebreaks were created.

73 A Portuguese league was equivalent to 6 kilometres, Jaime Cortesão, *Influencia dos Descobrimentos Portugueses na História da Civilização* (Lisbon: Casa da Moeda, 1993), 70–1.

74 AHU, CU, Reino, box 1a, folder 19.

75 AHU, CU, Reino, box 6, folder 34. The following lines are based on this document.

FIGURE 5.5 The firebreak stretching from the Lis River to Lagoa Limpa
SOURCE: DGT, IGP, CA-112, THE PINEWOOD OF LEIRIA IN 1769

Once some of the firebreaks were completed, in 1615 doctor Jerónimo de Souto gathered the *corregedor* of Leiria, the *guarda-mor* of the pinewood of Leiria and the junior guards (*couteiros*) of the pinewood with the proposal of reinstating them to their positions provided that each of them cleared 150 *palmos* (spans)[76] of firebreak. In return, the Crown would maintain their privileges and would increase the number of *couteiros* from 33 to 40 to address such a mammoth task. A significant number of the administrative staff of the pinewood resigned from their posts because of the huge amount of unpaid work entailed. Those who accepted the proposal signed an agreement laying down the arrangements made by the *desembargador* Jerónimo de Souto.

76 According to Filipe Vieira de Castro, 1 *palmo de vara* was equivalent to 22 centimetres, Filipe Castro, *The Pepper Wreck. A Portuguese Indiaman at the Mouth of the Tagus River* (College Station: Texas A&M University Press, 2005), 191.

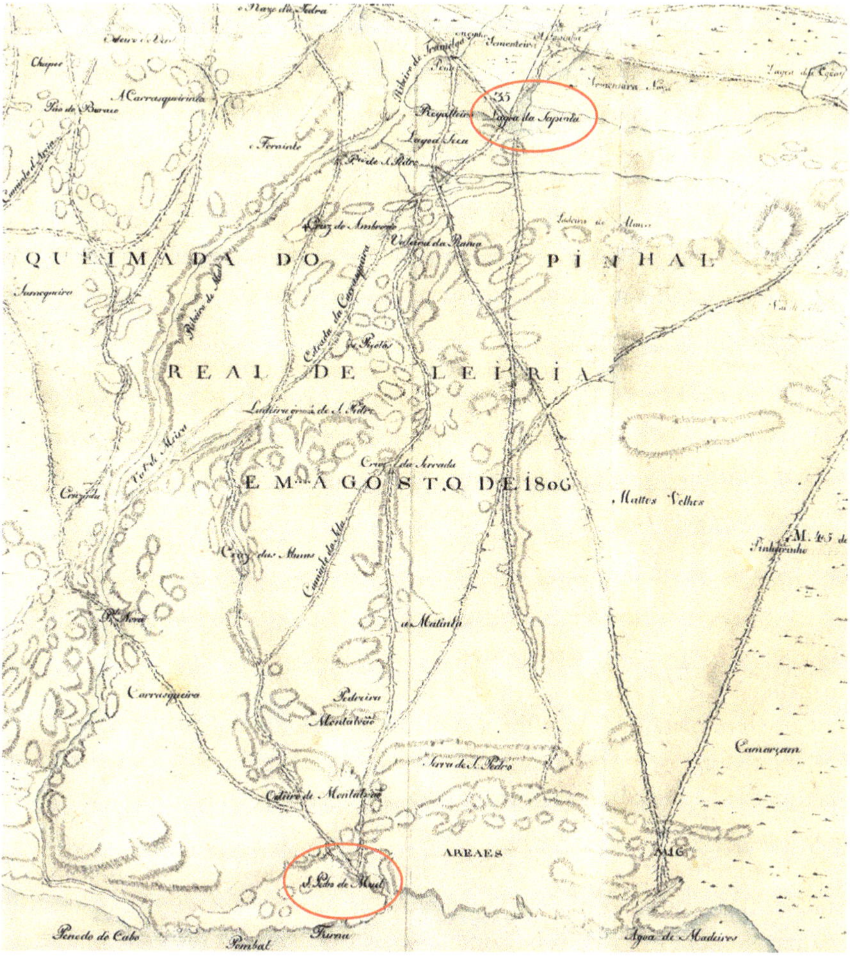

FIGURE 5.6 The firebreak extending from the Ermida of São Pedro de Muel (by the sea) to Lagoa Sa Pinha

SOURCE: DGT, IGP, CA-113, CARTA TOPOGRAFICA DO PINHAL DE LEIRIA IN 1807

In 1634, the Treasury Council (*Conselho da Fazenda*) confirmed that the claims made by the administrative staff of the pinewood of Leiria were true. It also turned to doctor Jerónimo de Souto, who asserted that the junior guards (*couteiros*) of the pinewood had efficiently worked on the firebreaks, which extended for two leagues, since 1615. Conversely, the Crown had not increased the number of the *couteiros* from 33 to 40.[77] On 24 October 1634, Cristóvão Silveira,

77 AHU, CU, Reino, box 6, folder 34.

who very likely held the position of *provedor* (purveyor) of the region of Leiria, backed the information provided by Jerónimo de Souto. The construction of the new firebreaks was still underway.[78] Although it is not possible to provide details of the physical location or width of the firebreaks, there is little doubt that it was the fires of 1613 that had triggered the Crown's efforts to establish policies for ensuring the future conservation of the pinewood of Leiria.

In this connection, the following pages deal with a report written by Gonçalo de Sousa in 1614 on the conservation of the pinewood of Leiria. It is very likely that the Crown had not only enlisted him to begin the inquiries but had also entrusted him with drafting a set of regulations for the conservation of the pinewood and the harvesting of shipbuilding timber. On 9 September 1614, King Philip III wrote a letter to the viceroy of Portugal acknowledging that he had seen the request from the Portuguese Treasury Council (*Conselho da Fazenda*) concerning the mission entrusted to doctor Gonçalo de Sousa regarding shipbuilding timber from Leiria. The king encouraged the *Conselho da Fazenda* to protect the royal forest and to perform plantings in the 'faldas' (foothills?) of the pinewood, which needed to be free of underbrush to avoid starting fires that could harm the pinewoods.[79]

Doctor Gonçalo de Sousa worked on these tasks for several weeks. The next mention is dated early January 1615, when the *Conselho da Fazenda* ordered an unknown minister (perhaps the *desembargador* Jerónomo de Souto) to examine the reports penned by Sousa in order to draft the regulations for the protection of the pinewood of Leiria and the harvesting of timbers because Sousa was transferred to the Island of Madeira.[80]

It is important to clarify that Sousa had not belonged to the administrative staff of the pinewood of Leiria, and that his proposals largely criticised the management of the pinewood conducted by the *guarda-mor* of the pinewood of Leiria and the administrative staff under his control. This document is of great importance for unravelling: 1) the Crown's perception of the pinewood of Leiria; 2) the policies it devised and implemented to ensure the conservation of the pinewood; and 3) the relations between the Crown and the local inhabitants not only in connection with the protection of the pinewood but, more importantly, the dynamics and interaction between the local inhabitants and the Crown when it came to harvesting and transporting shipbuilding timber.[81]

78 AHU, CU, Reino, box 6, folder 34.

79 AGS, SSP, lib. 1510, fol. 74r–v.

80 AGS, SSP, lib. 1512, fol. 6r.

81 ANTT, Ms. Livraria, 1113, doc. 16, fols. 198r–206r. The following lines are based on *ibid.*

This chiefly narrative section provides a good idea of the importance that wildfires had in the management and conservation of the pinewood of Leiria, both before they happened and after they took place. From at least the early fifteenth century, the Portuguese Crown established a permanent administration for the preservation of the pinewood, whose main duties were to create and to maintain firebreaks to prevent the spread of fires.[82] Wildfires also had a huge impact in their aftermath, as they constantly spurred the Portuguese Crown and foresters to seek ways to reduce their effects after they had broken out. Consequently, working out how to handle uncontrolled wildfires was an important trigger for forest management, a subject which requires in-depth research.

3.3 *Gonçalo Sousa's Draft of 1614 and the Aftermath Up to the Advent of the Braganza Dynasty*

Doctor Gonçalo de Sousa regarded the pinewood of Leiria as a strategic woodland for supplying pines for the royal navy. As it extended from north to south for 3 *leguas* (18 kilometres) along the coast,[83] it was large enough to meet all the needs of *pinho-bravo* (*Pinus pinaster*) for shipbuilding, as long as the Crown did not sell the timber and the pinewood was appropriately managed through *aceiros* (firebreaks).

Sousa had no qualms about blaming the Crown for the recent fires. He noted that the pinewood had been destroyed by the plantings and the enlargement of the royal pinewood carried out during the last years of Philip II's reign and the early years of Philip III's. It was important not only to restore the pinewood in order to supply timber for shipbuilding, but also to use the woodlands of Brazil and India as sources for constructing vessels in those territories, as this would give a respite to the pinewood of Leiria and the cork oaks, of which the kingdom of Portugal was running out.[84] This shows that, from the perspective of shipbuilding for the royal fleets, doctor Gonçalo de Sousa conceived the different regions or geographies of the empire as interconnected spaces.[85]

---

82 For example, the regulations of 1524 (BFDUL, Ms. 2–12–6) or 1751 (Melo, "Guerra, Imperios", 213–14).

83 According to the regulation of 1605, it stretched for 4 leagues, Silva, *Collecção Chronologica*, 120–21.

84 'que neste Rejno está acabada', ANTT, Ms. Livraria, 1113, fol. 198r.

85 This argument has been posited for the 18th century as well, Cristina Joanaz de Melo, Catarina Madureira Villamariz, Tânia Manuel Casimiro and Pedro Urbano, "Floresta de encanto," in *Como a Fénix Renascida – Matas, bosques e arvoredos (séculos XVI–XX): representações, Gestão, Fruição*, coord. by Cristina Joanaz de Melo (Lisbon: Colibri, 2020), 16. For the interconnected forest policies for Portugal and Brazil see Cristina Joanaz de Melo,

In addition, Sousa named alternative sources that could supply timber for the Lisbon shipyards more rapidly. The municipalities of Leiria, Alpedris, and Batalha contained large areas of pinewoods, which the Crown could acquire at very little cost. The system of relying on contractors, coupled with administrative mismanagement, posed a risk both to the conservation of the pinewood of Leiria, and to the private individuals from whom the monarchy had purchased timber. Over the past years, the mismanagement of the king's officers and the greed of contractors had compromised the availability of economic and easily accessible sturdy *Pinus pinaster* timber. The receipts issued by the king's officers to the contractors contained larger quantities of wood than the Crown really required.[86]

Moreover, the king's officers had not been supervising timber logging on private properties. They would instruct the owners of the pinewoods to return to their properties to await the arrival of the contractor and the officers, who, they were told, would come and check in person the quality of the pines and establish a reasonable price if they were of good quality. Owners had been known to have waited up to three months but nobody had shown up. Contractors entered private property to log the timber at any time and without notice. On plenty of occasions, the owners had not found out until the wood was in the port of Pederneira.[87]

In view of this systematic abuse, owners had decided to change the land use of their properties by transforming woodlands into areas of olive groves, vineyards or agricultural land, despite the fact that the soil of Leiria was not always appropriate for such economic activities. Rather, it was suitable for the natural growth of *Pinus pinaster* trees, which did not need a large administrative staff to carry out plantings and monitor their conservation and growth. It therefore made no sense to permit private contractors to break into private properties and cut their pines, as all this led local inhabitants to put their pinewoods to other uses, despite the low quality of the soil. Sousa asserted that no law would prevent private owners from destroying their pinewoods if the contractors and the king's officers continued with these illegal (and immoral) practices.[88]

"Floresta em movimiento: usar, regenerar, cuidar (séculos XIV–XIX)," in *Como a Fénix Renascida – Matas, bosques e arvoredos (séculos XVI–XX): representações, Gestão, Fruição*, coord. by Cristina Joanaz de Melo (Lisbon: Colibri, 2020), 99.

86 ANTT, Ms. Livraria, 1113, fol. 198r–v.

87 ANTT, Ms. Livraria, 1113, fol. 198v.

88 ANTT, Ms. Livraria, 1113, fol. 198r–v.

Sousa suggested ceasing to grant private contractors the legal right to chop down trees located on private properties. Instead, the Crown should buy the pines directly from the owners as this would be more economic than conserving the pines. In the event it continued to legally entitle contractors to cut privately owned pines, the Crown should force the contractors to deposit a security bail in Leiria to ensure that local inhabitants were paid for the logging and transportation of timber.[89]

Contractors' greed had diminished the canopy of *pinheiros-mansos*, especially the top-priority trees (the so-called 'paos reais') that produced highly important timber components such as lemes (*rudders*) and doublings (*calceces*). As contractors had no interest in the pinewoods beyond the duration of their contracts, they did not hesitate to cut the best trees to produce ship components of lesser quality such as deck beams (*latas*) and planks (*taboados*), even though the Crown could have used these timbers for *lemes* (rudders) if the trees had been given a few more years of growth. Furthermore, the poor condition of Leiria and other pinewoods was due to the mismanagement of the wardens (*couteiros*) of the pinewood. The main guard (*guarda-mor*), factors (*feitores*) and the *almoxarife* of the pinewood were supposed to go with the contractors to indicate the place where harvesting should be performed. This lack of control had resulted in reckless harvesting, and had led to seasoned and unseasoned timber being left in the woodlands with the subsequent waste of both funds and manpower.[90]

A large part of the document deals with the conservation of the pinewood. In this connection, the outer firebreaks (*aceiros*) were regarded as the chief means of ensuring its conservation. The firebreak was to stretch from the Lis River mouth all around the outer limits of the old pinewood as far as the beach located in São Pedro de Muel (see Figure 5.6). It was to be straight, with no bends (curves, *cotovelos*) – a firebreak with bends did not provide protection against external fires – in order to include all the pines and its clearing could be performed by burning vegetation.[91] The establishment of an administrative staff was intertwined with the creation and clearing of this firebreak, as the main duty of the 30 junior wardens (*mateiros*) was to clear and maintain this firebreak by burning. The firebreak was cleared on a yearly basis under the supervision of the *guarda-mor* and the *almoxarife* of the pinewood. To increase the effectiveness of these tasks, Sousa recommended entrusting the

89 ANTT, Ms. Livraria, 1113, fol. 198v.

90 ANTT, Ms. Livraria, 1113, fol. 198v.

91 ANTT, Ms. Livraria, 1113, fol. 199r.

corregedor of the district of Leiria with visiting the firebreaks with witnesses to report on their condition.[92]

Sousa pointed out the absurdity of Philip II's decision to establish the new pinewood in Camarção, which had eventually affected the conservation of the pinewood. This area was not only ill-suited to the growth of pines, but could have ensured the protection of the old pinewood, if controlled burning had been constantly practiced in this new area to prevent the growth of wood and underbrush that could easily spread fires. Moreover, the local inhabitants would have benefited from the absence of wolves and the abundance of grassland for their cattle.[93]

Sousa then listed in detail the activities that had had a direct impact on the conservation of the pinewood. Firstly, the Crown should grant permission to all hunters to enter the pinewoods to kill the deer (*veados*) now that the forest cover had been reduced. That way, there would be no further deer and hunters would not need to return to the pinewood, reducing the risks of fires.[94]

Contrary to the forestry legislation, Sousa regarded using the pinewood for shipbuilding timber as compatible with grazing, as 'experience shows that it [grazing] does not cause any damage, rather it is beneficial because oxen, goats, and sheep do not put pine in their mouths, and they [the pines] grow so close together that they resemble flax plants that eat each other.'[95] It was recommendable to allow cattle to enter as they helped the seeds spread naturally, and pine saplings were not harmed even when oxen trampled on them. This was particularly clear in the pinewoods owned by private individuals where cattle used to enter: the pines were so thickly spaced that private agents cut the small ones for firewood to permit the correct growth of the others.[96]

Sousa furthermore reported that cattle had usually grazed at the outer limits of the pinewood of Leiria without being penalized, but since the creation of the new royal forest the guards had extended their jurisdiction to raise more money through fines.[97] In contrast, Sousa was in favour of preventing farmers from allowing their herds of pigs to enter the pinewood, because they ate the pine seeds and got lost inside it, becoming wild.[98]

---

92 ANTT, Ms. Livraria, 1113, fol. 199r.

93 ANTT, Ms. Livraria, 1113, fol. 199v.

94 ANTT, Ms. Livraria, 1113, fol. 199v.

95 ANTT, Ms. Livraria, 1113, fol. 199v: 'a experiencia ensina que nenhum damno lhe faz antes proueito, porque o boj, cabra, e ouelha não toma na boca pinho e elles nascem tam iuntos que pareçem linho sameado e se comem huns a outros.'

96 ANTT, Ms. Livraria, 1113, fols. 199v–200r.

97 ANTT, Ms. Livraria, 1113, fol. 200r.

98 ANTT, Ms. Livraria, 1113, fol. 200r.

Conversely, during the early sixteenth century the inhabitants of Pederneira were allowed to take their herds of pigs into the pinewoods of the monastery of Alcobaça at certain times of year. It is more doubtful whether the inhabitants of Pederneira were permitted to do so with their herds of cows, sheep and goats as these species were regarded as harmful to young pines. However, Iria Gonçalves states that it is very likely that their cows and sheep grazed in the pinewoods.[99]

Sousa went one step beyond the usual understanding (and possibly the practice) of the management of the royal pinewoods in Portugal, as he supported letting farmers enter the pinewood to cut all the wood they required for their farm implements (*abegoaria*), as long as it was not from old pines (*pinho mor*). He suggested that farmers be allowed to use fallen pines that belonged to the *guarda-mor*, who could be compensated with something else. Permitting people to enter was beneficial and improved the condition of the pinewood, as it meant that they would turn up to help extinguish unexpected fires caused by lightning.[100]

Thus, Sousa not only supported letting cattle enter His Majesty's pinewood but held that the pinewood could be a space in which the gathering of firewood, cattle grazing, and the production of timber for shipbuilding were intertwined activities.

The third section of Sousa's report revolves around the administrative staff and is headed 'Officials' ('Officiais').[101] By far the most extensive, it includes a large array of elements that go beyond the category of the administrative staff, though conservation, administrative staff and shipbuilding were often interdependent realities. This makes senses if we take into account that the Portuguese Crown perceived restrictive legislation and the administrative staff entrusted with its enforcement as two key strategies for the conservation of the pinewood.

Sousa stated that the administrative staff entrusted with protecting the pinewood was made up of the main guard (*guarda-mor*), a clerk (*escrivão*), a bailiff (*meirinho*), and more than 30 wardens (*guardas*). The *guarda mor* was mainly responsible for ensuring that no fires were started in the pinewood, as well as for preventing people from stealing wood. The creation and clearing of the outer firebreak were regarded as the principal management techniques for avoiding wildfires. The *guarda-mor* was also responsible for indicating the places where wood should be cut. In this connection, Sousa claimed that there

99 Gonçalves, *O Património*, 473–74.

100 ANTT, Ms. Livraria, 1113, fol. 200r.

101 ANTT, Ms. Livraria, 1113, fols. 200r–209r.

was no law regulating the duties of the main guard – a statement which was untrue as the abovementioned regulations of the *guarda-mor* had been issued in 1524. In addition, the agents (*desembargadores*) dispatched from the court of Lisbon on extraordinary commissions could not oversee the main guard because they sojourned there for only short periods of time.[102]

After providing some general outlines on the administrative staff, Sousa went into detail about their posts, beginning with the head of the administration: the *guarda-mor*. The main guard had been responsible for appointing the junior guards. He stated that this had resulted in mismanagement because the main guard had chosen wealthy elderly people who lived far away from the pinewood. This had precluded a sound management of the pinewood as the guards were unable to come on time to perform their duties, as had been evident from the previous year's fire (1613). The Crown had enlisted guards who were blind and more than 80 years old, and others who lived too far away to carry out firefighting tasks. The wealthy inhabitants coveted these positions because of the privileges they entailed: not having their oxen seized by the Crown to transport timber (*jugada*).[103] This was a matter of great importance both for the Crown and for the local inhabitants. The former required human and animal workforces to harvest and to carry the wood from the pinewoods to the watercourses. For the latter, oxen were a crucial working tool on which they depended for their livelihoods.

The junior wardens (*couteiros*) were afraid of going against the main guard, as he was responsible for hiring and firing them. Therefore, Sousa believed that the Crown should delegate the function of hiring the main guards to the Treasury Council (*Conselho da Fazenda*) with the approval of the purveyor of Leiria (*provedor da Camara*). He also provided insights on the qualities and skills that the junior wardens should have. They should be men aged from 25 to 60 and dwelling solely in the houses of Torneiro and Mouta; the inhabitants of Marinha Grande, Graçia, Coucinheira, Vieira, Passagem, Ganderas, Carvide, and Monreal could also be included because the pinewood was visible from there. This would not only enable them to spot fires easily, but also to set about firefighting quickly as well to rapidly inform other dwellers if wildfires broke out.[104]

---

102 ANTT, Ms. Livraria, 1113, fol. 200r–v.

103 ANTT, Ms. Livraria, 1113, fol. 200v. Sousa also mentioned that wealthy local inhabitants paid the Master of Works of the monastery of Alcobaça to be given posts in the monastery because these entailed exemptions on having to lend their oxen for the transportation of timber, ANTT, Ms. Livraria, 1113, fol. 203r.

104 ANTT, Ms. Livraria, 1113, fol. 200v.

Furthermore, it was important to prevent the administrative staff of the pinewood from being involved in business activities related to the exploitation of the pinewood. None of the officials entrusted with protecting the pinewood should be allowed either to construct or to take part in the construction of ships in Pederneira and the areas surrounding the pinewood of Leiria, as there were high chances that the timber would come from the king's pinewoods. Moreover, the protection of the pinewood needed to be kept separate from the contracts to supply timber for the royal navy.[105]

Sousa furthermore recalled the importance of removing the main guard's capacity to grant permits to private individuals to make pitch from the pines of His Majesty's pinewood because it was a fire hazard. The main guard was not justified in complaining about this, as the Crown had just acknowledged his right to take 5% of the pitch made under each licence, and there were only three or four ovens (*fornos*) for making pitch in the area.

However, Sousa was aware that the combination of these measures (not being entitled to use fallen pines, not granting licences to make pitch, allowing poor farmers to gather firewood without cutting pines) were reason enough for the *guarda-mor* to complain. To avoid this, the author proposed giving him a small salary, as all the rights and payments in kind were not large, and nor were the duties of the *guarda-mor* substantial if the contracts were taken away.[106] It is very unlikely that the *guarda-mor* would have agreed with these observations due to the abundance of resources in the pinewood.

The need to regulate the office of the *guarda-mor* of the pinewood of Leiria was therefore clear. To ensure the conservation of the pinewood, to avoid the oppression of the poor local inhabitants, and to prevent mismanagement, it was important that the person entrusted with drafting the regulations should visit the whole pinewood in person to gather first-hand information.[107] The main guard should write down on paper all the regulations, royal orders, and dispatches he possessed in order for them to be included in the new regulations.[108]

After dealing with the main guard, Sousa mainly focused on the clerk (*escrivão*) of the pinewood to whom the Crown paid an income to prevent him from causing harm to the parties, to the timber contracts and the farmers. Sousa recalled that the current holder of the position had paid 10,000 reis to

105 ANTT, Ms. Livraria, 1113, fol. 201r.
106 ANTT, Ms. Livraria, 1113, fol. 201r.
107 ANTT, Ms. Livraria, 1113, fol. 200v.
108 ANTT, Ms. Livraria, 1113, fol. 201r.

'some servant of a minister'[109] for the post. Former clerks used to compile lists (referred to in the text as 'roes,' most likely *rols*) of the timber-related tasks assigned by the judges. Once the deadline set for providing timber ended, the clerk met the *guarda-mor* of the pinewood or the person in charge of supplying the timber to penalise those who had not fulfilled the designated quantities, but without filing complaints or initiating legal proceedings (*autos*). This ensured that they complied with their assignments, which benefited the Crown, and the farmers were not arrested: they were simply made to pay half or one *tostão* (1 *tostão* was equivalent to 10 *reis*) of the penalty. However, driven by greed, these ministers began to initiate proceedings and carry out arrests, as they received a small quantity for each, causing unnecessary harm to the farmers. Suddenly, Sousa asserted, the bailiff (*meirinho*) of the pinewood began to have plenty of work.[110]

At this point, Sousa questioned how it was possible that the local inhabitants had agreed to engage in cutting and transporting timber when the penalties for failing to fulfil their obligations were very small (1 *tostão* was to equivalent to 10 reis), yet they had lately been reluctant to perform these activities even though the penalties for not doing so were much higher, in some cases up to 2,000 *reis*. It is very likely that Sousa had exaggerated the situation. He did not blame the Crown and the local inhabitants for these issues, as the former paid for the work conducted by the local inhabitants, who were poor and, therefore, eager to earn money from these jobs. As usual, Sousa noted the two main reasons that explained the situation, both of which were related to the mismanagement of the officials of the pinewood, as well as to the transportation of timber.[111]

Firstly, officials' mismanagement affected the natural cycle of harvesting, sawing, and transportation of timber, causing major economic damage to the Crown and the local dwellers. Not only did it force the Crown to invest more funds but, at best, delays in the supply of timber hindered shipbuilding activity and, at worst, could either affect the quality of the construction or completely compromise the sailing of ships from Lisbon to India.

According to Sousa, the stages of working the timber should be as follows. The pines should be cut at the end of December and during January. Stripping off (*descascar*) the bark to prevent rotting and sawing work (*aserrar*) should be completed by March. Transportation of the timber should begin during this month, as it was the time when the grass grew and ensured that the oxen had

109 'deu dez mil reis a çerto criado de hum Ministro', ANTT, Ms. Livraria, 1113, fol. 201v.

110 ANTT, Ms. Livraria, 1113, fol. 201r–v.

111 ANTT, Ms. Livraria, 1113, fols. 201v–202r.

food on their way from the woodland to the port of shipment (Pederneira). Another reason for starting to transport the wood in March was that the cooler weather prevented the oxen from suffocating to death when crossing sandy areas. If any of the tasks began later, coercion became the only way of engaging the local inhabitants in the transportation of timber, as they were afraid of losing their oxen.[112] The Crown's demands may have interfered with local inhabitants' calendar of agricultural tasks, an element that requires further research.

Secondly, officials' greed had led them to establish extremely harsh legal proceedings and convictions because they received a percentage from each. Their avarice also influenced the allocation of timber-related jobs among the local population. As the staff of the pinewood aimed to make money, the *guarda-mor* and the clerk sold the wealthiest inhabitants of the area the privilege of not being included on the list of people to be assigned tasks of cutting, sawing, and transporting timber using their own animals. In addition, the main guard and the clerk established larger quantities of timber than were actually required, leading the poorest inhabitants to end up also working for the wealthiest inhabitants, who had previously paid the staff of the pinewood to be exempted from these obligations.[113]

Therefore, all the tasks required by the Crown – cutting, sawing, and transporting the timber – were allocated to widows and the poorest inhabitants. As a result, the work ran behind schedule, as the latter owned neither oxen nor carts, the essential tools for transporting the timber. In fact, they preferred to be imprisoned rather than attempt to meet the Crown's timber needs.[114]

The solutions to these issues lay in prohibiting the bailiff (*meirinho*) of the pinewood from filing complaints and initiating proceedings, as well as taking away his permission to arrest wives while their husbands were away working on timber transportation. Instead, offenders should be verbally cautioned and made to pay a maximum of one *tostão* as had traditionally been done. If legal proceedings and convictions were discontinued, the officers of His Majesty's pinewood would make sure that the processes of cutting, stripping the bark, sawing, and transporting timber were performed on time without clashing with the residents' agricultural duties.[115] On top of this, the pinewood officials' lists should reflect the Crown's real timber needs when it came to allocating the jobs to the *vintenas* to avoid overwhelming the local inhabitants.[116]

---

112 ANTT, Ms. Livraria, 1113, fols. 201v–202r.
113 ANTT, Ms. Livraria, 1113, fol. 202r–v.
114 ANTT, Ms. Livraria, 1113, fol. 202r–v.
115 ANTT, Ms. Livraria, 1113, fol. 202v.
116 ANTT, Ms. Livraria, 1113, fols. 202v–203r.

Sousa considered the clerk of the pinewood to be very cunning and clever, with considerable expertise on the pinewoods. As he was used to getting his own way, it was recommendable that the minister entrusted with dealing with him be highly knowledgeable on the matter.[117] The lists of timber-related jobs to be distributed among the inhabitants should not be handled by the clerk of the pinewood; rather, the Crown should entrust such an important task to the person in charge of managing the pinewoods (in this case the *guarda-mor*). When farmers and local inhabitants were considered to have completed their transportation duties it should be on the basis of their own testimonies, whereas, in practice, the clerk usually only did so as he saw fit. As a result, the locals were severely impaired as many of them depended on the wages paid by the Crown. Sousa also recommended that the king should request permission from the bishop of Leiria for the locals to work on Sundays and on the afternoons of Holy Days, so that they could fulfil their obligations as soon as possible and return to their homes.[118]

Sousa noted that the allocations of tasks were especially detrimental to the residents of the municipality of Leiria.[119] This was because of the criteria used by the officers to distribute the timber duties. Instead of establishing the municipality as a management unit, Sousa proposed taking the pinewood as the reference point for assigning the work. That way the areas of Coutos de Alcobaça (which belonged to the monastery of Alcobaça) and the municipalities of Porto de Mos and Pombal located nearer the pinewood would be included. Besides, the Crown should remove the privileges given to the inhabitants of Leiria of not having to transport timber from the pinewood to the port of Pederneira, as the residents of this area were not only plentiful but were also among the wealthiest in the area and owned means of transportation (oxen and carts).[120] Furthermore, the residents of Leiria had fulfilled their timber obligations by resorting to third parties who lived nearer the pinewood.[121]

According to Sousa, transportation issues continued up to the port of shipment. The Crown needed to ban the clerk of the pinewood from being present at the reception of timber in Pederneira.[122] In the event the Crown found itself running out of time to transport the timber, it was furthermore advisable to appoint a third person rather than relying on the clerk of the pinewood.[123] In

117 ANTT, Ms. Livraria, 1113, fol. 205v.
118 ANTT, Ms. Livraria, 1113, fol. 203r–v.
119 ANTT, Ms. Livraria, 1113, fols. 202r and 206r.
120 ANTT, Ms. Livraria, 1113, fols. 202–203r.
121 ANTT, Ms. Livraria, 1113, fol. 204r.
122 ANTT, Ms. Livraria, 1113, fol. 204v.
123 ANTT, Ms. Livraria, 1113, fol. 204v.

addition, the local inhabitants would benefit if His Majesty's clerks performed their duties appropriately. Firstly, the timber clerk at the port of Pederneira (*escrivão das madeiras da vila de Pederneira*) needed to be compelled to quickly hand over to the wood transporters (*carreteiros*) receipts confirming that they had fulfilled their timber assignments, and there was no need for the latter to make payments in kind (rabbits, hares, or partridges); otherwise, their oxen could starve to death. Secondly, the clerk of the pinewood and his family should be made to live in accommodation provided by the Crown in the pinewood, as this would reduce the time the local inhabitants needed to wait to be discharged from their assignments. These delays caused by the clerks had been used by the bailiffs to initiate judicial proceedings and to arrest the locals.[124]

It is very likely that the transportation of timber from the Leiria pinewood to Pederneira was hindered by the low density of road networks in the north part of the estate of the monastery of Alcobaça; the road that connected Pederneira with Pataias on the way to Leiria and perhaps two other minor roads are known. The first connected the city of Pederneira with the 'casais' (houses) of D. Bras. The second route ran from Pataias to Cós, and possibly from Paredes to Cos first by land and afterwards by water.[125] However, it is doubtful whether they were of any use for transporting timber from the pinewood to Pederneira. The route from the pinewood of Pederneira to the city's seaport remains unclear. Thus, it is not surprising that Sousa proposed checking whether the port of Paredes was suitable for transporting timber to Lisbon, as it was two leagues closer to the pinewood than the city of Pederneira. It would enable the Crown to save thousands of cruzados every year and the people would suffer less oppression from the authorities.[126]

From time to time the minister in charge of running the woods should either go in person or enlist someone to visit Pederneira to crosscheck the receipts for timber deliveries in Pederneira with the receipts for the timber transported by the carters from the pinewood, to avoid payments being made to people who had not actually transported the timber.[127] The timber clerk of Pederneira should be given the option to choose between performing all his duties appropriately or giving up the post, as he currently only worked when it suited him.[128]

124 ANTT, Ms. Livraria, 1113, fol. 203r–v.

125 Gonçalves, *O Património*, 380–86.

126 ANTT, Ms. Livraria, 1113, fols. 205v–206r.

127 ANTT, Ms. Livraria, 1113, fols. 203v–204r.

128 ANTT, Ms. Livraria, 1113, fol. 204r.

As has been noted, the Crown regarded the regulations as a key strategy for ensuring that its objectives were met. Sousa pointed out the need to regulate the position of the timber clerk of Pederneira (*escrivão das madeiras da vila de Pederneira*) by issuing a *regimento* or set of regulations for which he was to hand over all the royal orders he had concerning his office.[129] The regulations should contain a specific chapter, which would be used as a template form to record the timber delivered by the transporters in Pederneira. Sousa argued that these receipts should include specific details of the carters to easily differentiate them from others to avoid future confusion as people from the same *vintena* often had the same names.[130]

During the Early Modern Age, one of the main disputes between the Crown and local inhabitants arose over appraisals of the price the Crown paid for the transportation of timber for shipbuilding.[131] The writer pointed out that abusive practices took place when either the Crown or a contractor oversaw the transportation. In the latter case, contractors agreed with the appraisers to inflate the prices with the goal of sharing the profits between them, to the detriment of the local inhabitants.[132]

However, Sousa did not go into much detail about the malpractices and shortcomings when the Crown was in charge of transporting timber from the pinewood to Pederneira. Judging by similar cases in other geographical areas, it is very likely that the Crown not only intended to pay a lower price, but also fell behind on the payments.[133] He proposed taking the distance between the loading site and the place of delivery as the reference elements. The timbers would be marked, with each mark corresponding to the place of origin of the wood, which would be identified in Pederneira so that payment would be made according to the distance covered by the carter.[134] To avoid having to bring in assessors, the regulations would establish the locations of the pinewood and the surrounding areas from which timber was taken, as well as the

129 ANTT, Ms. Livraria, 1113, fol. 204r.

130 ANTT, Ms. Livraria, 1113, fol. 204r.

131 Álvaro Aragón Ruano, "Mar de árboles, vorágines de jurisdicciones. La complicada relación entre la Real Armada española y los bosques del Pirineo Occidental peninsular en el siglo XVIII," in *Árvores, barcos e homens na Península Ibérica* (*séculos XVI–XVIII*), coord. Rosa Varela Gomes and Koldo Trapaga-Monchet (Zaragoza: Pórtico Librerias, 2017), 47–8.

132 ANTT, Ms. Livraria, 1113, fol. 204v.

133 Aragón Ruano, "Mar de árboles," 45–8.

134 ANTT, Ms. Livraria, 1113, fol. 204v.

sum to be paid depending on the quality of the timber and the distance from those places to Pederneira.[135]

When harvesting was managed directly, not enough timber was produced from each tree. The solution lay in establishing fixed payments for the following elements: a day's work, cutting a tree, stripping bark from a tree, and sawing a *lata* (beck deam) and *taboa* (plank). That way, the workers would exert themselves more as their wages would depend on the amount of work they did, rather than on the number of days they worked.[136] Sousa did not realize that this method might lead to the production of low-quality timber components and a resulting decline in the quality of the ships, which might have detrimental effects in the long run.

Clearly, Sousa equated the pinewood of Leiria with the production of sturdy timber for the Lisbon shipyards. As a result, his recommendations were aimed at providing better-quality timber for the royal navy. As the Crown did not state the measurements (*vitolas*), that is, the length and width, of each of the timber components, contractors often supplied timbers that did not meet shipyards' requirements. This not only compromised the quality of the ships, but also harmed the carters as contractors sought to swindle them by lowering the value of the timbers. The contractors kept the best timbers for themselves either to sell them, or to construct ships.[137]

Sousa provided solutions to the different processes that led both to the loss of timber for shipbuilding and to the reduction of timber quality. He recommended that the regulations governing the duties of the people entrusted with transporting timber should include a clause requiring them to visit the pinewoods where the timber was to be harvested. If timber was left behind, they would be responsible for bringing it to Pederneira to prevent losses due to two causes. Firstly, the growth of the underwood (*mato*) covered timber. Secondly, rainfall rotted it. Special attention need to be paid to the 'paos' (sticks) and to waste timber called 'puntas.'[138]

Moreover, he cited incorrect stockpiling as one of the reasons that contributed to the rotting of timber. Once the timber arrived at the port of Pederneira, it was left on the beach (*praia*) without any supervision. The waves wetted the timber, and the sun caused it to rot. Leaving timber unprotected on the beach for even a whole year led to a significant reduction both in the amount of timber suitable for shipbuilding, and in the quality of that still regarded as usable.

---

135 ANTT, Ms. Livraria, 1113, fol. 204v.

136 ANTT, Ms. Livraria, 1113, fols. 204v–205r.

137 ANTT, Ms. Livraria, 1113, fol. 205r.

138 ANTT, Ms. Livraria, 1113, fol. 205r.

In addition, as the beach was an open area, local inhabitants and fishermen could steal the timber for shipbuilding and other uses. Thus, Sousa proposed building a warehouse to store the timber that needed to left from one year to the next.[139]

Some of the observations he made went beyond the Crown's pinewoods and included policies encompassing plantings in different land tenures. Sousa proposed on several occasions that plantings be conducted in the municipality and region (*comarca*) of Leiria, for which the Crown should acquire territories in the municipality of Leiria that were well suited to *pinhoes-mansos*.[140] The *corregedor* of Leiria could oversee these plantings in all the *sesmarias* (contracts made between the Crown and private individuals to work the land) by including a special article in the regulations governing this position. In general, throughout Portugal the *corregedores* were careless in their duty of conserving the 'mattas, e arvoredos' (woodlands and trees) and thoroughly negligent about conducting plantings. The seacoast of Leiria was filled with *charnecas* (moorlands), where pine trees could be planted.[141]

While with these draft regulations the Crown aimed to avoid another fire that could reduce the cover of the pines that provided timber components for shipbuilding, from 1614 to 1640 ships continued to be built in Lisbon and Oporto and the Crown constantly relied on the pinewoods of Leiria and Pederneira as a source of timber.[142] Yet it is very likely that the Crown had not implemented any of the measures proposed by Gonçalo de Sousa.

The documents generated by the royal administration mainly relate to issues that arose over the supply of timber for shipbuilding. In November 1616, the farmers of the city of Leiria and Coutos of Alcobaça submitted some notes (memoranda, or *apontamentos*) via their attorney, Manuel Esteves Serrão, concerning the pinewoods and the harvesting of timber for the royal fleets.[143] On 7 June 1618, the king asked Jorge da Silva, *guarda-mor* of the pinewood of Leiria, to report on the petitions the inhabitants of Leiria had made in 1616.[144]

139 ANTT, Ms. Livraria, 1113, fol. 205r–v.

140 ANTT, Ms. Livraria, 1113, fol. 199r.

141 ANTT, Ms. Livraria, 1113, fol. 199r.

142 Koldo Trápaga-Monchet, "Supplying Timber for his Majesty's Fleets: Forest Resources and Maritime Struggle in Portugal (1621–1634)," in *Maritime History and Archaeology of the Global Iberian World (15th–18th centuries)*, ed. Ana Crespo Solana, Filipe Castro and Nigel Nayling (Cham: Springer Nature, 2022), 215–48; Koldo Trapaga-Monchet, "*No es madera de vasallos, sino del rey*. Las políticas forestales de los Habsburgo en Portugal (1609–1640)," *Obradoiro de Historia Moderna* 28 (2019): 116–28.

143 AGS, SSP, lib. 1513, n.p. 16 November 1513.

144 AGS, SSP, lib. 1516, fol. 70r–v.

The conflicts with the local inhabitants continued during the following years. In November 1616, the king told the viceroy that the local farmers of Leiria had not carried out timber work because the Crown owed them money for transportation and wages.[145] In March 1623, the dwellers of Aranha claimed that the main guard of the pinewood of Leiria had forced them to transport to Pederneira timber to be used for the construction of ocean-going ships and galleons.[146]

As stated above, the Crown regarded the regulations as one of the main instruments for enforcing its objectives on the ground. There are a few mentions of how, in the end, it did not issue the regulations drafted by Gonçalo de Sousa. In November 1622, King Philip IV (1621–40) delivered to the viceroy of Portugal a copy of the regulations drawn up for the pinewoods and timber of Leiria and the Coutos of Alcobaça.[147] On 25 July 1625, a board was working on a set of regulations for the pinewood of Leiria, which had at least 83 articles.[148]

The following year, the king again issued regulations governing the pinewood of Leiria and Coutos of Alcobaça, which were to be used to draft regulations on the harvesting of timber in the Ribatejo area.[149] In November 1626, the Court of Madrid sent further regulations on the pinewood of Leiria and its timber, which were to be held in the Crown Archive situated in Lisbon (Torre de Tombo).[150]

In addition, the Crown had not raised the number of *couteiros* (junior wardens) to ensure the proper conservation of the pinewood. In 1636 and 1639, the *couteiros* belonging to the administrative staff of the pinewood of Leiria requested the Crown to increase the number of guards from 33 to 40, as had been promised in 1615. It is very likely that the Braganza dynasty had been on the throne when the Crown had agreed to this request on 23 March 1641.[151]

Finally, nor had the Crown invested in the construction of facilities (warehouses) to store the timbers that spent the winter on the beaches of Pederneira before embarkment. In March 1634, Alvaro Dias, an inhabitant of Pederneira, calculated that in the past two years alone not only had the Crown lost 6,000 cruzados worth of pine timber left in Pederneira during the winter, but that this timber would have been enough to meet the Crown's needs in 1634.[152]

145 AGS, SSP, lib. 1513, n.p., 30 November 1513.
146 AHU, CU, Reino, box 4, folder 26.
147 AHU, CU, Reino, box 3, folder 90.
148 AGS, SSP, lib. 1519, fol. 66r.
149 AHU, CU, Consultas do serviço real, códice 35, f. 236r, December 1623.
150 AGS, SSP, lib. 1520, fol. 125r–v.
151 BA, Ms. 51-VI-21, fols. 197v–201.
152 AHU, CU, Reino, box 6, folder 33.

4 Conclusions

By the early seventeenth century, the pinewood of Leiria spanned more than 24 kilometres along the coast, with *pinheiro-bravo* (*Pinus pinaster*) as the predominant species. This species adapts perfectly to poor soils, and it easily regenerates naturally with no need for human-conducted plantings.

Overall, during the period analysed here, the Portuguese Crown used a double interconnected strategy to ensure its material objectives for the pinewood of Leiria. Firstly, it passed restrictive legislation aimed at limiting the use of and accesses to the pinewood of Leiria. Secondly, it put in place an administrative staff to enforce the legislation. The pinewood of Leiria had a permanent administration for its conservation and management from the early fifteenth century. The main guard (*guarda-mor*) was the head of an administrative staff with two intertwined obligations. Firstly, he was responsible for policing tasks with the purpose of conserving the pinewood of Leiria to provide goods and commodities, especially related to the shipbuilding industry. Secondly, he was responsible for clearing the firebreaks to prevent the spread of fires, as they were viewed as the main natural hazard that endangered the conservation of the pinewood.

As for forestry legislation, the regulations governing the post of the *guarda-mor* of the pinewood of Leiria, issued in 1524, were somewhat short to be regarded as a thorough instrument for guaranteeing the sound management of the pinewood. There is no doubt that in the eyes of the Crown fires were the main risk to the conservation of the pinewood. Two out of the nine articles of those 1524 regulations directly dealt with prohibitions (hunters' access to the pinewood) and management techniques (firebreaks) to avert wildfires. The remaining clauses were more related to regulating certain uses of the pinewood, the duties and wages of the *guarda-mor*, and the systems of permits for using the pinewood. The Crown did not pass any further regulations on the conservation of the pinewood of Leiria during the remainder of the period addressed here. It was not until 1751 that the well-known regulations on the pinewood of Leiria were issued.[153]

Regarding the administrative staff, the 1524 regulations mentioned 16 junior wardens, a number which had been progressively raised to 33 by the end of the sixteenth century. The underlying reason was always the same: the number of guards was insufficient to ensure the conservation of such a large pinewood. The enlargement of the royal pinewood in 1597 and 1598 was a consequence of

153 For its study see, Melo, "Guerra," 199–220.

the Crown's objective of having a very large reserve of pinewoods for the royal fleets, and the number of guards was increased accordingly to 33.

Sometimes the administrative staff of the pinewood of Leiria, in particular the main guard, were also entrusted with supplying timber for royal shipbuilding. During the first decades of the seventeenth century, the Crown often turned to private contractors to meet the demand for timber for the construction of the ocean-going ships that connected Lisbon to India. The contracts entered into from 1601 to 1612 show a significant reduction in the prices paid by the Crown for specific pine timber commodities. It is very likely this indicates an abundance of these resources; otherwise, the Crown might have not been able to secure such a reduction.

The fire of 1613 marked a turning point in the history of the pinewood and its administration. In late September, two fires razed much of the pinewood, ushering in a disruptive period for Leiria. Although these were not the only fires in the history of the pinewood, they had far-reaching consequences. Soon after the fires, the Crown appointed doctor Gonçalo de Sousa to conduct an inquiry. His assignment was taken over in 1615 and finished by doctor Jerónimo Souto. The very extensive report written by Gonçalo de Sousa in 1614 was hitherto unknown.

At this point it is crucial to stress that fire was a key player in forest management in this Portuguese woodland, characterised by inflammable species that enabled fires to spread easily. Although Rego and Arala have noted the importance of fires in the management of the pinewood of Leiria, their history has been thoroughly disregarded.[154] This is clearly a topic that deserves a new historical approach.

From the very beginning of the report, there is no doubt that the Portuguese Crown regarded the pinewood of Leiria as a strategic resource for supplying timber for the royal fleets. In this connection, Sousa stated the pinewood was capable of largely meeting the royal fleets' demands for *pinheiro-bravo*. In addition, the report provides new insights into: 1) the administrative staff of the pinewood; 2) the relations between the Crown and the local inhabitants; and 3) the natural and human constraints that might have hindered the harvesting of timber.

Regarding the administrative staff, Gonçalo de Sousa was more concerned with the performance of the guards than with their number. The *guarda-mor* of the pinewood had been responsible for appointing the junior guards (*couteiros*), who were the Crown's eyes and hands on the ground. Their main duty was

154 Rego, *Florestas públicas*, 42; Pinto, *O Pinhal*, vol. 2, 272–310.

to police the pinewood to ensure its conservation, a task for which they needed to be based near the pinewood. Sousa complained about the mismanagement of the *guarda-mor,* who had appointed the wealthiest inhabitants of the areas surrounding Leiria, some of whom were over 80 years, blind, or simply lived too far away from the pinewood. They wanted the job because it exempted them from having to make their oxen available for the harvesting and transportation of timber. All of this became obvious when the fire of 1613 broke out.

It is important to bear in mind that the first point (administrative staff) is related to the second, an attempt to unravel the dynamics between the Crown and the local inhabitants both in the conservation of the pinewood and in the harvesting of timber. The junior wardens were appointed from among the local inhabitants surrounding the pinewood. As they did not receive a wage for their job, the Crown facilitated their livelihood by allowing them to use the forestry resources offered by the pinewood such as firewood, raw materials for construction, and the right to graze their cattle there. In addition, they were exempted from harvesting and transporting timber, the latter task being the one that triggered the most issues. The mismanagement of the *guarda-mor* over the choice of the junior guards meant that the burden of the timber transportation work was primarily borne by the poorest inhabitants, who did not own the required oxen.

As a result of all these factors, together with certain natural and social constraints, the tight timetable for the harvesting of timber was not met. Doctor Gonçalo de Sousa paid more attention to the social constraints, notable among which was the lack of warehouses to store the timber in Pederneira. The natural events were listed (rain, heat, sea tides) because they negatively affected the timber components that were left outdoors.

Gonçalo de Sousa's long, detailed, and impressive report therefore focuses more on the regulations and the administrative staff of the pinewood. This is not surprising considering that these were two of the main strategies used by the Crown to achieve its objectives. In this respect, Sousa's report resembles a draft for a set of regulations that the Crown did not finally issue during the Habsburg dynasty despite the constant references to it. In 1615, doctor Jerónimo de Souto took over from Gonçalo de Sousa and created two very large firebreaks all around the outer perimeter of the pinewood. The administrative staff of the pinewood of Leiria were offered to be reinstated to their positions if they agreed to take responsibility for clearing the firebreaks. As this entailed a huge amount of work, most of the guards resigned. The Crown offered those who stayed on to increase the number of guards from 33 to 40 to facilitate the correct management of the pinewood.

However, from 1615 to 1640 it is very unlikely that the Crown took any measures to increase the effectiveness of the management of the pinewood, aside from the work to finish all the firebreaks that had begun in 1615 and continued for almost two decades. Yet throughout this period the pinewood of Leiria and others situated in Pederneira and Batalha continued supplying timber for shipbuilding.

Abbreviations

AGS: Archivo General de Simancas
SSP: Secretarías Provinciales
AHU: Arquivo Histórico Ultramarino
CU: Conselho Ultramarino
ANTT: Arquivo Nacional Torre de Tombo
Ms. Livraria: Manuscritos da Livraria
BAHMOP: Biblioteca e Arquivo Histórico do Ministerio de Obras Púbricas
MMR: Montaria-mor do reino
BA: Biblioteca de Ajuda
BFDUL: Biblioteca da Faculdade de Direito da Universidade de Lisboa
DGT: Direcção-Geral do Territorio
IGP: Instituto Geográfico Português
CA: Cartografia Antiga

Acknowledgements

This contribution has benefited from the research project 'Protection, production and environmental change: the roots of Modern Environmentalism in the Iberian Peninsula (XVI–XVIIIth centuries)' of Gerda Henkel Stiftüng (project reference AZ 60/V/19)'.

I am grateful to Prof. Dr. Cristina Joanaz de Melo for reviewing this essay in depth and providing ideas and references for its improvement.

Bibliography

Appuhn, Karl. "Forests, Forestry, and State Power in Renaissance Venice." *The Journal of Modern History* 72, nº 4 (2000): 861–89. https://doi.org/10.1086/318548.

Aragón Ruano, Álvaro. "Mar de árboles, vorágines de jurisdicciones. La complicada relación entre la Real Armada española y los bosques del Pirineo Occidental peninsular en el siglo XVIII." In *Árvores, barcos e homens na Península Ibérica (séculos XVI–XVIII)*, coordinated by Rosa Varela Gomes and Koldo Trapaga-Monchet, 41–54. Zaragoza: Pórtico Librerias, 2017.

Baeta Neves, Carlos. *História Florestal, aquícola e cinegética. Colectânea de documentos existentes no Arquivo Nacional da Torre do Tombo* (Lisbon: Ministerio de Agricultura e Pescas, Instituto Florestal, 1980–1993), 6 vols.

Barbosa, Pedro. *Povoamento e estructura agrícola na Estremadura Central: Séc. XII a 1325*. Lisbon: Instituto Nacional de Investigação Científica, 1992.

Bluteau. Rafael. *Diccionario da Lingua Portugueza*. Lisbon: Officina de Thaddeo Ferreira, 1789, 2 vols.

Castro, Filipe Vieira de. *The Pepper Wreck. A Portuguese Indiaman at the Mouth of the Tagus River*. College Station: Texas A&M University Press, 2005.

Cortesão, Jaime. *Influencia dos Descobrimentos Portugueses na História da Civilização*. Lisbon: Casa da Moeda, 1993.

Devy-Vareta, Nicole, and António Monteiro Alves. "Os avanços e os recuos da floresta em Portugal – da Idade Média ao Liberalismo." In *Floresta e sociedade. Uma história em comum*, coordinated by Joaquím Sandes Silva, 55–76. Lisbon: Fundação Luso-Americana para o Desenvolvimento, 2007.

Devy-Vareta, Nicole. "Para uma geografia histórica da Floresta portuguesa. As matas medievais e a «Coutada velha» do Rei." *Revista da Faculdade de Letras – Geografia* 1 (1985): 47–67.

Devy-Vareta, Nicole. "Para uma geografia histórica da floresta portuguesa. Do Declínio das matas medievais à política florestal do Renascimento (séc. XV e XVI)." *Revista da Faculdade de Letras – Geografia* 1 (1986): 5–37.

Fernow, Bernhard E. *History of forestry in Europe, the United States and other countries* (Toronto: Toronto University Press, 1911).

Freire Costa, Leonor. *Naus e galeões na ribeira da Lisboa. A construção naval no século XVI para a Rota do Cabo*. Cascais: Patrimònia, 1997.

Gonçalves, Iria. *O Património do Mosteiro de Alcobaça nos séculos XIV e XV*. Lisbon: Universidade Nova de Lisboa, 1989.

Grossi, Paolo. *El orden jurídico medieval*. Madrid: Marcial Pons, 1992.

Grossi, Paolo. *Mitología jurídica de la Modernidad*. Madrid: Trotta, 2003.

Hespanha, António Manuel. *As vésperas do Leviathan. Instituições e poder político. Portugal (séc. XVII)*. Lisbon: Pedro Ferreira, 1987, 2 vols.

Labrador Arroyo, Félix. *La Casa Real en Portugal (1580–1621)*. Madrid: Polifemo, 2009.

Leite, Carlos. "Os trabalhos de Reinaldo Oudinot em Leiria nos finais do século XVIII: um plano global de ordenamento hidráulico, agrícola e florestal." Master Thesis: University of Coimbra, 2016.

Luxán Meléndez, Santiago. "El control de la hacienda portuguesa desde el poder central: la Junta de Hacienda de Portugal (1602–1608)." In *Política y Hacienda en el Antiguo Régimen*, edited by José Ignacio Fortea Pérez and Carmen Mª Cremades Griñán, 377–88. Murcia: Universidad de Murcia, 1993, vol. 1.

Maria Silva, Francisco, and Caetano Maria Batalha. *Memoria sobre o Pinhal Nacional de Leiria. Suas madeiras e productos rezinosos*. Lisbon: Imprensa Nacional, 1859.

Martínez Gutiérrez, Diego J. *La Junta de Hacienda de Portugal*. Pamplona: Newbook Ediciones, 1996.

Melo, Cristina Joanaz de. "Guerra, impérios e Corte Joanina nas Coutadas de Caça: Alavancas de Regeneração Florestal em Portugal, em meados do século XVIII." *Manuscrits. Revista d'Història Moderna* 42 (2020): 199–220. https://doi.org/10.5565/rev/manuscrits.296.

Melo, Cristina Joanaz de. "Menos coutadas melhores pinhais: imperio, inundações, fisiocracia, guerra e especialização das matas reais em Portugal (1777–1824)." *Tiempos Modernos. Revista Electrónica de Historia Moderna* 9 nº 39 (2019): 456–87.

Melo, Cristina Joanaz de. "Floresta em movimiento: usar, regenerar, cuidar (séculos XIV–XIX)." In *Como a Fénix Renascida – Matas, bosques e arvoredos (séculos XVI–XX): representações, Gestão, fruição*, coordinated by Cristina Joanaz de Melo, 79–130. Lisbon_ Colibrí, 2020.

Melo, Cristina Joanaz de, Catarina Madureira Villamariz, Tânia Manuel Casimiro, Pedro Urbano. "Florestas de encanto." In *Como a Fénix Renascida – Matas, bosques e arvoredos (Séculos XVI–XIX). Representações, Gestão, Fruição*, coordinated by Cristina Joanaz de Melo, 7–17. Lisbon: Colibrí, 2020.

Observatorio Técnico Independiente. *Recuperação da Mata Nacional de Leiria após os incêndios de outubro de 2017*. Lisbon: Assambleia da República, 2020.

Rakcham, Oliver. *Woodlands*. United Kingdom: Harper Collins, 2006.

Pinto, António Arala. *O Pinhal do Rei – Subsídios*. Alcobaça: Oficina de Oliveira Junior, 1938, 2 vols.

Rego, Francisco. *Florestas públicas*. Lisbon: Direcção Geral das Florestas, 2000.

Rivero Rodríguez, Manuel. *La edad de oro de los virreyes. El virreinato en la Monarquía Hispánica durante los siglos XVI y XVII*. Madrid: Akal, 2011.

Silva, José Justino. *Colecção chronologica da legislação portuguesa compilada e anotada (1603–1612)*. Lisbon: Imprensa de J. J. A. Silva, 1854.

Trapaga Monchet, Koldo. "*No es madera de vasallos, sino del rey*. Las políticas forestales de los Habsburgo en Portugal (1609–1640)." *Obradoiro de Historia Moderna* 28 (2019): 105–34. http://doi.org/10.15304/ohm.28.5915.

Trápaga Monchet, Koldo. "Supplying Timber for his Majesty's Fleets: Forest Resources and Maritime Struggle in Portugal (1621–1634)." In *Maritime History and Archaeology of the Global Iberian World (15th–18th centuries)*, edited by Ana Crespo Solana, Filipe Castro and Nigel Nayling, 215–48. Cham: Springer Nature, 2022.

Trapaga-Monchet, Koldo, and Romero-Calcerrada, Raúl. "Forest policies, administration, and management of the Leiria pinewood in Portugal (13th–18th centuries)." *Management & Organizational History* 17, nº 3–4 (2022): 138–65; https://doi.org/10.1080/17449359.2022.2153142.

Trapaga-Monchet, Koldo. "A destruction that preserves": Maritime warfare, empirical forestry and sustainability in Portugal (13th–17th centuries)." In *Roots of Sustainaibility in the Iberian Empires: shipbuilding and forestry, 14th-19th centuries*, edited by Koldo Trapaga-Monchet, Álvaro Aragón-Ruano and Cristina Joanaz de Melo, 183–208. United Kingdom: Routledge, 2023.

Varnhagen, Friderico. *Manual de instrucções praticas sobre a Sementeira, cultura e corte dos pinheiros*. Lisbon: Typografia da Academia, 1836.

Vaz Correia, Alexander, Ângelo Carvalho Oliveira and António Fabião, "Biologia e ecologia do pinheiro-bravo." In *Pinhais e eucaliptais. A floresta cultivada*, edited by Joaquim Sande Silva, 17–34. Lisbon: Fundação Luso-American para o Desenvolvimento, 2007.

Warde, Paul. "Fear of Wood Shortage and the Reality of the Woodland in Europe, *c.* 1450–1850." *History Workshop Journal* 62 (2006): 28–57. https://doi.org/10.1093/hwj/dbl009.

CHAPTER 6

Forestry Capability and Sustainability in the *Monteiro-mor* Ordinance of 1605 Study Area

Raúl Romero-Calcerrada and Koldo Trapaga-Monchet

1 Introduction

In the early fifteenth century, the Portuguese Monarchy began to expand beyond the Iberian Peninsula, first to North Africa with the conquest of Ceuta in 1415, from where they sailed to all over Africa and reached India (1498) and Brazil (1502). The first ships were caravels, Mediterranean-style vessels weighing around 50–150 tonnes. However, as the decades progressed, fleets became larger, more complex and permanent. Maritime warfare forced the Iberian Empires to invest more sizeable sums in the construction and maintenance of ocean-going vessels, which were paramount not only to keeping the maritime empires connected, but also to ensuring the arrival of income for the always needy Royal Treasury.[1] This encouraged the Iberian Empires to take steps not only to protect home-grown trees, but also to ensure the importation of the required wood through merchant timber networks.[2]

Such large warships and ocean-going vessels were highly complex technological creatures, and they required very specific types of wood, especially for the structural timbers and masts. Consequently, the Portuguese Crown took several steps to ensure the construction of these ships. Among other elements,

1 Leonor Freire Costa, *Naus e galeões na ribeira de Lisboa: a construção naval no século XVI para a Rota do Cabo* (Cascais: Patrimonia Historica, 1997). Francisco Contente Domingues, *Os navios do mar oceano: teoria e empiria na arquitectura naval portuguesa dos séculos XVI e XVII* (Lisbon: Centro de História da Universidade de Lisboa, 2004); Jan Glete, *Warfare at Sea, 1500–1650: Maritime Conflicts and the Transformation of Europe* (New York: Routledge, 2000); Augusto Alves Salgado, "Portugal e o Atlântico: organização militar e acções navais durante o período filipino: (1580–1640)" (PhD diss., Universidade de Lisboa, 2009); Bartolomé Yun Casalilla, *Iberian world empires and the globalization of europe 1415–1668* (Basingstoke: Palgrave Macmillan, 2019), 51–98.

2 Ana Crespo Solana and Nigel Nayling, "ForSEAdiscovery. Forest resources for Iberian Empires: Ecology and Globalization in the Age of Discovery (16th–18th centuries)," in *Actas del V Congreso Internacional de Arqueología Subacuática* (IKUWA V) (Madrid: Ministerio de Educación, Cultura y Deporte, 2016), 896–904; Ana Crespo Solana, "ForSEAdiscovery: la construcción naval y el comercio de la madera del siglo XVI al XVIII," *Revista PH* 96 (2019): 114–41.

DOI:10.1163/9789004745698_008

the Monarchy built a cluster of facilities near the royal palace in Lisbon, known as the *Ribeira das naus*, which was already being run by Crown officials in 1524.[3]

In addition, it was necessary to guarantee a steady supply of high-quality timber to the shipyards. The tree species most commonly used at the Lisbon shipyards were *Quercus suber*, *Pinus pinea* and *Pinus pinaster.*[4] Consequently, these species turned out to be strategic for the Early Modern Portuguese Monarchy, which issued an extensive range of administrative documentation concerning their protection, availability, production and scarcity. Early Modern policymakers and shipwrights deliberately crafted the argument about the scarcity of wood and timber to legitimise new forest policies.[5] Therefore, it is not surprising that claims of timber shortages were directly related to the availability of these tree species.

Numerous authors have studied Portuguese forestry legislation on hunting and shipbuilding.[6] In Portugal, the royal forests began to increase in number and extension in the late thirteenth century. The *montaria-mor* was the main institution entrusted with protecting the royal forests for hunting and shipbuilding from at least the 1300s to the 1700s.[7] As some authors have correctly pointed out, the *monteiro-mor* ordinance or regulation issued in 1605 noted a scarcity of timber for shipbuilding which was endangering vassals' well-being and the survival of the Portuguese Monarchy.[8]

3 Leonor Freire Costa, "Aspectos empresariais da construção naval no século XVI: o caso da Ribeira das Naus de Lisboa," *Análise Social* 31, no. 136/137 (1996): 295–304.

4 Filipe Castro, *The Pepper Wreck: A Portuguese Indiaman at the Mouth of the Tagus River*, Texas: Texas A&M University Press, 2005), 114–61; Costa, *Naus e galeões*, 305–33; Domingues, *Os navios, passim.*

5 Karl Appuhn, *A Forest on the Sea: Environmental Expertise in Renaissance Venice*, (Baltimore: Johns Hopkins University Press, 2009); Alfredo José Martínez González, *Las Superintendencias de Montes y Plantíos (1574–1748): derecho y política forestal para las armadas en la Edad Moderna*, Valencia: Tirant lo Blanch, 2015; Paul Warde, "Fear of Wood Shortage and the Reality of the Woodland in Europe, c. 1450–1850," *History Workshop Journal* 62, no. 1 (2006): 28–57; John T. Wing, *Roots of Empire. Forests and State Power in Early Modern Spain, c. 1500–1750*, Leiden: Brill, 2015.

6 Félix Labrador Arroyo, *La Casa Real en Portugal, 1580–1621* (Madrid: Polifemo, 2009), 215–41; Koldo Trápaga-Monchet, "El estudio de los bosques reales de Portugal a través de la legislación forestal en las dinastías Avis, Habsburgo y Braganza (ca. 1435–1650)," *Philostrato: Revista de Historia y Arte* 1 (2017), 5–27.

7 Labrador Arroyo, *La Casa Real*, 225–41; Cristina Joanaz de Melo, *Coutadas reais (1777–1824): Privilégio, poder, gestão e conflito* (Lisbon: Montepio Geral, 2000); Cristina Joanaz de Melo, *An Analysis of the Royal Preserves in Portugal. Issues of privilege, power, management and conflict* (Sheffield: Wildtrack, 2015); Trápaga Monchet, "El estudio," 5–27.

8 Nicole Devy-Vareta, "Para uma geografia histórica da floresta portuguesa: do declínio das matas medievais à política florestal do Renascimento (sécs. XV e XVI)," *Revista da Faculdade de Letras* 1 (1986): 24–5, 33–4; Labrador Arroyo, *La Casa Real*, 238–41.

However, the idea of forest destruction dated back at least to the nineteenth century. Since the mid-nineteenth century, scholars on Portuguese maritime history have constantly churned out the argument that imperial shipbuilding had been the main reason for the destruction or decline of the Portuguese woodlands, with this destruction reaching its peak in the eighteenth century.[9] Nevertheless, this literature is missing a contribution that makes numerical calculations on the capacity of Portuguese land to produce trees belonging to the species used for shipbuilding during the Early Modern Age.

For this reason, the objective of this contribution is a) to create maps showing the limits of the management and protection of the forest area according to the *monteiro-mor* ordinance of 1605; b) to determine the potential distribution of *Quercus suber*, *Pinus pinaster* and *Pinus pinea* (established through Integrated Suitability for Tree Species); and c) to use different scenarios to provide an approximation of the number of trees available in the jurisdictional territory of the *monteiro-mor* ordinance of 1605.

2 Materials and Methods

Cartography is a handy geographic tool for understanding human decisions and ascertaining the spatial dimension of human activity in a territory. Unfortunately, prior to the late eighteenth century, no historical maps reflect systematic information about management areas or existing resources in the spaces studied here.

2.1 *Cartographic Approach to the Limits of the* Monteiro-mor *Ordinance or Regulation of 1605*

The legislative compilation carried out by the judge of the royal forests (*juiz das coutadas*) Heitor Botelho in 1584, during the Union of the Crowns, constitutes one of the most crucial documentary corpuses of Portuguese forestry legislation, together with the law or regulation of the *monteiro-mor* of 1605.[10] The aforementioned compilation of 1584 includes legislation issued throughout

9 Among others José Bonifacio Silva, *Memoria sobre a necessidade e utilidade do plantio de novos bosques em Portugal* (Lisbon: Na Typographia da Academia Real das Sciencias, 1815), 16–22; James Duffy, *Shipwreck & Empire. Being an Account of Portuguese Maritime Disasters in a Century of Decline* (Harvard: Harvard University Press, 1955); Charles R. Boxer, *The Portuguese Seaborne Empire (1415–1825)* (London: Hutchinson, 1977); Costa, *Naus e galeões*, 191–92.

10 Labrador Arroyo, *La Casa Real*, 233–41; Nicole Devy-Vareta and António A. Monteiro Alves, "Os avanços e os recuos da floresta em Portugal – da Idade Média ao Liberalismo," in

the sixteenth century concerning: a) new and old regulations for the protection of the royal preserves and b) officers who worked exclusively in the royal preserves. Especially noteworthy are the rules issued in 1560 to regulate the office of bailiff of the royal forests (*juiz das coutadas*).[11]

The *monteiro-mor* ordinance of 1605 governed the obligations, duties and privileges of the *monteiro-mor* and the staff under his supervision. The *monteiro-mor* was responsible for the *montaria-mor*, which was the main institution entrusted with the protection and preservation of the royal forests from the fourteenth to the early nineteenth century in Portugal.[12] The ordinance of 1605 constituted a benchmark in Portuguese forestry legislation because it gathered a legislative effort that the Crown had carried out since the fifteenth century. Moreover, this legislation detailed hundreds of private woodlands that contained suitable (in the present or in the future) trees to produce sturdy timber for naval construction.[13] It was one of the most important Portuguese forestry laws until the mid-1700s.

The ordinance indicates the limits of the forests and forest districts based on: 1) natural landscapes without precisely specifying the boundaries; 2) human elements that are difficult to trace today; and 3) a literal description of the landscape based on ancient toponymy and natural landscapes that have been modified over time, as well as manmade elements that no longer exist, making their exact georeferencing difficult. These were the same difficulties that Nicole Devy-Vareta faced when she tried to locate the woodlands and royal forests of the fifteenth-century Óbidos forest district.[14]

It may be concluded from an analysis of the 1584 compilation of forestry legislation and the law of 1605 that: a) the main population centres that acted as the "capital or main centres" of the forest districts, including those of Colares, Sintra and Sierra de Sintra, were already established because these districts were still royal forests, despite not being included in the ordinance of 1605, though they were regulated by several provisions listed in the compilation of 1584; and b) as for the limits of the forest areas included in the ordinance, in

---

Floresta e sociedade. Uma história em comum, ed. Joaquim Sande Silva (Lisbon: Fundação Luso-Americana, 2007), 64–5.

11 Biblioteca de Ajuda (BA), Ms. 44-XIII-61.

12 Labrador Arroyo, *La Casa Real*, 205–41; Joanaz de Melo, *An Analysis*, *passim*.

13 José Justino de Andrade Silva, *Collecção chronologica da legislação portugueza (1603–1612)* (Lisbon: Imprensa de J. J. A. Silva, 1854), 112–21; Labrador Arroyo, *La Casa Real*, 238–41; Trápaga Monchet, "El estudio," 18–21.

14 Nicole Devy-Vareta, "Para uma geografia histórica da floresta portuguesa: as matas medievais e a "coutada velha" do Rei," *Revista da Faculdade de Letras* 1 (1985): 62–3.

the north (Coimbra) we established the boundary at twenty kilometres north of the mouth of the Mondego river. In contrast, there is more detailed information about the Coruche forest district, making it possible to establish the territorial delimitation in the Montargil mountain range at the southeast end of the map. Further south (Alcácer do Sal forest district), the information provided by the 1605 regulation is scarce: it merely indicates that there were many trees on both banks of the Sado River, so we chose Alcácer do Sal as the southern limit. The boundaries can likely be extended further to the south as the law stipulated that the entire Sado River fell within the scope of the regulation.

However, in this essay, we preferred not to push the boundaries further to the south for two reasons: a) the Portuguese Monarchy did not appoint new keepers to safeguard such a large area, whose protection it would have been impossible to ensure; and b) although the historical documents on timber supplies from 1605 to 1640 include references to logging timber (pine) from this area (Alcácer do Sal) to the shipyards of Lisbon, it is not known if there was a permanent administrative staff in place to protect it.[15] Furthermore, we included the boundaries of the Palmela forest district, located on the left margin of Tagus River from around Samora Correia up to the Atlantic Ocean and Setúbal, even though the regulation of 1605 specified the disbandment of this royal forest. We made this decision for two reasons: 1) it was not the first time that the Crown decided to dismantle this forest;[16] and 2) this area had supplied timber for shipbuilding for centuries.[17] The legal situation of the royal forest of Montemor-o-Novo, which had been disbanded in 1498 and again in 1605, remains unclear.[18]

The new maps we generated show our first approximation of the forestry limits of the scope of jurisdiction of the *monteiro-mor* regulation of 1605, based on current cartography and using the historical sources on forestry legislation.

The current maps used are the Administrative Limits (Direção-Geral do Território (Portugal), 2018) "Geographical information given by the

15 Biblioteca e Arquivo Histórico do Ministerio de Obras Púbricas (BAHMOP), Montaria-Mor do Reino (MMR), núcleo 8; Trapaga Monchet, "El estudio," 114; Koldo Trapaga-Monchet, "Supplying Timber for his Majesty's Fleets: Forest Resources and Maritime Struggle in Portugal (1621–1634)," in *Heritage and the Sea: Volume 1: Maritime History and Archaeology of the Global Iberian World (15th–18th centuries)*, ed. Ana Crespo Solana, Filipe Castro and Nigel Nayling (Cham: Springer International Publishing, 2022), 227–31.

16 Devy-Vareta, "Para uma geografia," 1986, 21–3.

17 Devy-Vareta, "Para uma geografia," 1986, 10.

18 Silva, *Collecção chronologica*, 123; Devy-Vareta, "Para uma geografia," 1986, 21–4.

Direção-Geral do Território," maps of towns, cities, etc. of Portugal,[19] a Digital Elevation Model (DEM),[20] the watershed lines,[21] and the River Network,[22] relating to Mainland Portugal. The main population centres of the forest districts under the forestry legislation issued up to 1584 were generated after consulting maps of Portuguese towns, cities, etc.[23]

In addition to the population centres and the hydrographic network, we chiefly used the watershed lines[24] and the Digital Elevation Model (MDE) to generate digital maps. With these two elements, the hydrographic basins of rivers were determined from the 1st to the 3rd levels[25] and, in some cases, other watersheds were used to subdivide hydrographic basins.

We attempted to generate maps that "interpret" the meaning of the values reported in the historical documents. These maps provide an approximation by taking into account the geographical limits and natural elements that were probably used by the officers of the Portuguese Monarchy to create, select and establish the limits of the royal forests.

Topography was the primary element we used to define the geographical space covered by the *monteiro-mor* ordinance or regulation of 1605, as it would have facilitated the mobility and extraction of forest resources. Given the technology, infrastructure and resources of the time, we considered that the summit line is a natural border that would have covered this extension and/or would have fallen within the area of influence of the "capital or main headquarters" of the forest districts. We used ArcGIS Desktop 10.8 and the Coordinate System ETRS_1989_Portugal_TM06 (EPSG:3763). Basic spatial analysis operations were carried out (selection by attributes, area of influence or spatial overlap).

19 Frederik Ramm, "Portugal: OpenStreetMap Data in Layered GIS Format." Data/Maps Copyright 2018 Geofabrik GmbH and OpenStreetMap Contributors. Map tiles: Creative Commons BY-SA 2.0 Data: ODbL 1.0, 2019.

20 European Digital Elevation Model (EU-DEM), version 1.1." European Environment Agency (EEA) under the framework of the Copernicus programme, 2016.

21 João Ferreira Silva, Manuela Raposo Magalhães and Natália Cunha, "Linhas de Festo de Portugal Continental," Lisbon: LEAF/ISA/ULisboa, 2013, http://epic-webgisportugal.isa.utl.pt/.

22 João Ferreira Silva, Manuela Raposo Magalhães and Natália Cunha, "Classificação hierárquica e toponímica das Linhas de Água de Portugal Continental," Lisbon: LEAF/ISA/ULisboa, 2013, http://epic-webgis-portugal.isa.utl.pt/.

23 Ramm, "Portugal."

24 Silva, Raposo and Cunha, "Linhas de Festo."

25 Silva, Raposo and Cunha, "Classificação hierárquica."

3 Estimating the Forestry Area and the Number of Trees Available Per Year within the Territorial Scope of the *Monteiro-mor* Ordinance or Regulation of 1605

The main Portuguese species used for crafting vessels in Lisbon shipyards according to shipbuilding treatises, archival sources and material remains from archaeological sites were *Pinus pinea*, *Pinus pinaster* and *Quercus suber*.[26]

Undoubtedly, establishing the location of forest resources is fundamental in determining availability and the load capacity that human activity can exert on forest spaces. Unfortunately, there are no precise data on the extension (area of the forestry species being exploited) and characteristics (number of trees or volumes per species) for the historical period studied here. We examined different scenarios and assumptions to achieve an approximate quantification of the forest area in question and the number of trees available per year (for example, for shipbuilding or other activities that required this resource).

We calculated the approximate area of influence of the limits of the *monteiro-mor* ordinance or regulation of 1605 using the maps generated in the previous section (Figure 6.1).

To determine the potential distribution and attempt a quantification of the forest area, we carried out an Integrated Suitability assessment for *Quercus suber*,[27] *Pinus pinea*[28] and *Pinus pinaster*.[29]

Using Figure 6.1, we calculated the Integrated Suitability for these species within the proposed territorial limits of the *monteiro-mor* ordinance or regulation of 1605.

Our analysis made several assumptions: a) it was considered that 10% of the total forest area resulting from the Integrated Suitability for Tree Species consisted of good-quality, productive forest masses usable for shipbuilding. In addition, b) these productive forest masses were taken to have measured

26 Castro, *The Pepper Wreck*, 114–41; Costa, *Naus e galeões*, 305–31; Domingues, *Os navios*, *passim*; Koldo Trapaga-Monchet, "No es madera para vasallos, sino del rey. Las políticas forestales de los Hasburgo en Portugal (1609–1640)," *Obradoiro de Historia Moderna* 28 (2019): 116–29.

27 Manuela Raposo Magalhães, Ana Müller and João Ferreira Silva, "Aptidão Integrada ao Sobreiro (*Quercus suber* L.) para Portugal Continental," LEAF/ISA/ULisboa, 2015a, http://epic-webgis-portugal.isa.utl.pt/.

28 Manuela Raposo Magalhães, Ana Müller and João Ferreira Silva, "Aptidão Integrada ao Pinheiro-manso (*Pinus pinea* L.) para Portugal Continental," LEAF/ISA/ULisboa, 2015b, http://epic-webgis-portugal.isa.utl.pt/.

29 Manuela Raposo Magalhães, Ana Müller and João Ferreira Silva, "Aptidão Integrada ao Pinheiro-bravo (*Pinus pinaster* Aiton) para Portugal Continental," LEAF/ISA/ULisboa, 2015c, http://epic-webgis-portugal.isa.utl.pt/.

densities of 25 trees per hectare (trees/ha) for *Quercus suber* and 100 trees/ha for *Pinus pinea* and *Pinus pinaster*. Finally, c) a cutting rotation of 100, 125 and 150 years was considered for *Quercus suber*, and 50, 75 and 100 years for *Pinus pinea* and *Pinus pinaster* to determine the number of trees available annually.

The criteria established in the design of our scenario were pretty restrictive in order to approximate the minimum quality of the wood used for shipbuilding at the time and, especially, to work with the lower threshold to determine the forestry capacity and the sustainability of the forest management.

4 Results and Discussion

Mainland Portugal is located in the west part of the Iberian Peninsula. The northern part of the country is mountainous, an extension of the Galician Massif with the Serra da Estrêla and foothills of the Central System as its southern limit. The Douro River, embedded in the sharp relief, crosses this northern part of Portugal.

The Tagus River can be considered the natural element that marks the transition towards small mountain ranges and low plateaus and mainly through plains and broad valleys that characterize the centre and south of Portugal. A prominent feature of this area is the Alentejo Plateau, surrounded in its southern part by the elevations of the Algarve.

The whole of Portugal is located in the Mediterranean climatic zone with Atlantic influence. Temperatures increase as we move south, while rainfall decreases to aridity.

4.1 *Cartographic Approach to the Limits of the* Monteiro-mor *Ordinance or Regulation of 1605*

The *monteiro-mor* ordinance or regulation of 1605 stemmed from the need to structure and manage the territory. It was aimed at regulating and protecting a resource that was valuable for energy (domestic and industrial use), shipbuilding, food for people and cattle, shade for livestock, construction, and multiple uses for Portuguese dwellers. This regulation is of great relevance to understanding forest management at that historical moment; unfortunately, determining its spatial scope of action is a complex task, given the imprecision of many of its limits, the presence of ancient toponymy and natural landscapes that are difficult to locate today and constructions unoccupied by humans.

As an initial approximation, we drew up a proposal for the geographical space enclosed by the limits of the forests and forestry districts or the territorial scope of application of the *monteiro-mor* ordinance of 1605. The cartographic

TABLE 6.1 Total area (in hectares) of the jurisdictional scope of the *monteiro-mor* ordinance of 1605

	Total area (in hectares)
Monteiro-mor ordinance of 1605	1 918 584.67

proposal should be interpreted as a starting point for carrying out initial quantitative estimates (Table 6.1) that provide a rough idea of its size. In subsequent studies, an attempt will be made to create a more detailed and accurate map of some forest districts.

Forest harvesting areas for shipbuilding were mainly located near the coastline or navigable rivers due to the wood-producing areas' lack of access and transport infrastructure. We interpreted that the spirit of the ordinance stemmed from natural (hypsometry and morphology of the land, among other things) and spatial (maximum reference distances) criteria. In some cases, specifically in the Southeast area, we included hydrographic basins that it is logical to think would have fallen within the area of influence, although there is no absolute certainty. Therefore, this map should be considered an initial approximation, which can be improved on.

The cartographic proposal (Figure 6.1) is based on an analysis of the population centres that acted as the "capitals" of the forest districts and the topographical elements that provided precise references (for example, 20 kilometres north of the mouth of the Mondego river; or in the forest district of Coruche the territorial delimitation in the Montargil mountain range). We made an initial approximation by applying criteria for choosing limits similar to the interpretation made for the map of the Cork Oak Law of 1546.[30]

In the analysis carried out, in most cases we observed a certain correspondence between the Hydrographic Basin and the "capital" of the forest districts. For this reason, we considered that landforms and natural elements (lines of mountain peaks or hydrographic basins) could provide a valid approximation of the administrative limits. There is a sound basis for this assumption since, at the time, the accessibility of the land, the availability of important watercourses and the optimal location of the forest management centres

30 Raúl Romero-Calcerrada and Koldo Trapaga-Monchet, "La ley del alcornoque de 1546 y la ordenanza del Monteiro-Mor de 1605," in *Historia, Sociedad y Medio ambiente: la sostenibilidad*, coord. Koldo Trapaga-Monchet and Luis Alberto Polo Romero (Madrid: Sílex, 2022), 143–79.

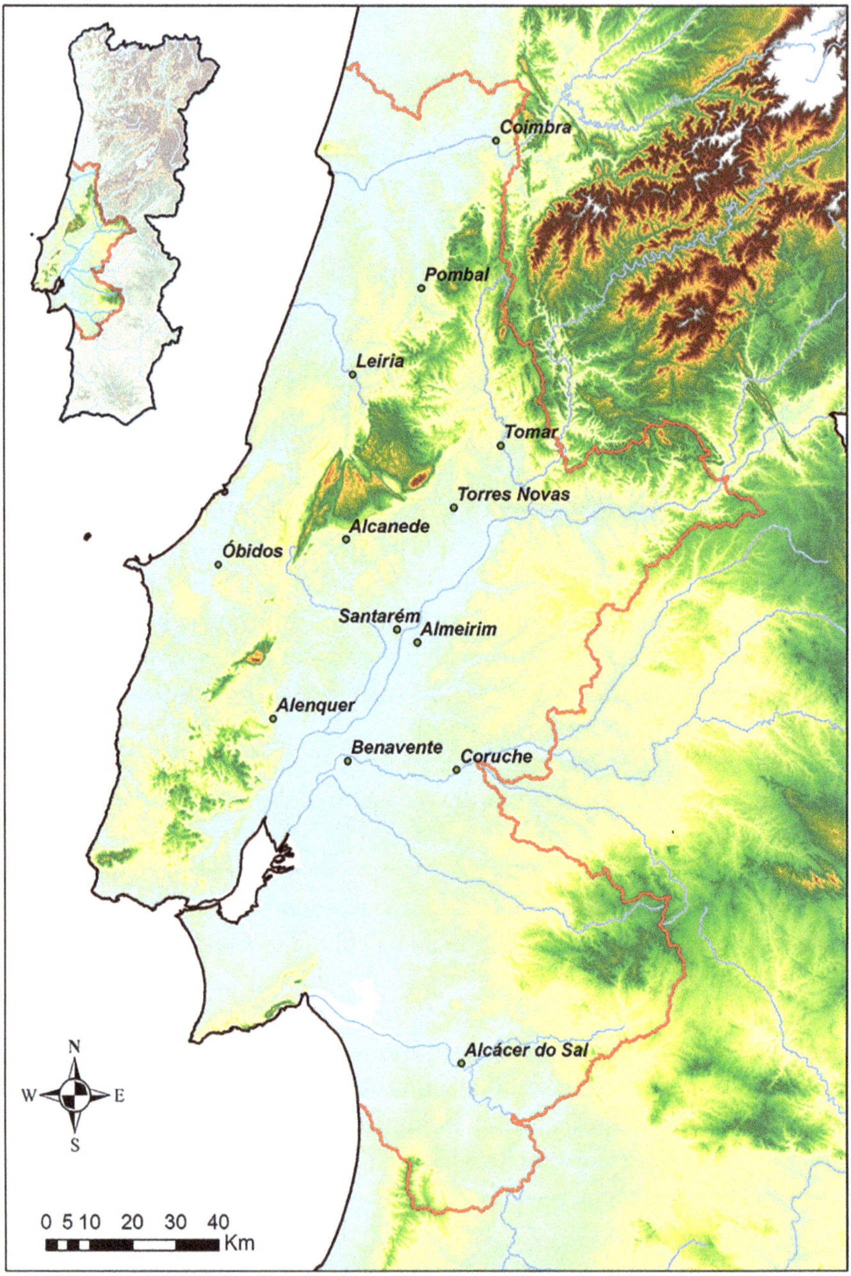

FIGURE 6.1 The geographic space of the *monteiro-mor* ordinance or regulation of 1605
SOURCE: AUTHORS' OWN ELABORATION, EUROPEAN DIGITAL ELEVATION MODEL (2016) AND RAMM, F. (2019)

were significant factors.[31] It would not make sense to ascend hills with large volumes or weights of wood only to descend again on the other slope because this would have required a larger workforce (with the consequent economic outlay). In other words, logic indicates that whenever it was possible the law of least effort or the law of gravity would have prevailed instead of having been fought against. These elements would have generally underlain the definition of the management zones or forest districts.

4.2 *Estimating the Forestry Area and Number of Trees Available per Year for the Territorial Scope of the* Monteiro-mor *Ordinance or Regulation of 1605*

The location of forest resources is an important factor in any production process, especially when there are technical and logistical limitations, as there were in the Early Modern Age when it came to harvesting timber for shipbuilding. Both the distance between the resource (the tree) and the place of processing (e.g., sawmill or shipyard) and, mainly, the area available for exploitation were very relevant to determining the supply of resources. In general, it was intended that the exploited forest masses and the shipyards should be as close as possible.[32] This was because forests were cleared by dragging logs, with the aid of oxen and sometimes even on men's shoulders, to waterways that allowed them to be transported by flotation.[33]

There is no precise documentation or cartography on forests (location, surfaces, etc.) and the activities related to their management in the kingdom of Portugal during the period studied here. Only general references can be found on the location of mountains, areas of forestry activity, etc., in historical records (Devy-Vareta 1985). At best, there are narrative maps of the limits of each of the forest districts, such as the compilation carried out in 1584 and the regulation of 1605, which have made it possible to generate the scope of this regulation (Figure 6.1).

For this reason, establishing the potential distribution of the forest species of *Quercus suber*, *Pinus pinaster*, and *Pinus pinea* helps ascertain the location and the exploitable area (Figure 6.2, Tables 6.2 and 6.3).

31 Jorge Gaspar, "Os portos fluviais do Tejo," *Finisterra* 5, no. 10 (1970): 153–215.

32 Appuhn, *A forests*, 47; Paul W. Bamford, *Forests and French Sea Power 1660–1789* (Toronto: University of Toronto Press, 1956), 49–50; Martínez González, *Las Superintendencias*, 381–3; Wing, *Roots of Empire*, 166–9.

33 Gaspar de Aranda y Antón, *La carpintería y la industria naval en el siglo XVIII* (Madrid: Instituto de Historia y Cultura Naval, 1999).

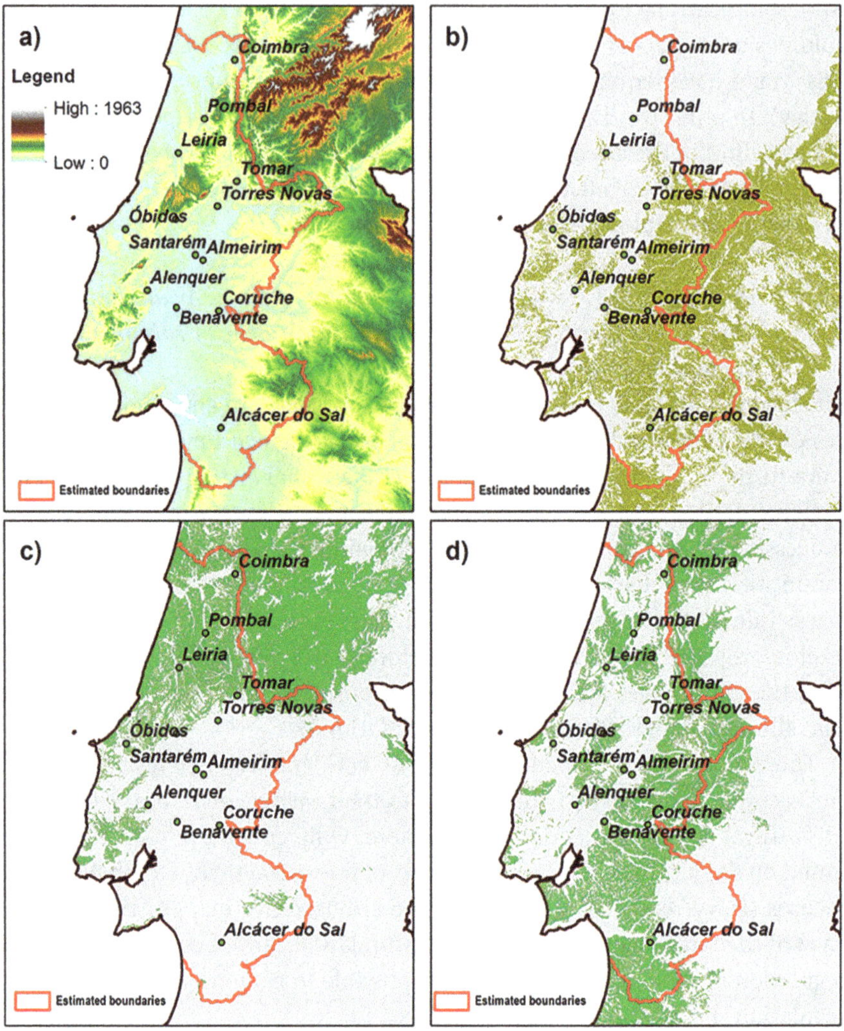

FIGURE 6.2 a) Hypsometry (in metres) of the geographic space of the *monteiro-mor* ordinance or regulation of 1605 (line in red). Integrated Suitability for b) *Quercus suber*, c) *Pinus pinaster* and d) *Pinus pinea*
SOURCE: AUTHORS' OWN ELABORATION, EUROPEAN DIGITAL ELEVATION MODEL (2016), MAGALHÃES ET AL (2015A), MAGALHÃES ET AL (2015B) AND MAGALHÃES ET AL (2015C)

Although these data are approximate and based on scenarios, this is a first step towards being able to evaluate the management of resources and the effects of human activity on them. As stated earlier, there is no detailed information on the potential distribution of the forest species of interest. At this point, the following question arises: how can this issue be solved?

TABLE 6.2 Estimation of forested areas (in hectares) for *Quercus suber, Pinus pinaster* and *Pinus pinea*

	Quercus suber	*Pinus pinaster*	*Pinus pinea*
Monteiro-mor ordinance of 1605	607,661.63	420,719.88	627,647.31

TABLE 6.3 Estimating the number of trees available in Figure 6.2 with the indicated density (trees/ha) considering each species to account for 10% of the forest area

		25 trees/ha	100 trees/ha
Quercus suber	Total number of trees	1,519,154	
	Number of trees available each year (silvicultural rotation every 100 years)	15,191	
	Number of trees available each year (silvicultural rotation every 125 years)	12,153	
	Number of trees available each year (silvicultural rotation every 150 years)	10,128	
Pinus pinea	Total number of trees		6,276,473
	Number of trees available each year (silvicultural rotation every 50 years)		125,529
	Number of trees available each year (silvicultural rotation every 75 years)		83,686
	Number of trees available each year (silvicultural rotation every 100 years)		62,765
Pinus pinaster	Total number of trees		4,207,199
	Number of trees available each year (silvicultural rotation every 50 years)		84,144
	Number of trees available each year (silvicultural rotation every 75 years)		56,096
	Number of trees available each year (silvicultural rotation every 100 years)		42,072

We chose to use Integrated Suitability for Tree Species as a valid means of solving these shortcomings. Integrated Suitability for Tree Species, in our case *Quercus suber*, *Pinus pinaster* and *Pinus pinea* (Figure 6.2), is the result of integrating the Soil and Morphological Suitability to Silviculture map and the Bioclimatic Suitability to each of Tree Species and Pastures map. Combining the results/data of both provides a more accurate location for each tree species, depending on the ecological value of the soil, the land morphology and the slope.

These data (Figure 6.2, Tables 6.2 and 6.3) make it possible to establish the location and the theoretical area where these species grew. Although this starting point is approximate and based on scenarios, it is the first step towards evaluating the sustainability of resource management and human activity's effects on these resources.

In addition, resource management is relevant to the productive process if its continuity or sustainability is desired. Wood was a strategic natural resource for human societies during the Middle and Early Modern ages[34] and even until the end of the nineteenth century. Wood is renewable if adequately managed, providing goods (fuel, tools, building materials, etc.) and services to human societies since the dawn of history.[35] These characteristics made it an essential resource and support for the activity and development of any country, especially in the Middle and Early Modern ages. Its importance reached such heights that some authors have called the Modern Age the "Wooden Age."[36]

Forest management ensured that the felling of forest masses yielded the types and qualities of wood that were required for each use. Felling could be selective (aimed at specific pieces for shipbuilding where quality prevailed), but clear-cutting was also carried out to obtain raw timber.[37] As stated, wood was multifunctional and those types that were not suitable for shipbuilding

34 María Jesús Melero Guilló, "A la mar madera": La madera en la arquitectura naval española," in *Andalucía, América y el mar: Actas de las IX Jornadas de Andalucía y América*, ed. Bibiano Torres Ramírez (Sevilla: Diputación de Huelva, 1991), 145–57.

35 Bill Slee, "Landscape goods and services related to forestry land use," in *Multifunctional Land Use: Meeting Future Demands for Landscape Goods and Services* (Berlin–Heidelberg: Springer, 2007), 65–82.

36 John A. Perlin, *A Forest journey: The Story of Wood and Civilization* (Woodstock, Vermont: Countryman Press 2005); Michael Williams, *Deforesting the Earth: From Prehistory to Global Crisis. An Abridgment.* Chicago: University of Chicago Press, 2010.

37 Aranda y Antón, *La carpintería, passim.*

were put to other uses.[38] For this reason, many fellings were made to obtain firewood and charcoal for household and industrial use.[39] For these purposes, the felling of trees was not selective and could lead to an entire forest mass being cut down.

The map generated (Figure 6.1) enabled us to calculate the total jurisdictional area (Table 6.1), and, based on the proposed limits, the area (Table 6.2) occupied by forest species. Figure 6.1 and Table 6.2 provide an approximate idea of the potential distribution of *Quercus suber, Pinus pinaster* and *Pinus pinea*.

Based on these data, we made the assumption that 10% of the available forest areas yielded good-quality trees suitable for forest uses (wood, charcoal, etc). This analysis, first and foremost, enabled us to estimate the forest cover in areas defined in our scenarios as regulated by the *monteiro-mor* ordinance of 1605 and, therefore, subject to forest management.

Considering non-overlapping areas, there would have been 60,766 ha of *Quercus suber* (3.2% of the jurisdictional area of the *monteiro-mor* ordinance of 1605), 42,072 ha of *Pinus pinaster* (2.2%) and 62,765 ha of *Pinus pinea* (3.3%). As can be seen, the figure of 8.6% of usable forest area is well below the minimum percentage of forest area in the 1995–2015 period:[40] 35% in 2010 in mainland Portugal, 21% in 2010 in the NUTS II Lisbon Region, 37% in 2010 in the NUTS II Center Region and 42% in 1995 in the NUTS II Alentejo Region.

Table 6.3 estimates the number of available trees based on their density: 25 trees/ha for *Quercus suber* and 100 trees/ha for *Pinus pinaster* and *Pinus pinea*. These densities include lower thresholds than those found nowadays.[41] Quercineas (e.g., *Quercus suber*) tend to have a lower density than pine forests, typical figures for this species being between 25 trees/ha and 75 trees/ha. In addition, silvicultural rotation is usually longer, in the region of 100 or 150 years, although it could be done from 75 years. In the case of pine forests (e.g., *Pinus pinaster*), the density could vary from 100 trees/ha to 200 trees/ha, and optimal silvicultural rotation was between 50 and 100 years.

---

38 Félix Labrador Arroyo and Koldo Trápaga-Monchet, "La configuración del espacio y la explotación forestal de un enclave singular: el Real Sitio del Soto de Roma durante la dinastía Hasburgo," *Studia Histórica. Historia Moderna* 39, no. 3 (2017): 310–22.

39 Fernando Reboredo and João Pais, "Evolution of forest cover in Portugal: A review of the 12th–20th centuries," *Journal of Forestry Research* 25, no. 2 (2014b): 249–56.

40 José Sousa Uva, *6.º Inventário Florestal Nacional* (*IFN6*). Instituto da Conservação da Natureza e das Florestas (ICNF), 2019; José Sousa Uva and Sónia Pacheco Faias. *6.º Inventário Florestal Nacional* (*IFN6*). *Anexo Técnico. Versão 1.0 I*: Instituto da Conservação da Natureza e das Florestas (ICNF), 2019.

41 Uva, *6.º Inventário*; Uva and Pacheco, *6.º Inventário*.

The above assumptions define scenarios of extreme or unfavourable conditions that are expected to give rise to reflection on the sustainability of forestry activity and open new avenues of research on the problem of human–nature interaction. The purpose is not to accurately reproduce or forecast what existed in the period studied but to understand, from a spatial and quantitative approach, what resources these territories could provide in the Early Modern Age in order to shed some light on the processes of deforestation or pressure on the natural environment that are reported in the literature of various fields of study.[42]

The data and assumptions in Tables 6.2 and 6.3 seem conclusive for determining potential stocks and making estimates of resource availability. Acting on less than 9% of the jurisdictional area of the *monteiro-mor* ordinance of 1605, between 15,191 and 10,128 *Quercus suber* trees aged between 100 and 150 years could be extracted annually and sustainably. For *Pinus pinea*, between 125,529 and 62,765 trees aged between 50 and 100 years could be cut annually without compromising their viability; and for *Pinus pinaster*, between 84,144 and 42,072 trees under the same conditions as *Pinus pinea*.

In this case, with scenarios that pose extreme conditions, the results show that a sustainable extraction of 22,486,400 trees every 100 years (*Quercus suber*/*Pinus pinea* and *Pinus pinaster*, respectively) and 11,496,500 trees every 100 years (*Quercus suber*/*Pinus pinea* and *Pinus pinaster*, respectively) could have been carried out. It is important to stress the idea that forest masses are living systems. Forest cover not only decreases with logging but also grows back and expands with astonishing speed once human action is relaxed,[43] or through human actions such as reforestation.

Some assumptions have been made here, and the results of some of the possibilities have been provided. Reducing silvicultural rotation or modifying density would indeed have affected the volume of wood and, in some cases, its quality for shipbuilding. However, under similar environmental conditions, the viability of the forests would not have been impacted, nor would deforestation processes have been triggered. Nevertheless, further investigation is required to establish whether there were repeated efforts to exploit certain local areas. Undoubtedly, these figures should spark reflection on the

42 Marta Domínguez-Delmás, *et al.*, "Tree-rings, forest history and cultural heritage: Current state and future prospects of dendroarchaeology in the Iberian Peninsula," *Journal of Archaeological Science* 57 (2015): 184–5; Reboredo and Pais, "Evolution of forest," 2014b, 250–4.

43 Michael Williams, "Dark ages and dark areas: Global deforestation in the deep past," *Journal of Historical Geography* 26, no. 1 (2000): 28–46.

causes that generated the deforestation processes or pressure on the natural environment.[44]

The number of trees available annually and every ten years, resulting from maintaining the abovementioned silvicultural rotation, would have provided the necessary woods for maritime expansion and also ensured the sustainability of the forest and the maintenance of the forest resource of interest. It is highly challenging to establish how much timber was required in this period for the construction of an ocean-going ship. Archival information on timber supplies offers some insights into this matter. From 1621 to 1634, the amount of cork oak timber required for the construction of a single ocean-going ship ranged from around 2,200 to 3,200 cork oak trees, with an exceptional case in 1630 when 4,500 cork oaks were requested.[45]

It is more challenging still to assess pine needs. In December 1622, Roque da Silveira (purveyor of the warehouses for shipbuilding) stated that 3,960 *pinheiro-mansos* (*Pinus pinea*) were needed to build two ocean-going ships. In November 1632, the Portuguese government requested the *monteiro-mor* for the amounts of cork oak, stone and maritime pine listed in table 4 for the construction of a new galleon. It asked for 3,200 cork oaks, 1,400 *pinheiros-mansos* and a range of timber components which do not allow us to calculate how many *Pinus pinea* were necessary for their construction. This pine would have come from the area of Ribatejo. From Pederneira, it requested 1,006 *pinheiro-bravos* (*Pinus pinaster*) and an assortment of timber components for which it is not possible to calculate the number of trees required. A rough estimate would be 2,000 pines of each type (*Pinus pinea*, *Pinus pinaster*) to build a single vessel.

According to Reboredo & Pais,[46] during the 1400s and the 1500s, maritime expansion was responsible for the destruction of all *Quercus sp.* between the Duero and Tagus rivers and, in general, between four and five million trees were felled (approximately 2 to 2.5 million trees in 100 years), predominantly *Pinus sp.* and *Quercus sp.* According to these authors, deforestation was chiefly motivated by shipbuilding needs, although industrial activities and population growth were also influential. Furthermore, they go so far as to indicate that

---

44 Domínguez-Delmás et al., "Tree-rings," 184–5; Fernando Reboredo and João Pais, "Evolution of Forest Cover in Portugal: From the Miocene to the Present," in *Forest Context and Policies in Portugal. Present and Future Challenges*, ed. Fernando Reboredo (Cham: Springer International Publishing, 2014a), 11–23.

45 BAHMOP, MMR, núcleo 9; Trápaga Monchet, "Supplying Timber," 227–29.

46 Reboredo and Pais, "Evolution of forest," 2014a, 14.

TABLE 6.4 Amounts of cork oak and stone and maritime pine specified for the construction of a new galleon in November 1632

Required timber	Ship components
3,200 cork trees	Stem posts, doublings (*coisses*, here understood as *calçês*), keels, frames, first futtocks, *aposturas*, clamps or beam shelves, breasthooks, waterways, weatherdeck knees, *curvas de reves*, bilge stringers, deck-support knees, and other necessary items not detailed
Stone pine (*Pinus pinea*) from Ribatejo area	
1,000 trees	Wales, filler timbers, and *meias latas* (half deck beams)
400 trees	Stanchions
80 dozens	Lateral planks
40 dozens	*Dalcaixa* planking
140	Stern planks
2	*Madres de leme* (rudders)
4	*Azafroes*
2	Doublings
10	*Pinçoes* (here understood as *pinção*, whipstaff)
200 maritime pine trees	*Armaçãos*
Maritime pine from the area of Pederneira	
140 dozen	Deck planks
140 dozen	Ceiling planks
1,000 trees	Deck beams, bilge stringers, carling
6	*Asafroes*
12 tabuas	Channels
24	*Apostiças* (possibly *aposturas*, if so top timbers)
6 trees	Pumps
6	*Asafrões mansos*

SOURCE: BAHMOP, MMR, NÚCLEO 9, NOVEMBER 1632. PUBLISHED IN TRAPAGA MONCHET, "SUPPLYING TIMBER," 229

maritime expansion was the main factor responsible for the near-destruction of the oaks between the Duero and Tagus rivers.

Published historical information, such as the *monteiro-mor* ordinance of 1605, refers to numerous pine and cork oak forests and some oak groves.[47]

47 Silva, *Colecção chronologica*, 115–21.

Furthermore, Domínguez-Delmás *et al.* indicate that the Spanish Empire in the Age of Discovery needed some six million trees (approximately three million trees in 100 years) for its fleet between the sixteenth and seventeenth centuries.[48] In both cases, it can be seen that the figures given by these authors are below those we obtained (Tables 6.2 and 6.3) for the territories under the jurisdiction of the *monteiro-mor* ordinance of 1605. All these data indicate that it is necessary to look for other causes, or carry out a multi-causal analysis, to understand the dynamic processes of land occupation and the idea of deforestation.

Judging by the archival information on timber supplies for shipbuilding – that 3,200 cork oaks, 2,000 stone pines, and 2,000 maritime pines were required to construct a single ocean-going ship – Portuguese forest cover would have been able to meet the needs of the shipyards of Lisbon without great difficulty. Therefore, this experimental and quantitative approach provides a basis for questioning and interpreting territorial processes and deforestation in the Early Modern Age from a different perspective. Undoubtedly, it would be interesting to examine whether the pressure on forest masses was very localized, regional or national, or the result of the combined action of more complex processes linked to models of economic development (agrarian expansion, industrial demand for raw materials and energy sources) and a change in the property regime (a shift from communal and royal forests to private ownership) that led to a reduction in the forest area.

The main Portuguese shipyards demanded quality raw materials continuously so as not to interrupt the production chain. Having nearby forest areas would have made it easier to meet these requirements. However, the Viana do Castelho shipyard had difficulties obtaining raw materials from early on,[49] a possible indication at least that the necessary forest masses were not located nearby.

The underlying idea that takes on particular importance is that a deficiency in the logistics chain affecting the availability of forest resources, or the pressure on forests "close" to the places of demand, could have created a false sense of high prices and an increase in prices. Therefore, deforestation – possibly understood in this historical context as the pressure on forest masses above their carrying capacity – would not have been a widespread issue but localized and concentrated around spaces with high demand (in our case, the shipyards). The logistical limitations of the time, local pressure on the forest

48 Delmás et al., "Tree-rings," 184.

49 Fernando Reboredo and João Pais, "A construção naval e a destruição do coberto florestal em Portugal – Do Século XII ao Século XX," *Ecologi@* 4 (2012): 34.

masses generated for shipbuilding and the multiple uses which were given to forests and which all (though not only) European maritime monarchies tried to restrict should be the object of specific studies. Moreover, the shipyards were located in the main cities of Portugal, which required large amounts of wood for heating. Firewood was by far the socio-economic activity that consumed the most sizeable quantity of wood.[50]

The results raise new and suggestive questions about the persistence of the concern about deforestation.[51] The construction of the so-called Early Modern State notably influenced the emergence of increasingly invasive legislation issued by the territorial princes, which included use of and access to natural resources. A rhetoric was required to justify the approval of such measures and among the given reasons were fears of deforestation or the disappearance of forest masses (there is an extensive bibliography: see, for example, the case-studies of Venice and Württemberg, or the overview provided by Paul Warde for most of Europe[52]).

Therefore, deforestation processes are complex with multiple causes and effects and directly related to the economy and population growth.[53] The processes of forest dynamics,[54] the availability of forest resources and deforestation spark controversy. Despite the control and repopulation advocated by the authorities, especially the Iberian monarchies, the perception and transmission of the idea of scarcity of forest resources, unsustainability of human activity, and deforestation resulting from cutting down forest masses at a faster rate than they could regenerate have been present since ancient times.[55]

---

50 Warde, "Fear of wood," 34–6.

51 J. Donald Hughes and J.V. Thirgood, "Deforestation, Erosion, and Forest Management in Ancient Greece and Rome," *Journal of Forest History* 26, no. 2 (1982): 60–75; J. Donald Hughes, "Ancient Deforestation Revisited," *Journal of the History of Biology* 44, no. 1 (2011): 43–57.

52 Appuhn, *A Forest*, 1–8; Warde "Fear of wood," 34–55; Paul Warde, *Ecology, Economy and State Formation in Early Modern Germany* (Cambridge: Cambridge University Press, 2006), 161–223.

53 Sumit Chakravarty, S. K. Ghosh, C.P. Suresh, A.N. Dey, and G. Shukla, "Deforestation: Causes, Effects and Control Strategies," in *Global Perspectives on Sustainable Forest Management*, ed. Okia Clement Akais (Rijeka: IntechOpen, 2012), 3–27; Robert M. Ewers, "Interaction effects between economic development and forest cover determine deforestation rates," *Global Environmental Change* 16, no. 2 (2006): 161–69.

54 Pere Serra Rúiz, Xavier Pons Fernández and David Saurí Pujol, "Land-cover and land-use change in a Mediterranean landscape: A spatial analysis of driving forces integrating biophysical and human factors," *Applied Geography* 28, no. 3 (2008): 189–209.

55 William E. Harris (ed.), *The Ancient Mediterranean Environment between Science and History* (Leiden: Brill, 2013); Milena Holmgren and Marten Scheffer. "To Tree or Not to Tree: Cultural Views from Ancient Romans to Modern Ecologists," *Ecosystems* 20, no. 1

Our study, which examines scenarios and restrictive assumptions for the case of Portugal, points to the existence of important forest resources within the jurisdictional area of the *monteiro-mor* ordinance of 1605. The results allow us to venture that timber reserves existed and were sufficient to cater to demand. It is very likely that there were local limitations or restrictions[56] and possibly numerous problems of accessing them. These aspects should be studied in detail in future research.[57]

5 Conclusions

Judging by the elements described in the historical texts, the limits of the forest protection and management zones seem to be based on geographical aspects. The consideration of watersheds as the main element of the boundaries established in the forest legislation provides cartography with limits consistent with the technical and accessibility limitations of the Early Modern Age.

The analysis carried out makes it possible to establish a potential distribution of *Quercus suber*, *Pinus pinea* and *Pinus pinaster*. According to our calculations, which are based on quite restrictive assumptions, these species would have occupied an area of 150,000 hectares, or 8.7% of the scope of jurisdiction of the *monteiro-mor* ordinance of 1605 (*Quercus suber*: 3.2%; *Pinus pinaster*: 2.2% and *Pinus pinea*: 3.3%). In an area of this size, the sustainable extraction of 22,486,400 and 11,496,500 trees of all three species could have been carried out every 100 years. Our analysis indicates that there would have been sufficient good quality trees (that met the requirements for shipbuilding) for minimum annual removals of 10,000 *Quercus suber*, 60,000 *Pinus pinea* and 40,000 *Pinus pinaster*.

These data show that there was a sufficient forestry capacity to address the wood needs for shipbuilding reflected in literature and archival sources without compromising the viability and future sustainability of this area. The results highlight the need for more complex and interdisciplinary analyses to

(2017): 62–8; J. Donald Hughes, "How the Ancients Viewed Deforestation," *Journal of Field Archaeology* 10, no. 4 (1983): 435–45; Brian F. Synder, "The genetic and cultural evolution of unsustainability," *Sustainability Science* 15, no. 4 (2020): 1087–99.

56 Koldo Trapaga-Monchet, "Guerra y deforestación en el reino de Portugal (siglos XVI–XVII)," *Tiempos Modernos: Revista Electrónica de Historia Moderna* 9, no. 39, (2019): 396–425.

57 Álvaro Aragón Ruano, "Ríos de madera." Recursos forestales e hídricos para la Real Armada durante el siglo XVIII en Guipúzcoa y Navarra," *Tiempos Modernos: Revista Electrónica de Historia Moderna* 9, no. 39 (2019): 426–55.

enable us to understand the Portuguese navies' decline during the seventeenth century. According to the provisional data presented here in terms of the productive capacity of the land, from an environmental point of view there does not seem to have been any reason for the Portuguese authorities to be forced to use 'green woods' or young pines/cork oaks that did not meet the necessary conditions for shipbuilding.

These results should encourage us to contemplate reality. The causes of the problem, as reported in the literature, suggest new hypotheses involving issues related to location, the logistics chain, resource accessibility, price fluctuations and other factors. These factors are considered pivotal in the processes of local deforestation. They also drive the need for a more comprehensive definition of forest masses, including their distribution and characteristics, and the identification of key actors and pressures that shaped the state and evolution of forested areas during the Modern Age. Additionally, they prompt a reassessment of the origins and widespread legislative regulations concerning forest resources, with a focus on the aspects previously highlighted in this study.

Undoubtedly, these data encourage reflection on human-environment interactions. Even though there were sufficient forest resources at the time, the Portuguese Monarchy created a legislative body to conserve and protect the woodlands. It is particularly interesting to reflect on why the idea of deforestation or reduction of forest area took shape. If the reductions in forest masses were so significant and there was awareness of the loss of ecosystem services and its impacts (e.g., loss of soil or biodiversity), how could these monarchies have subsisted? How is it possible that forestry masses have recovered by over 20% today, and high-value natural spaces can be found in these areas? Undoubtedly, these issues are worth reflecting on and deserve to be examined and analysed in detail.

Abbreviations

BA: Biblioteca de Ajuda
BAHMOP: Biblioteca e Arquivo Histórico do Ministerio de Obras Púbricas
MMR: Montaria-Mor do reino

References

Appuhn, Karl. *A Forest on the Sea: Environmental Expertise in Renaissance Venice*. Baltimore: Johns Hopkins University Press, 2009.

Aragón Ruano, Álvaro. "Ríos de madera". Recursos forestales e hídricos para la Real Armada durante el siglo XVIII en Guipúzcoa y Navarra." *Tiempos Modernos: Revista Electrónica de Historia Moderna* 9, no. 39 (2019): 426–55.

Aranda y Antón, Gaspar de. *La carpintería y la industria naval en el siglo XVIII*. Madrid: Instituto de Historia y Cultura Naval, 1999.

Bamford, Paul W. *Forests and French Sea Power 1660–1789*. Toronto: University of Toronto Press, 1956.

Boxer, Charles R. *The Portuguese seaborne Empire (1415–1825)*. London: Hutchinson, 1977.

Castro, Filipe. *The Pepper Wreck: A Portuguese Indiaman at the Mouth of the Tagus River*. Texas: Texas A&M University Press, 2005.

Chakravarty, Sumit, S.K. Ghosh, C.P. Suresh, A.N. Dey, and G. Shukla. "Deforestation: Causes, Effects and Control Strategies." In *Global Perspectives on Sustainable Forest Management*, edited by Okia Clement Akais, 3–27. Rijeka: IntechOpen, 2012.

Costa, Leonor Freire. "Aspectos empresariais da construção naval no século XVI: o caso da Ribeira das Naus de Lisboa." *Análise Social* 31, no. 136/137 (1996): 295–312.

Costa, Leonor Freire. *Naus e galeões na ribeira de Lisboa: a construção naval no século XVI para a Rota do Cabo*. Cascais: Patrimonia Historica, 1997.

Crespo Solana, Ana and Nigel Nayling. "ForSEAdiscovery. Forest resources for Iberian Empires: Ecology and Globalization in the Age of Discovery (16th–18th centuries)." In *Actas del V Congreso Internacional de Arqueología Subacuática (IKUWA V)*, 896–904. Madrid: Ministerio de Educación Cultura y Deportem 2016.

Crespo Solana, Ana, Filipe Castro, and Nigel Nayling. *Heritage and the Sea. Volume 1: Maritime History and Archaeology of the Global Iberian World (15th–18th centuries)*. Switzerland: Springer Cham, 2022, https://doi.org/10.1007/978-3-030-86460-6.

Crespo Solana, Ana. "ForSEAdiscovery: la construcción naval y el comercio de la madera del siglo XVI al XVIII." *Revista PH* 96 (2019) 114–41.

Devy-Vareta, Nicole and António A. Monteiro Alves. "Os avanços e os recuos da floresta em Portugal – da Idade Média ao Liberalismo." In *Floresta e sociedade. Uma história em comum*, edited by Joaquim Sande Silva, 55–76. Lisbon: Fundação Luso-Americana, 2007.

Devy-Vareta, Nicole. "Para uma geografia histórica da floresta portuguesa : do declínio das matas medievais à política florestal do Renascimento (sécs. XV e XVI)." *Revista da Faculdade de Letras* 1 (1986): 5–37.

Devy-Vareta, Nicole. "Para uma geografia histórica da floresta portuguesa: as matas medievais e a "coutada velha" do Rei." *Revista da Faculdade de Letras* 1 (1985): 47–73.

Domingues, Francisco Contente. *Os navios do mar oceano: teoria e empiria na arquitectura naval portuguesa dos séculos XVI e XVII*. Lisbon: Centro de História da Universidade de Lisboa, 2004.

Domínguez-Delmás, Marta, R. Alejano-Monge, S. Van Daalen, E. Rodríguez-Trobajo, I. García-González, J. Susperregi, T. Wazny, and E. Jansma. "Tree-rings, forest history and cultural heritage: Current state and future prospects of dendroarchaeology in the Iberian Peninsula." *Journal of Archaeological Science* 57 (2015): 180–96.

Duffy, James. *Shipwreck & Empire. Being an Account of Portuguese Maritime Disasters in a Century of Decline.* Harvard: Harvard University Press, 1955.

European Environment Agency (EEA). "European Digital Elevation Model (EU-DEM), version 1.1." European Environment Agency (EEA) under the framework of the Copernicus programme, 2016.

Ewers, Robert M. "Interaction effects between economic development and forest cover determine deforestation rates." *Global Environmental Change* 16, no. 2 (2006) 161–69.

Gaspar, Jorge. "Os portos fluviais do Tejo." *Finisterra* 5, no. 10 (1970): 153–215.

Glete, Jan. *Warfare at Sea, 1500–1650: Maritime Conflicts and the Transformation of Europe.* New York: Routledge, 2000.

Harris, William E. (ed.). *The Ancient Mediterranean Environment between Science and History.* Leiden: Brill, 2013.

Holmgren, Milena and Marten Scheffer. "To Tree or Not to Tree: Cultural Views from Ancient Romans to Modern Ecologists." *Ecosystems* 20, no. 1 (2017): 62–8.

Hughes, J. Donald and J.V. Thirgood. "Deforestation, Erosion, and Forest Management in Ancient Greece and Rome." *Journal of Forest History* 26, no. 2 (1982): 60–75.

Hughes, J. Donald. "How the Ancients Viewed Deforestation." *Journal of Field Archaeology* 10, no. 4 (1983): 435–45.

Hughes, J. Donald. "Ancient Deforestation Revisited." *Journal of the History of Biology* 44, no. 1 (2011): 43–57.

Labrador Arroyo, Félix and Koldo Trápaga-Monchet. "La configuración del espacio y la explotación forestal de un enclave singular: el Real Sitio del Soto de Roma durante la dinastía Hasburgo." *Studia Histórica. Historia Moderna* 39, no. 3 (2017): 293–327.

Labrador Arroyo, Félix. *La Casa Real en Portugal (1580–1621).* Madrid: Polifemo, 2009.

Magalhães, Manuela Raposo, Ana Müller and João Ferreira Silva. "Aptidão Integrada ao Sobreiro (*Quercus suber* L.) para Portugal Continental." http://epic-webgis-portugal.isa.utl.pt/: LEAF/ISA/ULisboa, 2015a.

Magalhães, Manuela Raposo, Ana Müller and João Ferreira Silva. "Aptidão Integrada ao Pinheiro-manso (*Pinus pinea* L.) para Portugal Continental." http://epic-webgis-portugal.isa.utl.pt/: LEAF/ISA/ULisboa, 2015b.

Magalhães, Manuela Raposo, Ana Müller and João Ferreira Silva. "Aptidão Integrada ao Pinheiro-bravo (*Pinus pinaster* Aiton) para Portugal Continental." http://epic-webgis-portugal.isa.utl.pt/: LEAF/ISA/ULisboa, 2015c.

Martínez González, Alfredo José. *Las Superintendencias de Montes y Plantíos (1574–1748): derecho y política forestal para las armadas en la Edad Moderna.* Valencia: Tirant lo Blanch, 2015.

Melero Guilló, María Jesús. "A la mar madera": La madera en la arquitectura naval española." In *Andalucía, América y el mar: Actas de las IX Jornadas de Andalucía y América*, edited by Bibiano Torres Ramírez, 145–57. Sevilla: Diputación de Huelva, 1991.

Melo, Cristina Joanaz de. "Menos coutadas melhores pinhais: império, inundações, fisiocracia, guerra e especialização das matas reais em Portugal (1777–1824)." *Tiempos Modernos. Revista electrónica de Historia Moderna* 9, no. 39 (2019): 456–87.

Melo, Cristina Joanaz de. *An Analysis of the Royal Preserves in Portugal. Issues of privilege, power, management and conflict.* Sheffield: Wildtrack, 2015.

Melo, Cristina Joanaz de. *Coutadas reais (1777–1824): Privilégio, poder, gestão e conflito.* Lisbon: Montepio Geral, 2000.

Perlin, John. *A Forest journey: The Story of Wood and Civilization.* Woodstock, Vermont: Countryman Press, 2005.

Ramm, Frederik. "Portugal: OpenStreetMap Data in Layered GIS Format." Data/Maps Copyright 2018 Geofabrik GmbH and OpenStreetMap Contributors. Map tiles: Creative Commons BY-SA 2.0 Data: ODbL 1.0, 2019.

Reboredo, Fernando and João Pais. "A construção naval e a destruição do coberto florestal em Portugal – Do Século XII ao Século XX." *Ecologi@* 4 (2012): 31–42.

Reboredo, Fernando and João Pais. "Evolution of Forest Cover in Portugal: From the Miocene to the Present." In *Forest Context and Policies in Portugal. Present and Future Challenges*, edited by Fernando Reboredo, 1–37. Cham: Springer International Publishing, 2014a.

Reboredo, Fernando and João Pais. "Evolution of forest cover in Portugal: A review of the 12th-20th centuries." *Journal of Forestry Research* 25, no. 2 (2014b): 249–56.

Romero-Calcerrada, Raúl and Koldo Trapaga-Monchet. "La ley del alcornoque de 1546 y la ordenanza del Monteiro-Mor de 1605." In *Historia, Sociedad y Medio ambiente: la sostenibilidad*, edited by Koldo Trapaga-Monchet and Luis Alberto Polo Romero (cords.), 143–79. Madrid: Sílex, 2022.

Salgado, Augusto Alves. "Portugal e o Atlântico:organização militar e acções navais durante o período filipino: (1580–1640)". PhD diss., Universidade de Lisboa, 2009.

Serra Rúiz, Pere, Xavier Pons Fernández and David Saurí Pujol. "Land-cover and land-use change in a Mediterranean landscape: A spatial analysis of driving forces integrating biophysical and human factors." *Applied Geography* 28, no. 3 (2008): 189–209.

Silva, João Ferreira, Manuela Raposo Magalhães, and Natália Cunha. "Classificação hierárquica e toponímica das Linhas de Água de Portugal Continental." Lisbon: LEAF/ISA/ULisboa, 2013, http://epic-webgis-portugal.isa.utl.pt/.

Silva, João Ferreira, Manuela Raposo Magalhães, and Natália Cunha. "Linhas de Festo de Portugal Continental." Lisbon: LEAF/ISA/ULisboa, 2013, http://epic-webgisportugal.isa.utl.pt/

Silva, José Bonifacio. *Memoria sobre a necessidade e utilidade do plantio de novos bosques em Portugal.* Lisbon: Na Typographia da Academia Real das Sciencias, 1815.

Silva, José Justino de Andrade. *Collecção chronologica da legislação portugueza (1603–1612)*. Lisbon: Imprensa de J. J. A. Silva, 1854.

Slee, Bill. "Landscape goods and services related to forestry land use." In *Multifunctional Land Use: Meeting Future Demands for Landscape Goods and Services*, 65–82. Berlin-Heidelberg: Springer, 2007.

Snyder, Brian F. "The genetic and cultural evolution of unsustainability." *Sustainability Science* 15, no. 4 (2020): 1087–99.

Trapaga Monchet, Koldo. "El estudio de los bosques reales de Portugal a través de la legislación forestal en las dinastías Avis, Habsburgo y Braganza (ca. 1435–1650)." *Philostrato: revista de historia y arte* 1 (2017) 5–27.

Trapaga Monchet, Koldo. "Guerra y deforestación en el reino de Portugal (siglos XVI–XVII)." *Tiempos Modernos: Revista Electrónica de Historia Moderna* 9, no. 39, (2019): 396–425.

Trapaga Monchet, Koldo. "No es madera para vasallos, sino del rey. Las políticas forestales de los Hasburgo en Portugal (1609–1640)." *Obradoiro de Historia Moderna* 28 (2019): 105–34.

Trapaga Monchet, Koldo. "Supplying Timber for his Majesty's Fleets: Forest Resources and Maritime Struggle in Portugal (1621–1634)." In *Heritage and the Sea: Volume 1: Maritime History and Archaeology of the Global Iberian World (15th–18th centuries)*, edited by Ana Crespo Solana, Filipe Castro and Nigel Nayling, 215–48. Cham: Springer International Publishing, 2022.

Trapaga Monchet, Koldo. "Who protected portuguese forests? Safeguarding and preserving royal and private forests in Portugal (1605–1640)." In *Árvores, barcos e homens na Península Ibérica (séculos XVI–XVIII)*, edited by Rosa Varela Gomes and Koldo Trápaga-Monchet, 135–48. Zaragoza: Pórtico Librerías, 2017.

Uva, José Sousa and Sónia Pacheco Faias. *6.º Inventário Florestal Nacional (IFN6). Anexo Técnico. Versão 1.0 I.*: Instituto da Conservação da Natureza e das Florestas (ICNF), 2019.

Uva, José Sousa. *6.º Inventário Florestal Nacional (IFN6)*. Instituto da Conservação da Natureza e das Florestas (ICNF), 2019.

Warde, Paul. "Fear of Wood Shortage and the Reality of the Woodland in Europe, *c.* 1450–1850." *History Workshop Journal* 62, no. 1 (2006): 28–57, https://doi.org/10.1093/hwj/dbl009.

Warde, Paul. *Ecology, Economy and State Formation in Early Modern Germany*. Cambridge: Cambridge University Press, 2006, doi:10.1017/CBO9780511497230.

Williams, Michael. "Dark ages and dark areas: Global deforestation in the deep past." *Journal of Historical Geography* 26, no. 1 (2000): 28–46.

Williams, Michael. *Deforesting the Earth: From Prehistory to Global Crisis. An Abridgment*. Chicago: University of Chicago Press, 2010.

Wing, John T. *Roots of Empire. Forests and State Power in Early Modern Spain, c. 1500–1750*. Leiden: Brill, 2015, https://doi.org/10.1163/9789004261372.

Yun-Casalilla, Bartolomé. *Iberian world empires and the globalization of europe 1415–1668.* Basingstoke: Palgrave Macmillan, 2019, doi:https://doi.org/10.1007/978-981-13-0833-8.

CHAPTER 7

Woodland in the Ensenada Cadastre: Distribution, Composition and Role in the Peasant Economy

Concepción Camarero Bullón, Ángel Ignacio Aguilar Cuesta and Ana Luna San Eugenio

One of the concerns of Don Zenón de Somodevilla y Bengoechea, the first Marquis of Ensenada and minister of finance, war, the Navy and the Indies from 1743 to 1754, was the management and use of woodlands. He was well aware that woodlands were a fundamental asset in the peasant economy as they provided grazing for livestock and game animals for local inhabitants and ensured a supply of firewood, timber and charcoal for villages, towns and cities. He also realized that they needed to be improved, properly managed and protected against misuse as a *sine qua non* for implementing his ambitious shipbuilding programme. Without quality timber, there can be no ships. And they had to be managed in such a way as to prevent clashes between the traditional uses of woodlands and their increase and conservation for naval purposes, as well as between the agents involved: the councils, the *corregidores* and the Navy. The forest ordinances enacted in 1749, during his term as minister, encompass both lines of action.[1] Many places refer to them in their cadastral declarations when they speak of new plantings or plantings made according to "the new ordinances."

In a way, accomplishing the two goals set by the minister amounted to squaring the circle, and Ensenada must have been fully conscious that this called for thorough and reliable information about the size, composition, location and uses of wooded areas. That is why, at first sight, it is surprising to find that the cadastral document providing the most comprehensive, albeit less precise, information about the localities, the *Interrogatorio de la letra A* – the questionnaire which gave rise to the well-known *Respuestas generales* (general answers) – contains no specific questions dedicated solely and exclusively to

1 *Ordenanza que Su Magestad, Dios le guarde, manda observar para la cría, conservación, plantíos y cortas de los montes, con especialidad los que están inmediatos a la mar y ríos navegables: método y reglas que en esta materia deben seguir los intendentes de marina, establecidos en los tres departamentos de Cádiz, Ferrol y Cartagena*, of 31 January 1748, and *Real Ordenanza sobre aumento de plantíos y conservación de montes*, of 7 December of the same year.

woodlands.[2] Indeed, in this respect it differs from most of the questionnaires on which it was modelled: those of Philip II's surveys, the *Relaciones de Felipe II*, for Spain and the Americas,[3] which refer to them very directly. For instance, in the 1575 questionnaire for Spain, questions 18 and 24 request this information in connection with satisfying villages' needs for timber, firewood, grazing land and game:[4]

> 18. Whether the land is plentiful, or lacking in firewood, and from where they gather provisions, and if it is mountainous, what kind of woodland and trees, and what animals, game and wild species breed and are found in it. [...]
> 24. The grazing land and wood-pasture there is within the confines of the aforesaid village, with the forests and any hunting and fishing grounds there are, and which they are and what they are worth.
>
> 18. Si es tierra abundosa, o falta de leña, y de donde se proveen; y si montañosa, de qué monte y arboleda, y qué animales, cazas y salvaginas se crían y hallan en ella. [...]
> 24. Los pastos y dehesas señaladas que en términos del sobredicho pueblo hubiese, con los bosques y cotos de caza y pesca que asimismo hubiese, y cuales son y lo que valen.

Years after the Cadastre, the geographer Tomás López included in the questionnaire[5] given to parish priests of the villages and towns a question that directly requested information as accurate as possible about wooded areas. This is only logical, as he sought to gather the maximum amount of geographical data as specific and accurate as possible to draw up maps. Given the resources available then, maps had to been compiledfrom existing cartographic records as,

2 AHN, Consejos, lib. 1510.
3 For Spain they are called *Relaciones Topográficas* and for the Americas *Relaciones Geográficas*.
4 The complete questionnaire can be found in Nadezda Konyushikhina, "Los cuestionarios para las Relaciones topográficas de Felipe II y las Relaciones geográficas de Indias de los años 1570," *CT Catastro* 89 (2017), 25–8.
5 Tomás López used his 15-point questionnaire for just over three decades, making only slight changes. The first documented variation dates from 1763, though it was still not fully defined them. The best-known model is that of the 1780s. It was appended in manuscript form to the letters he sent to the villages. The surviving replies are held in the Biblioteca Nacional and are available at the Biblioteca digital hispánica. A reference book on Tomás López's work is Antonio López Gómez y Carmen Manso Porto, *Cartografía del siglo XVIII. Tomás López en la Academia* (Madrid: Real Academia de la Historia, 2006).

unlike France, Spain had not begun to triangulate its territory and lacked astronomical measurements:[6]

> 6 What forests, woodlands and groves the place has; with what stands it is populated, what they are called, in what cardinal point they are located and how far they stretch.
>
> 6ª Qué bosques, montes o florestas tiene el lugar; de qué matas poblado, cómo se llaman, a qué aire caen y cuánto se extienden.

Years later, in 1790, the court (*audiencia*) of Extremadura was established. It immediately set about inspecting the territory of the region and drew up a questionnaire to be answered in the various localities with the aid of the inspectors, consisting of 57 questions on a broad variety of subjects. No fewer than six of the questions requested information about woodlands and their uses.[7] Specifically:

> XLIV. If there are woodlands populated with trees or shrubs, their species, destination and uses; and if it is known whether any medicinal herbs are produced, or others which can be used to manufacture any goods, such as soap, dyes or others; whether firewood, charcoal or timber can be harvested without causing them to deteriorate, and to what use the latter can be put; whether these woodlands are public, or to whom they belong; whether they are well or poorly kept, and the causes that influence this.
> XLV. If there are woodlands impenetrable to livestock, which only serve as shelter for wild animals, which it would be appropriate to clear, and by what means this can be achieved.
> XLVI. If the woodlands are customarily burned, and for what purpose: what damage results from this, and how this excess is usually punished.
> XLVII. If the woodlands are stripped of bark, and whether stripping bark spoils them, or whether it is carried out by law; to whom the price of the

6 Concepción Camarero Bullón and Ángel Ignacio Aguilar Cuesta, "La cartografía, instrumento para conocer el territorio, planificar y gestionar las reformas en la España del siglo XVIII," *Manuscrits. Revista d'Història Moderna* 42 (2020), 151 and ff.

7 On the *Interrogatorio de la Audiencia de Extremadura*, see: Miguel Rodríguez Cancho, "Interrogatorios del siglo XVIII. Estudio comparativo," *Norba* 2 (1981), 221–32. The replies from the villages to the *audiencia*'s questionnaire are published in Miguel Rodríguez Cancho and Gonzalo Barrientos Alfageme, eds. *Interrogatorio de la Real Audiencia Extremadura a finales de los tiempos modernos* (Mérida: Asamblea de Extremadura, 1993–1996), 11 vols.

bark belongs; if it is to the local residents, how much each *arroba* produces for them, and what price is paid by tanners.
XLVIII. If, on the pretext of cultivating and planting trees in some lands or plots, they have been closed pursuant to royal orders, and in fraud of the latter and to the public detriment they are kept as woodlands to take advantage of grazing, depriving the other residents of them, and their extension or size.
XLIX. If there are any wood-pastures, their number, and to whom they belong; if they are for grazing and farming and if so whether they have been reduced to grazing only, and their size.
L. The plantings or sowings carried out pursuant to royal orders shall be acknowledged, and their state.

XLIV. Si hay montes poblados de árboles ó arbustos, su especie, destino y utilidades; y si se tiene noticia de que produzcan algunas yerbas medicinales, ú otras que puedan beneficiarse en alguna fabrica, como para jabón, tintes, ú otras: si se puede sacar sin deteriorarlos leña de ellos, carbón, ó madera, y qué uso se puede hacer de éstas: si estos montes son públicos, ó á quién pertenecen: si están bien, ó mal cuidados, y las causas que influyen en esto.
XLV. Si hay montes impenetrables al ganado, que solo sirven al abrigo de fieras, que sea conveniente desmontar, y por qué medios se puede conseguir.
XLVI. Si se suelen quemar los montes, y para qué fines: qué perjuicios se siguen de esto, y cómo se suele castigar este exceso.
XLVII. Si los montes se descascan, y si de los descasques resulta su ruina, ó se descascan á ley; a quién pertenece el precio de la casca; si es de propios, quánto produce para éstos cada arroba, y á qué precio sale á los Curtidores.
XLVIII. Si á pretexto de cultivar y arbolear algunas tierras ó terrenos, se han cerrado con motivo de las Reales órdenes, y en fraude de éstas y perjuicio público se conservan de monte, y para aprovecharse de los pastos, privando á los demás vecinos de éstos, y su extensión, ó cabida.
XLIX. Si hay dehesas, su número, y á quién pertenecen; si son de pasto y labor, y si siéndolo se han reducido á solo pasto, y su extensión.
L. Se reconocerán los plantíos, ó semilleros executados en virtud de Reales órdenes, y su estado.

This shows that the court of Extremadura was interested in all aspects of woodlands and their possible uses – be they in a "natural" state, wood-pasture, thinned, cleared of underbrush, liable to be cleared, or newly planted. In other words, it wished to know about the situation of the woodlands and any related aspects that could trigger conflicts.

In view of these antecedents and consequences, does the absence from the cadastral questionnaire of a direct question focused solely and exclusively on wooded areas mean that the woodlands were left out of the major survey promoted by Ensenada or that they were of no interest to the minister? Not at all – quite the opposite, in fact: they are there, but it is necessary to know where to look for them and for information about their floral composition, uses, area and value, bearing in mind that the Ensenada Cadastre is not a single record but a set of documents with different territorial levels, different purposes and differing degrees of data aggregation. As far as those relevant to this essay are concerned, some are local and others are provincial; and some are collective and others are individual. Furthermore, the approach taken to the subject by Tomás López's questionnaire and, above all, that of the Royal Court gives the impression that they (especially the latter) specifically address all aspects of the cadastral information gathered on woodlands, including the impossibility of putting some areas to any use because their topography or thickness made them inaccessible.

On another note, a point needs to be made before continuing with the information on woodlands provided by the cadastral documentation. When dealing with the Cadastre, besides being aware at all times of the level of the documents they are working with, researchers need to accept that the documentation is highly homogeneous as a whole but heterogeneous detail, as will be demonstrated by a few examples discussed below. This heterogeneity, which is acknowledged by the creators of the Cadastre and yields acceptable results, stems from several factors: the geographical diversity of the vast territory surveyed (just over 372,000 km2), the types of settlement, the types of wooded areas, the characteristics of the social groups and population centres, the economic organization, the human factor (cadastral teams, intendant ...). etc. And this heterogeneity means that in many cases the information on forests that appears in the replies given to the various questions of the *Interrogatorio general* varies in degree of detail depending on the areas, the type of forest, its uses, its importance to the local economy, the cadastral teams (courts) and the intendancy to which they were responsible, among other factors. In contrast, the information gathered in another cadastral document, the *Libro de lo real* (record of real estate) is more homogeneous and, certainly, more detailed and accurate, as will be seen later.

We will now take a closer look at the different local cadastral documents in which information can be found on the woodlands (*montes* is the term usually employed in them). It should be noted that where this information is clearly described and differentiated is in the local documents, both collective and individual. In the provincial documents, information about woodlands is more diluted and distorted by the different land uses. This is due to the level of data aggregation and the criteria adopted, since the purpose was to establish the value of the assets and income with a view to determining the locality's base for levying the single tax in accordance with the different branches of wealth. In other words, if a unit of wooded land had the same value as one of, say, grazing land and one of third-class vineyard, the areas of all three were added together to give a single figure which could not be broken down into different types of use. Information on woodlands can occasionally also be found in the correspondence between the Royal Board of the Single Tax (*Real Junta de Única Contribución*) and the provincial heads of the Cadastre: intendants, commissaries and accountants.

The local cadastral documentation is made up of the *Respuestas generales*, the *Memoriales*, the *Libros de lo real*, the *Libros de cabezas de casa*, the *Estados locales*, the *Autos y diligencias* to which the enquiry gave rise, and the *documentos probatorios* submitted as proof of the replies to various questions. These had to be copied literally and are included among the *autos y diligencias*.[8] For the subject studied here, the main, but not the only, sources are the first three. Let us now take a closer look at these documents to find out what information they can provide about the woodlands.[9]

1 The *Respuestas Generales*: Textual and Cartographic Information

As their name indicates (general answers), the *Respuestas generales* are the result of the replies to a 40-point questionnaire (*Interrogatorio de la letra A*) on a wide range of aspects of the locality: name, boundaries, size, jurisdictional situation, population and settlement, economic activities, industrial facilities,

8 Concepción Camarero Bullón and Miguel C. Vivancos: "Con 'letras antiguas y en latín': la copia de los privilegios antiguos en el Catastro de Ensenada," in *La Corte de los Borbones: crisis del modelo cortesano*, coord. José Martínez Millán, Concepción Camarero Bullón and Marcelo Luzzi Traficante (Madrid: Ediciones Polifemo, 2015), vol. 1, 77–119.

9 For the documentary structure of the Ensenada Cadastre, see Concepción Camarero Bullón, "Vasallos y pueblos castellanos ante una averiguación más allá de lo fiscal: el Catastro de Ensenada, 1749–1756," in *El Catastro de Ensenada. Magna averiguación fiscal para alivio de los vasallos y mejor conocimiento de los reinos*, dir. Ignacio Durán Boo and Concepción Camarero Bullón (Madrid: Ministerio de Hacienda, 2002), 194.

ecclesiastical and healthcare institutions, among others.[10] They provide an overview of the locality and its people. Two copies of this document were made at the time: one was sent to the Board of the Single Tax in Madrid between 1755 and 1756, which is the one currently kept in the Archivo General de Simancas, and the second was submitted to the councils in January 1760 together with a copy of the *Libro de lo real* and another of the *Libro de cabezas de casa*.[11] The original was deposited in the provincial counting houses (*contadurías*) along with the rest of the local documentation.[12]

The copy of the *Respuestas* kept in the Archivo General de Simancas became more accessible a couple of decades ago when it was made available through the Spanish Ministry of Culture's PARES portal.[13] The ease with which data can be retrieved and handled, given its structure of direct questions on a whole battery of specific topics, makes it the cadastral document on which most work has been carried out since 1957, when Matilla Tascón first brought to attention the extensive group of documents of the Ensenada Cadastre in his book *La única Contribución y el Catastro de la Ensenada*.[14] It was not until several decades later that work gradually began on the *Libros de lo real* and the *Libros de cabezas de casa* and subsequently, albeit less intensively, on the *Memoriales*.

The use to which researchers – chiefly historians and geographers – have put the Cadastre has gone through different stages and has experienced ups and

10 Concepción Camarero Bullón, "El Catastro de Ensenada, 1749–1759: diez años de intenso trabajo y 80.000 volúmenes manuscritos," *CT Catastro* 46 (2000), 61–88 (Spanish), 141–53 (English). Alejandro Vallina Rodríguez and Nadezda Konyushikhina, "Los interrogatorios de los Catastros españoles de la Edad Moderna: fuentes geohistóricas para conocer los paisajes y las sociedades," CT *Catastro* 91 (2017), 39–63.

11 The Archivo General de Simancas houses a complete copy of the *Respuestas generales* that were sent to the board. It is only missing the *Respuestas generales* of Madrid, Villa y Corte, which were not drawn up, and those of the municipality of Madrid, which are held in the Archivo Histórico Nacional. They are published, together with the *Estados locales* of the Villa y Corte, in Concepción Camarero Bullón, *Madrid y su provincia en el Catastro de Ensenada. I. La Villa y Corte, II. Los pueblos* (Madrid: Ediciones del Umbral, 2001–2005), 2 vols.

12 The local documentation in the Ensenada Cadastre that was deposited at the provincial counting houses in the capitals of the former provinces are now housed in the Archivos Históricos Provinciales, except for those of Madrid, which are in the Archivo Histórico Nacional, Burgos, which are in the Archivo de la Diputación de Burgos, and Coruña, which are in the Archivo del Reino de Galicia. The provincial documentation is distributed between the Archivo General de Simancas and the Archivo Histórico Nacional.

13 They are available at the PARES portal. https://pares.mcu.es/Catastro/servlets/ServletController?ini=0&accion=0&mapas=0&tipo=0.

14 Antonio Matilla Tascón, *La Única Contribución y el Catastro de la Ensenada* (Madrid: Ministerio de Hacienda, 1957).

downs. At times they have highly valued its information, while at others they have asked more of it than it was able to give, consequently judging it to be of little use as a source, generally due to a lack of knowledge of the criteria used to gather and process the information. In other cases, they have systematically questioned the information provided, deeming that as a document compiled for tax purposes it would necessarily have entailed a systematic and high level of concealment of assets and income. The scholars who have taken such an approach to the Cadastre seem to have turned a blind eye to the systems established by the cadastral authorities for verifying the declarations. On the whole, these systems guaranteed a high level of accuracy in the data, and although there may have been some concealment, it is more the exception than the rule.

In any case, scholars from very different branches of knowledge are increasingly turning to the documents of the Ensenada Cadastre to aid them in their research on very different subjects and with very different methodologies and objectives, as a result of which the range of subjects studied has been enormously broadened. This has gone hand in hand with a greater knowledge of this large ensemble of cadastral documentation and of cadastres in general – which are increasingly better catalogued – the ease of consulting them in archives and reproducing them, the possibility of making part of this documentation available on the internet, and current developments in computer software that facilitate the processing of large volumes of data and old maps. All this has made the Ensenada Cadastre in particular and cadastres in general primary geohistorical sources.

It is worth stressing two issues before going on to examine the *Respuestas generales*: firstly, over the past two decades, researchers from different fields of knowledge, some far removed from the humanities, have been handling the Castilian cadastral documentation; and secondly, work is gradually starting to be carried out on all the documentary levels of the Ensenada Cadastre, albeit with varying intensity and incorporating new topics of study, as in the case of the woodlands.[15]

---

15 The following are examples of research on woodlands for which the Cadastre was used, in some cases without a very clear idea of its documentary structure and the criteria for gathering the information. Jesús Bravo Lozano, *Montes para Madrid. El abastecimiento de carbón vegetal a la villa y corte en los siglos XVII y XVIII* (Madrid: Caja de Madrid, 1993); Inocencio Cadiñanos Bardeci, "Los Montes de Sanabria a fines del siglo XVIII y comienzos del XIX," *Anuario del Instituto de Estudios Zamoranos Florián de Ocampo* 27 (2010), 237–54; Vicente Casals Costa, "Conocimiento científico, innovación técnica y fomento de los montes durante el siglo XVIII," In *El Siglo de las luces: de la industria al ámbito agroforestal*, coord. Manuel Silva Suárez (Zaragoza: Diputación Provincial, 2005), 453–500; Luis Javier Coronas Vida, "Montes y arbolado en los pueblos de la jurisdicción de Burgos

Let us now return to the *Respuestas generales*, the first cadastral document that we will address here. It is necessary to bear in mind that the designer of the survey was not aiming for accuracy of data in this document, but rather for an overall picture of the locality. Accuracy is provided by the information gathered in the declarations (*Memoriales*), after all the statements were reviewed and verified by the experts appointed for the purpose, and in the certifications requested, among others. Indeed, those responsible for the Cadastre were conscious of this when they accepted, without asking for further clarification, answers of the type given by the council and experts of Santo Domingo de Silos and its hamlets (Burgos) to the first question.[16] This reply, which is similar to those provided by many other villages, makes reference to the basic documentation of the Cadastre, what is declared in the *Memoriales*, the information resulting from the latter and the verification of what is declared in them:[17]

durante el siglo XVIII," *Boletín de la Institución Fernán González* 232 (2006), 179–222; Concepción Diego Liaño and Juan Carlos García Codrón, "La Corona y los pueblos en la explotación de los montes de Cantabria: deforestación y gestión del bosque en la segunda mitad del siglo XVIII," *Cuadernos de la Sociedad Española de Ciencias Forestales* 16 (2003), 215–20; Rafael Fernández Aldana, "Evolución de los hayedos en las cuencas de los ríos Leza, Jubera y Cidacos entre los siglos XVIII y XX, a partir del Catastro de Ensenada, de las Relaciones de Tomás López, del Diccionario de Madoz y de la clasificación y el catálogo de los montes públicos," *Zubía* 13 (2001). 113–138; Concepción Fidalgo Hijano and Juan Antonio González, "El entorno de las Lagunas de Ruidera en el siglo XVIII a la luz del catastro de Ensenada y la cartografía de la época," *CT Catastro*, 77 (2013), 43–66; Gloria Sanz Sanjosé, "Dinámica de las masas forestales en el territorio de Riofrío (Segovia)," *Cuadernos de la Sociedad Española de Ciencias Forestales* 16 (2003), 273–78; Jorge Mongil Manso and Javier Álvarez Martínez, "El Catastro de Ensenada y sus aplicaciones en trabajos sobre el medio natural," *Medio Ambiente en Castilla y León* 17 (2002), 43–8; Jorge Mongil Manso and Javier Álvarez Martínez, "Análisis de algunos factores determinantes de la superficie forestal de las provincias de Segovia y Soria en el siglo XVIII," *Cuadernos de la Sociedad Española de Ciencias Forestales* 16 (2003), 221–25; José Ramón Moreno Fernández, *El monte público en La Rioja durante los siglos XVIII y XIX: aproximación a la desarticulación del régimen comunal* (Logroño: Gobierno de La Rioja, 1994); Emilio Pérez Romero, *Patrimonios comunales, ganadería trashumante y sociedad en la Tierra de Soria. Siglos XVIII–XIX* (Salamanca: Junta de Castilla y León, 1995); Ofelia Rey Castelao, *Montes y política forestal en la Galicia del antiguo régimen* (Santiago de Compostela: Universidad de Santiago de Compostela, 1995); Ofelia Rey Castelao, "Montes, bosques y zonas comunales aprovechamientos agrícola-ganaderos, forestales y cinegéticos," in *VII Reunión Científica de la Fundación Española de Historia Moderna*, coord. Francisco José Aranda Pérez (Ciudad Real: Universidad de Castilla-La Mancha, 2004), vol. 2, 907–66; Hortensio Sobrado Correa, "Rozas, estivadas y pan de vedro: el cultivo temporal del monte en la Galicia de la Edad Moderna," *Historia agraria, Revista de agricultura e historia rural* 89 (2023), 61–94.

16 "1. What the village is called" (1ª Cómo se llama la Población").

17 ADBU, CE, leg. 1912 and AGS, DGR, 1ª remesa, lib. 21.

> They said that this village is called Santo Domingo de Silos, that its limits encompass, without division, the hamlets of Peñacoba, Ynojar and Hortezuelos; together with this village they all form a Council for the appointment of posts, in both political and economic governance, and in the tax roll for royal taxes. However the urban area and hamlets have the use and exploitation of some woodlands and places for private grazing; which ones they are and in what form is stated in the declaration [*Memorial*] of the council.
>
> Dijeron que esta población se llama Sto. Domingo de Silos, que comprehende en su término, sin divisa [sic] de el, las aldeas de Peñacoba, la de Ynojar y Hortezuelos, haziendo todas con esta villa un Concejo así para el nombramiento de ofizios, en el govierno político, como económico, y en encavezamiento de tributos Reales, sin embargo que el Casco de villa y Aldeas tienen el uso y aprovechamiento en algunos montes y sitios destinados para pastos privativos a cada una, que lo que son y en la forma que es consta con expresión en el Memorial del Concejo.

The council's *Memorial*, or declaration, is referred to again in connection with the shared use of various woodlands and preserves with neighbouring villages. They state, for instance, that the village and its hamlets share with Contreras (Burgos):

> grazing within their limits from sunrise to sunset, which produces no income for the community. With the bordering places on the other three sides […] it shares grazing and gathering of firewood in the woodlands and areas which are recorded in this village's declaration [*Memorial*], with a specification of their sizes.
>
> comunidad en los pastos de sus términos de sol a sol, sin utilidad a sus comunes. Item. con los lugares de las otras tres confrontaciones […] tiene comunidad en pasto y rozo de leña en los montes y términos que, con expezificación de sus cavidas, constan en el Memorial desta Villa.

Although, as we have just seen, the answers to the first three questions may contain references to woodlands, it is generally the answer to question 4 that provides the first specific reference to wooded areas. This question is designed to gather information about land types and uses in the village limits, and woodland is specifically included as one of them.[18] This is the only question in which direct

18 "4. What kinds of land are there in the village limits: irrigated, rainfed, specifying whether they are of vegetables, arable, vines, pastures, forests, thickets, woodlands and others

reference is made to woodland and it appears in the context of the different land uses in the locality – no doubt an indication that the authors of the Cadastre were aware that it was one of several aspects of peasants' use of the land and its resources. The amount of information gathered with respect to woodland varies considerably from one locality to another, although the answers always refer at least to the existence of wooded areas and, more often than not, provide information on the dominant species in them, their condition, legal situation and uses. Continuing with the example of Silos, besides reporting on the forest areas, the respondents draw a distinction between those used exclusively by the village and those whose use is shared with its hamlets, stating that there are[19]

> shared woodlands, of high and low juniper, and its own, of holm oak, oak and pine, which are stated in the declaration [*Memorial*], from which neither is charcoal made nor is firewood harvested to sell in or outside this village. And only in the years when its own woodlands of holm oaks produce abundant acorns are pigs brought in from outside to graze and eat the acorns, and in a five-year period this earns the urban area, which exclusively exploits them, 200 *reales de vellón*.
>
> montes comunes, de enebro alto y bajo, y propios, de encina, roble y pino, los que se hallan expresados en el Memorial, de los que no se fabrica carbon ni se saca a vender leña fuera ni en esta villa, y sí solo en los años que hay abundancia de grana en los de encina, propios de ella, traen de fuera ganado de cerda a hervajar y comer la grana, y en un quinquenio dexa de utilidad al Casco de Villa, por serla este aprovechamiento privativo, 200 reales de vellón.

El Escorial, then in the province of Segovia and now part of Madrid, declared that it had high forests of ash trees, low woodlands of Pyrenean oaks and ash, low rockrose woodland with pasture, and black poplars.[20]

---

there may be, explaining whether there are any which produce more than one harvest per year, those that yield only once, and those which need to rest for an intermediate year." ("4ª Qué especies de tierra se hallan en el término; si de regadío, y de secano, distinguiendo son de hortaliza, sembradura, viñas, pastos, bosques, matorrales, montes, y demás, que pudiera haber, explicando si hay algunas que produzcan más de una cosecha al año, las que fructificaron solo una, y las que necesitan de un año de intermedio de Descanso.").

19 ADBU, CE, leg. 1912 and AGS, DGR, 1ª remesa, lib. 21.

20 The town of El Escorial should not be confused with the Royal Site of San Lorenzo de El Escorial, which was not recorded in the cadastre because the royal sites where the king

The answers given to this question generally also provide information about thinning the woodlands for cultivation, as in the cases of the parishes of San Juan de Arroxo (Lugo) and Salas de los Infantes (Burgos). The experts and council of the former stated that[21]

> the types of land there are in the abovementioned parish and within its limits are [...] riparian woodlands populated with chestnut trees, woodlands belonging to private owners, woodlands that are shared for grazing, and useless by nature; [...] with respect to both the woodlands that are privately owned and those shared for grazing, they produce rye and are cleared of trees every 15 years for those of first-class quality, every 24 for those of second-class quality, and every 36 for those of third-class quality.
>
> las especies de tierra que hay dentro de la referida feligresía y su término son [...] sotos poblados de castañales, monte de particulares, montes comunes quanto al pasto, e inútiles por su naturaleza; [...] por lo que les respecta a los montes tanto de particulares como comunes quanto al pasto, producen centeno y se rompen, siendo de primera calidad, de 15 en 15 años, de segunda, de 24 en 24, y de tercera, de 36 en 36.

The respondents of the village in Burgos report the existence of lands now cleared – which most likely were originally wooded – located within its confines and in some of the areas shared with other neighbouring villages, and comment on their low quality and consequent long crop rotation period.

---

spent seasonal sojourns and the newly created Royal Site of San Fernando were excluded from the process. This is not true of the rest of the royal sites and the king's properties, which were recorded along with other localities, for example Soto de Roma, Gózquez and Aceca, among others. See Félix Labrador Arroyo "Protection and Production: Soto de Roma in the seventeenth century," in *Árvores, barcos e homens na Península Ibérica (séculos XVI–XVIII)*, eds. Rosa Varela Gomes and Koldo Trápaga-Monchet (Zaragoza: Pórtico Librerías, 2017), 1–12; Félix Labrador Arroyo and Koldo Trápaga-Monchet, "La configuración del espacio y la explotación forestal de un enclave singular: el Real Sitio del Soto de Roma durante la dinastía Habsburgo," *Studia historica. Historia moderna* 39, no. 2 (2017), 293–327; Concepción Camarero Bullón and Ángel Ignacio Aguilar Cuesta, "Sitios Reales menores y Sitios del Rey en el Catastro de Ensenada del Reino de Granada," in *De reinos a naciones: espacios, territorios y mentalidades*, coords. Juan Jimémez Castillo and Manuel Rivero Rodríguez, *De reinos a naciones: espacios, territorios y mentalidades* (Madrid: Polifemo, 2021), 155–92; Concepción Camarero Bullón and Laura García Juan, "Geografía histórica de los espacios reales: Alóndiga, Aceca y Barciles, despoblados del rey en la vega del Tajo," *Estudios Geográficos* 284 (2018), 209–35.

21 AGS, DGR, 1ª remesa, lib. 546 (San Juan de Arroxo) and ADBU, CE, lib. 1740 (Salas de los Infantes).

They also state the composition of the flora and the various uses to which the woodlands are put[22]:

> Rainfed lands suitable for growing cereal [*tierras de pan llevar*], those which commonly yield crops with an interval of one year, though there are some which are poorer in nature and do so with four years' rest, and they are those which are called cleared land [*rozas*] [...] And in the remainder and centre, as well as in private and common land, there are different areas of barren land, shrubland, scrub, thickets, high forests and low woodlands, all of which neither this village nor its residents put to any uses other than for cutting firewood for their homes, and using them for their livestock, except for the five hundred and seventy-nine *reales* the community receives from the shared farmland they lease in common, and the income usually obtained from leasing some riparian woodland for hunting, which will be recorded in the declaration [*Memorial*] submitted by the procurator.

> tierras de pan llebar de secano, las que por lo común fructifican con un año de yntermedio, aunque existen algunas que, por ser de peor naturaleza, lo hacen con quatro años de descanso, y son las que llaman rozas [...] Y que en el resto y centro, así del término particular como de los comunes expuestos, se encuentran diferentes tierras herías, estepales, vreñas, matorrales, montes altos y bajos, de todo lo qual no tiene esta villa ni sus vezinos otro aprobechamiento que la leña que cortan para la fogata de sus casas y el que tienen con sus ganados, a exzepción de los quinientos setenta y nueve reales que percibe el Común del término que en común arriendan de la Campiña y lo que suele producir el arrendamiento de la caza de unos sotos, que constará en el memorial que por parte del procurador está presentado.

Neila (Burgos) also refers to the thinning of woodland in the reply to this question, as it reports having[23]

> thinned woodland with stands of oak, beech, pine juniper, holly and underbrush, which is not used at all by the community except for grazing and the firewood needed to heat their fireplaces.

22 ADBU, CE, lib. 1740 and AGS, DGR, 1ª remesa, lib. 21.

23 ADBU, CE, lib. 1184 and AGS, DGR, 1ª remesa, lib. 13.

> monte hueco con matas de roble, aya, pino enebro, acevo y matabaja, que en nada se aprovecha el común, más que de los pastos y la leña necesaria para la calefacción de sus chimeneas.

A water-powered sawmill is owned by a resident of the town, Joseph Bariola, due to the abundance of timber in the area. However, it only operates for three months – March, April and May – because the river is unfit for this use during the rest of the year due to the shortage of water in summer and ice in winter. For this reason, its owner only obtains a meagre annual profit of 250 *reales*.[24]

In their replies to this question, villages sometimes also refer to new plantings made in compliance with the provisions of the ordinance referred to above, although this information can also appear in answers to other questions too. For example, Robledo de Chavela and its districts of Peralejo, La Nava, El Espino, La Paradilla, Santa María, Las Serreras, Robledondo, la Cereda and la Oya, now in Madrid and then part of Segovia, declared that they had[25]:

> Wood-pasture for grazing, newly planted [...], in the meadows there is a portion of scattered oak and beech trees and different black poplars in the grazing wood-pasture, woodlands with the same species ... pollarded holm oak woodlands and kermes oak woodlands with some broom that is not cut because it is so ordered, with Scots and mountain pine, and in the naturally unfruitful land, a small quantity of old oak woodland.

> Dehesa boyal, nuevo plantío [...], en los prados hay una porción de árboles de roble y fresno dispersos y diferentes álamos negros en la dehesa boyal, montes de las mismas especies ... monte de encina olibado y chaparral con algo de retama que no se corta por estar así ordenado, con pinar albar y negral, y en la ynfructífera por naturaleza, monte de roble viejo en corta cantidad.

---

24 When the Cadastre was drawn up in 1753, according to the *Respuestas generales* Neila had 233 lay *vecinos* (heads of household), including two disabled people and twelve widows, and five clergy, two or them belonging to minor orders. There are also four inhabitants. The key to its economy is transhumance, an activity in which a very significant portion of the male population was engaged. The village's merino sheep breeders were part of the sheep owners' guild ("comprehendida en la Real Cabaña de Merinas").

25 AGS, DGR, 1ª remesa, lib. 546.

References to new plantings also appear fairly often in the answers to question 6.[26] For example, the aforementioned village of Salas de los Infantes reports that, in addition to the trees that make up the woodlands, there are various poplars and black poplars in the vicinity of the village, planted by royal order. And the respondents of Roa de Duero (Burgos) state that in its limits there are various poplars, elms and willows on the banks of the river Duero, and that the same species has been planted in the past few years on His Majesty's orders. They regret that these have not taken root properly, and many have been lost despite the care devoted to them by the wardens of the area and the local council.

The other question that elicited the most information about woodlands is number 10,[27] as it specifically enquired about the areas given over to different uses in the limits of the locality. This question is closely linked to the previous one, number 9, which requested information about the unit or units of measurement of area used in the locality,[28] and to number 12, asking for the value of the woodland products to be specified when appropriate.[29] As can be seen from the examples provided, these questions are complementary and offer researchers an insight into the use and valuation of the woodlands. For instance, after declaring that there are some 2,475 hectares of land in the

26 "6. If there are any plantings of trees in the lands they have declared, such as fruit, mulberry, olive, fig, almond trees, grapevines, carob trees, etc." ("6ª Si hay alguno plantío de árboles en las tierras que han declarado, como frutales, moreras, olivos, higueras, almendros, parras, algarrobos, etc.").

27 "10. What number of measures of land there are in the village limits, specifying those of each species and quality; for example: how many *fanegas*, or the name of the unit of measurement for arable land, of best quality; how many of medium quality, and how many of low quality; and the same for the other species they have declared" ("10ª Qué número de medidas de tierra havrá en el término, distinguiendo las de cada especie y calidad; por exemplo: tantas fanegas, o del nombre que tuviese la medida de tierra de sembradura, de la mejor calidad; tantas de mediana bondad, y tantas de inferior; y lo propio en las demás especies que huvieren declarado").

28 "9. What number of measures of land is used in that village; of how many paces or Castillian square *varas* it is composed; what is the quantity of each species of grain which is harvested in the village limits, how much is sown in each one" ("9ª Qué número de medidas de tierra se usa en aquel pueblo; de quántos pasos o varas castellanas en quadro se compone; qué cantidad de cada especie de granos, de los que se cogen en el término, se siembra en cada una").

29 "What quantity of produce each species yields, from year to year, with ordinary cultivation, a measure of land of each species and the quality of those there are in the village limits, not including any produce from trees" ("12ª Qué cantidad de frutos de cada género, unos años con otros, produce, con una ordinaria cultura, una medida de tierra de cada especie y calidad de las que hubiere en el término, sin comprender el producto de los árboles que hubiese").

locality, El Escorial states that 51 are high ash forest, considered to type of highest quality woodland, 225 are Pyrenean oak and low ash, considered to be of second-class quality in the same species, 28 are low rockrose with pasture, of third-class quality, and 2 are of black poplar, considered to be of second-class quality. In the answer to question 12, the town provides further information about its woodlands, stating that an *obrada* of black poplars, depending on their cutting, should be valued at 35 *reales*, one of rockrose, for its grazing, at one *real*, one of high ash forest, cut for charcoal making, at 41 *reales* and 14 *maravedíe*s, and one of Pyrenean oak and low ash at 50 *reales*.

Bélmez (Jaén) states in the answer to question 9 that the unit of measurement used locally is the *fanega*, consisting of 8,760 Castilian square *varas*, amounting to 666 *estadales* and two-thirds; in turn, one *estadal* is equivalent to three *varas* and five *octavos*, which is the standard *fanega* in Córdoba. The respondents add that, after clearing and burning woodland, in mountainous areas each *fanega* is sown with nine *celemines* of wheat or twelve of barley. This information is expanded on in the answer to question 10: they estimate the whole area in the village limits to be 29,441 *fanegas*, of which 8,720 are low woodland, whose pastures serve the village's residents and those of the localities that are part of the same community. The residents pay nothing for this use. However, they stress that[30]:

> were they to be sold each *fanega* would be worth thirty-four *maravedíes* and, with the extraordinary practice of clearing and burning the woodland, wheat, barley and rye are sown in them by the residents at no cost.
>
> si se vendieran valdría cada fanega treinta y quatro maravedíes y en ellas con la extraordinaria cultura de rozar y quemar el monte se siembra trigo, zebada y zenteno por los vezinos sin pagar cosa alguna.

Of particular interest in connection with the question of woodland are the places where it had a high economic value. For example, the economy of the area called Tierra de Pinares de la Demanda in the provinces of Soria and Burgos was based on exploiting the woodland and on cart transportation.[31] Its villages provide a wealth of information about its forests. For instance, in the reply to question 10, Canicosa de la Sierra reports that they have not measured

30 AGS, DGR, 1ª remesa, lib. 323 and AHPJA, CE, lib. 7654.

31 Pedro Gil Abad, *Junta y Hermandad de la Cabaña Real de Carreteros Burgos-Soria* (Burgos: Diputación Provincial de Burgos, 1983).

the limits, but give a conservative estimate of the *arrobas*[32] of different types of land – noting that greater accuracy can be found in the information resulting from the declarations (*Memoriales*) provided by residents and outsiders, following the reviewing and checking of all the statements. They reckon that within the exclusive confines of the village itself (that is, not including common land shared with neighbouring villages), the area[33]

> of grazing land populated with pines, oaks and a few holm oaks, known by the name of Sertil, whose grass and wood are made use of, is composed of 6,600 measures. Of woodland populated with the abovementioned trees, called Mata del Horno, whose wood is made use of, 9,110 *arrobas*. Of wooded land, useless, 550. Of heath with no use, either common or private, 1,000 *arrobas*. Of rocky and craggy land, 2,000.
>
> de pastos poblada de pinos, robres y algunas encinas, conocido con el nombre del sertil, cuya yerba y madera se beneficia, se compondrá de 6.600 medidas. De monte poblado de esos expresados árboles, llamado Mata del Horno, cuya madera se beneficia, 9.110 arrobas. De tierra montuosa, inútil, 550. De verezales sin utilidad ni en común ni en particular, 1.000 arrobas. De peñascos y riscos, 2.000.

The reply to question 12, which describes the income they obtain from the woodlands, supplements all this information:

> the limits and land and woodland and pastureland of this village, which consists of 16,000 *arrobas* of land more or less, half of which is populated with pine, 2,000 with tall oaks, and the rest with barren land for grazing, produces 3,500 *reales* in grass, which the community receives from its leases, and as for the wood which the community sells to residents and outsiders, it produces and receives 500 *reales* yearly, with no further income.
>
> el término y tierra y monte y pastos de esta villa, que se compone de 16.000 arrobas de tierra poco más o menos, la mitad poblada de pino, 2.000, de robres altos y las restantes de tierra yerma para pastos, produce por lo correspondiente a yerbas, 3.500 reales, que recibe el Común de sus

32 In this locality an *arroba* of land was equivalent to 202 feet.

33 ADBU, CE, lib. 396 and AGS, DGR, 1ª remesa, lib. 12.

> arriendos y, por lo que mira a la madera que vende el Común a vecinos y forasteros produce y recibe anualmente 500 reales, sin otro útil.[34]

Quintanar de la Sierra (Burgos), a village where the pinewood and cart transportation were the backbone of the local economy,[35] provides very interesting information about these woodlands, both those located within the limits of the village itself and those existing in the common land of Revenga, measuring a league and a half all around, of which it enjoys one-third together with the villages of Regumiel and Canicosa. The respondents state in their reply to question 10 that the limits and territory measure six leagues in circumference. Of them, one, circular shaped, is given over to orchards and meadows. They also claim to have wood-pasture one and half leagues in circumference which is fenced off for neighbourhood cattle to graze in, and is top-quality woodland[36]

> due both to the good grazing and to the cutting of pines the council does there and sells to both residents and outsiders, whose profits go to the community; the area of all this is four thousand five hundred *haces* of land, more or less, each composed of one thousand and eighty-nine square paces. Another four thousand five hundred *haces* of second-class woodland, of this species, which amounts to another league and a half in respect of three thousand *haces* per league, of which the Council similarly makes use; and two leagues that are of third-class quality as tree cutting cannot be used by the community and privately because it is rough

34 The Cadastre records about a hundred carts and some 80 individuals who earned a living from transportation. According to the *Vecindario de Ensenada* (census) the town had 111.5 lay heads of household (*vecinos*) and 2 who were secular clergy. Widows counted as half a *vecino*. Using the data gathered from the cadastral survey, in 1759 a census was drawn up – a document not envisaged in the Instructions – listing the *vecinos* of all the surveyed localities after checking and correcting the information provided in the declarations. That of all the provinces except those of Jaén and León-Asturias still survives. It was transcribed and published in Colección Alcabala del Viento, Concepción Camarero Bullón and Jesús Campos, *Vecindario de Ensenada, 1759* (Madrid: Centro de Gestión Catastral y Cooperación Tributaria y Tabapress, 1991), tomo 1, 96).

35 The local carters had no fewer than 1,023 oxen for transporting and 273 untamed young bulls and cows as replacements. The number of carts registered in the Cadastre was more than 350 and a hundred or so men were engaged in this activity. According to the *Vecindario de Ensenada*, the town had 155 lay heads of household and 3 who were secular clergy, see Camarero Bullón and Campos, *Vecindario de Ensenada*, tomo 1, 102; Pedro Gil Abad, introduction, *Quintanar de la Sierra 1753, según las Respuestas generales del Catastro de Ensenada* (Madrid: Centro de Gestión Catastral y Cooperación Tributaria y Tabapress, 1992).

36 ADBU, CE, lib. 1497 and AGS, DGR, 1ª remesa, lib. 14.

> and brambly, and only pasture is leased by the village for the grazing of transhumant merino flocks.

> así por el buen pasto como por la corta de pinos que en ella haze el Conzejo y enajena así a vezinos como forasteros, cuyo útil será cargado al Común, que su cavida de toda ella es de quatro mil y quinientos azes de tierra, poco más o menos, compuesto cada uno de los mil y ochenta y nueve pasos en quadro. Otros quatro mil y quinientos azes de tierra de monte de segunda calidad, en esta espezie, que compone otra legua y media al respecto de tres mil azes por legua, que en la misma forma se aprovecha el Concejo; y dos leguas de tercera calidad por no poderse utilizar en común y particular de la corta de árboles por lo áspero y breñoso y solo el pasto se arrienda por la villa para herbaje de ganados merinos trashumantes.

In this case, it is a village where logging is so important that in the reply to question 13, on fruit trees,[37] the respondents state that trees of this kind do not exist within its limits owing to the coldness of the place. This does not come as a surprise given the continental climate of the area and the altitude of the locality: the population centre is at 1,100 metres above sea level and much of its territory lies at as much as 1,900 metres, though

> of the pines that are regularly cut and sold by this village from the 4,500 *haces* of wood-pasture land, as they are of first-class quality in this species, 400, sold at one and a half *reales* apiece, fetch 600 *reales de vellón* every year, and in the 4,500 *haces* of woodland of second-class quality, every year 273 pines are cut which, at that price, total 490 *reales*. Grazing is not taken into account as it is enjoyed by the neighbourhood livestock. And this village leases third-class woodland at 5,000 *reales* per year for the grazing of transhumant merino sheep; and as for pines, the community does not use them for anything because they are in a brambly and rough place; and in the one-third that belongs to this village within the limits of Revenga nothing is used because the livestock of the community enjoy grazing and its firewood is consumed for their homes.

---

37 "What are the trees there are considered to yield per measure of land, according to the form in which they are planted, each in its species" ("13ª Qué producto se regula darán por medida de tierra los arboles que hubiere, según la forma en que estuviese hecho el plantío, cada uno en su especie").

> de los pinos que regularmente se cortan y venden por esta villa de los 4.500 azes de tierra que comprehende la dehesa, por ser de primera calidad en esta especie, 400 que, vendidos a real y medio cada uno, producen 600 reales de vellón cada un año y, en los 4.500 hazes de tierra que comprehende el monte de segunda calidad se cortan en cada un año 273 pinos que, a dicho precio, suman 490 reales, y que del pasto no se haze consideración por disfrutarlo los ganados del vezindario, y que el monte de terzera le tiene arrendado esta villa en 5.000 reales cada año para herbaje del ganado merino trashumante; y por lo tocante a pinos, no se utiliza el Común en cosa alguna por lo breñoso y áspero de su sitio; y que en la terzera parte que a esta villa pertenece en el término Rebenga no se utiliza en cosa alguna por disfrutar el pasto los ganados del común y su leña para el consumo de sus casas.

The importance of timber exploitation in the village is also revealed by the existence of two sawmills for sawing wood, one at the river Arlanza, owned by Juan Pérez, which they reckon yields a yearly profit of 1,500 *reales*, and the other at a stream they call Piavanares (probably the Rialares stream), owned by Antonio Medrano Cuesta, whose taxable base the experts calculate at 900 *reales* (question 17). A great deal of timber and a large expanse of woodland were needed to construct and repair so many carts, to load them, and to supply the two industrial facilities.

In some cases, the data supplied in writing in the replies to the questionnaire is accompanied by interesting visual information, as together with the necessary literal information the third question of the *Interrogatorio* requests a "figura" – that is, a drawing of the area. This gave rise to an interesting group of non-technical maps of a fair number of localities[38]:

> What territory do the village limits cover, how far does it stretch from east to west and from north to south, and what is its circumference, in hours and leagues, what boundaries or borders it has; and what its appearance is, including a drawing in the margin.

> Qué territorio ocupa el término, cuánto de levante a poniente y del norte al sur, y cuánto de circunferencia, por horas, y leguas, qué linderos o confrontaciones; y qué figura tiene, poniéndola al margen.

38 AHN, Consejos, lib. 1510.

The fact that the respondents were asked to include a drawing in the margin of the document indicates that it needed to be small, probably schematic, and with little detail. The authors of the Cadastre were aware from the outset that they would have to make do without technical maps of the villages' limits and measurements of the millions of parcels of land, as there were insufficient surveyors to carry out such a mammoth task. Nevertheless, we will see how there are provinces – for instance, Jaén and La Mancha – which measured all their localities,[39] but not the parcels of land. We will also see that whereas some drawings are highly schematic, generally no more than rough outline sketches, others take up a whole page or most of one and include more detail. Some are even land-use maps and indicate how the space was perceived by the people of the time: the experts, the court clerks who drew them, and the residents themselves, whose statements and comments made during the fieldwork no doubt led the draughtsmen to focus on certain land features.[40] The province of La Mancha[41] and the kingdom of Granada[42] are notable for the quantity and quality of their maps. In the other provinces there are a few maps that can be considered good – though they are few and far between – which it would be interesting to locate and catalogue systematically.

When these maps exist, a detailed analysis of the elements included in them often enables us to identify the presence of wooded areas – on the basis how the relief is represented, because they are actually drawn, because they

39 Concepción Camarero Bullón, Amparo Ferrer Rodríguez and Juan Gámez Navarro, "El proceso de elaboración del Catastro de Ensenada en el Reino de Jaén," *CT Catastro* 43 (2001), 19–50. Ángel Ignacio Águilar Cuesta, *Catastrar las Castillas: racionalidad frente a despilfarro. El coste de la realización del catastro de ensenada en el Reino de Jaén* (PhD dissertation: Universidad Autónoma de Madrid, 2021).

40 Concepción Camarero Bullón, "La cartografía en el Catastro de Ensenada, 1750–56," *Estudios Geográficos* 231 (1998), 245–83.

41 Eduardo Rodríguez Espinosa, Mª Ángeles Rodríguez Doménech and Concepción Camarero Bullón, "La representación cartográfica de los municipios manchegos en el XVIII. El Catastro de Ensenada," *Anales de Geografía de la Universidad Complutense* 40, no. 2 (2020), 499–540; Eduardo Rodríguez Espinosa and Mª Ángeles Rodríguez Doménech, *Mapas mentales y realidad en la Intendencia de la Mancha a mediados del XVIII. Superficie, población y croquis municipales del Catastro de Ensenada* (Valencia: Tirant humanidades, 2023).

42 Mª José Ortega Chinchilla, "Cartografía del espacio vivido: los croquis del Catastro de Ensenada y del Diccionario Geográfico de Tomás López desde el enfoque de la Geografía de la Percepción," *CT Catastro* 95 (2019), 9–44. Ana Luna San Eugenio, "Una propuesta para la sistematización y la difusión de la cartografía del catastro de Ensenada," in *Presentar, divulgar, conocer y valorar el patrimonio: Propuestas de trabajo transdisciplinares*, dirs. Félix Labrador Arroyo and Pablo Osma Rodríguez (Madrid: Dykinson, 2023), 79–94.

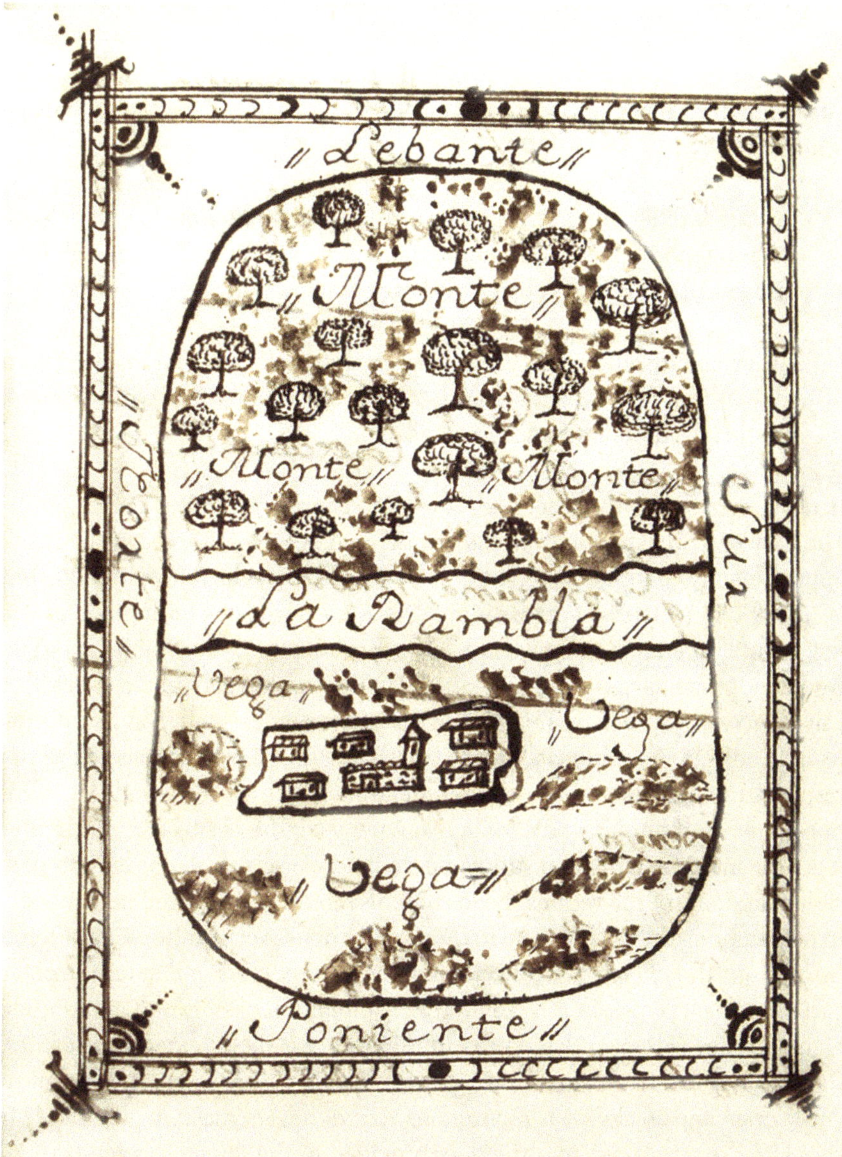

FIGURE 7.1 Map of the village of Albox included in the *Respuestas generales*
SOURCE: ARCHIVO HISTÓRICO PROVINCIAL DE GRANADA, CE, LIB. 962

are labelled "monte," or a combination as in the case of Albox (now in Almería, then part of Granada) (see Figure 7.1).

Albox is a village whose river – simply referred to as 'La rambla' (watercourse) and identified as Rambla de Horda on the 1:50,000 scale map – divides it into two areas with different characteristics: woodlands and water meadows. In the drawing the woodlands are depicted with tall trees with patches of dense vegetation that appear to suggest that they are interspersed with scrub. The answer to question 4 states, after listing all the uses of the land in the village limits, that[43]:

> there are different grazing lands and high forest and some low woodland, and different pines and some Iberian holm oaks in the first, though they are few in number, scrubland and useless rocky land.
>
> se hayan distintos pastos y monte alto y algún monte vaxo, y en el primero diferentes pinos y algunas carrascas, aunque estas son de corto número, matorral y tierra inútil de peñascos.

It is likewise stated that there is no wood-pasture land in the village limits. The answer to question 10 expands on and qualifies the information, as of the approximately 1,600 *fanegas* of land, 700 are high forest and low woodland, that is, 45.75%. Some twelve *fanegas* more or less are populated with pines, as the few existing Iberian holm oaks are scattered throughout the arable land. The woodland must have been significant and important to Albox's economy if it was depicted with such prominence in the non-technical map. This hypothesis would appear to be confirmed by the fact that in the answer to question 32 the respondents state that there are different lumberjacks who do a trade with one or two beasts of burden and earn 15 *reales* per year; this seems to indicate that it is a secondary activity that supplemented these peasants' economy. It is possible that many of the *vecinos* (heads of household) who engaged in this activity were among the 35 who lived in the country houses or farmhouses scattered throughout the village. There were also three carpenters and several master cartwrights, all professions linked to the exploitation of firewood and timber. A total of 637 *vecinos* provided the information for the *Respuestas generales*, though after the verifications were carried out they were found to amount to 788 laymen and 26 clergy, as reported in *Vecindario de Ensenada* (census).[44] It should be remembered that the data provided in the *Respuestas* tends to be approximate. Greater accuracy is found in the *Memoriales* and in the verification conducted by the experts.

As the maps are non-technical, they vary in style of execution and content, depending on different variables. Whereas Albox is an example of maps

43 (AGS, DGR, 1ª remesa, lib. 277 y AHPGRA, CE, lib. 962).

44 Camarero Bullón and Campos, *Vecindario de Ensenada*, tomo 2, 522.

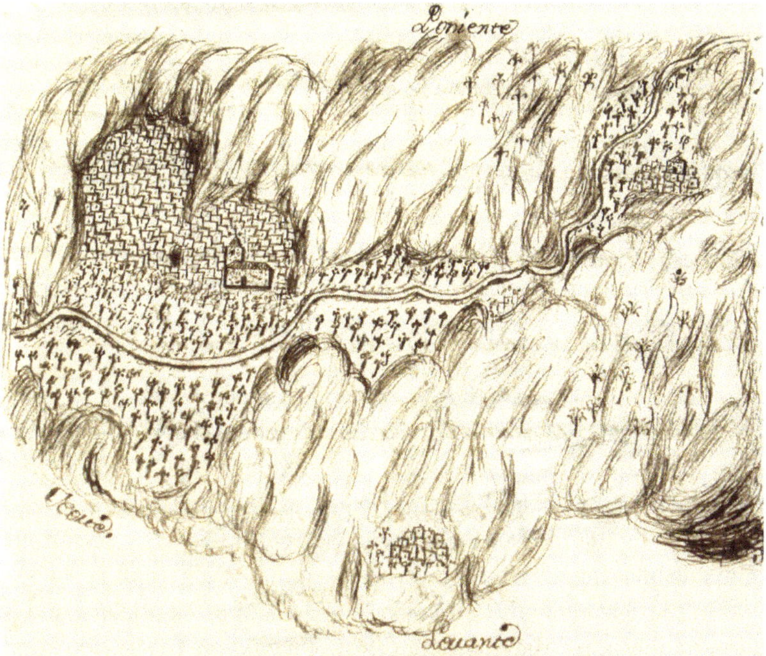

FIGURE 7.2 Map of Alboloduy included in the *Respuestas generales*
SOURCE: ARCHIVO HISTÓRICO PROVINCIAL DE GRANADA, CE, LIB. 955

in which woodland is represented and labelled, Alboloduy (see Figure 7.2) belongs to the type that depict woodland and topography.[45] The draughtsman enables viewers to grasp the land use and the presence of woodland in certain areas at first sight, as well as the rugged topography that may indicate the existence of natural vegetation of some kind, of rocky hills, and of spaces that by nature are unsuited to any uses. This visual information is borne out by the textual data contained in the *Respuestas generales*. Alboloduy, located in the Sierra Nevada nature park between the Mongenegro and Mencal massifs, has a rugged terrain which the draughtsman of the cadastral map depicts with loose strokes and a few trees at specific points in its midst. The population centre, represented by the church and the tiers of houses, is located on the right bank of the river Nacimiento. The area along the river and beside the population centre is wooded; much of it is probably riparian forest. Lastly, the map shows two smaller population centres: the one in the north is referred to in the records as "farmsteads called del Nacimiento" ("Cortijos llamados del Nacimiento") and

45 It is spelled Alboludoi in both the local and the provincial cadastral information.

beside it is a wooded area; the second is not correctly named but must be Las Alcubillas Altas. In the data provided about the village in the answer to question 21, it is stressed that the town has 450 lay *vecinos*, 330 of whom dwell in the main population centre and the other 120 in the farmhouses and country houses under its jurisdiction in the village limits. Seven clergymen, three of them belonging to minor orders, also live in the village. After checking the statements made, the *Vecindario de Ensenada* adjusts the data to 461 lay heads of household and 5 clergy.[46] They reside in 329 houses located in the main population centre and in 101 "farmhouses scattered throughout its jurisdiction" ("casas cortijo repartidas en el todo de su jurisdicción"). Lastly, it should be noted that the map is oriented to the west.

The information provided by the experts and council of Alboloduy in the *Respuestas* allows us to gain a better understanding about the woodlands represented on the map: they state that there are different areas of high forest and low woodland in the village limits, as well as various areas of grazing land, which are shared by the residents of this village and those of the neighbouring Sta. Cruz de Marchena. In their reply to question 6, they list as species of flora found in those woodlands holm oaks, poplars and black poplars, and in their answer to question 7 they state that the holm oaks are in the woodland. Given the water requirements of poplars and black poplars, we may deduce that these are the species shown lining the course of the river Nacimiento in the map, forming a gallery forest. The answer to question 13 provides further detailed information about the holm oak groves[47]:

> as for the oaks there are in the high forest, they have declared that they cannot assign a particular yield to each tree [...] because most of them are kermes oaks and Iberian holm oaks, which produce nothing [...], though what can be said is that all of them will produce 300 *fanegas* of acorns yearly. And as for income from firewood, they cannot produce it either as cutting them is prohibited by the community and naval ministers, who have recognised this.

> por lo que haze a las enzinas que ay en el monte alto que dexan declarado no pueden asignarle producto especial a cada árbol, [...] porque la maior parte de ellas son chaparros y carrascas, que nada produzen, [...] que lo que sí pueden decir, es que producirán el todo de ellas 300 fanegas de

46 Camarero Bullón and Campos, *Vecindario de Ensenada*, tomo 2, 524.

47 AGS, DGR, 1ª remesa, lib. 279 y AHPGR, CE, lib. 955.

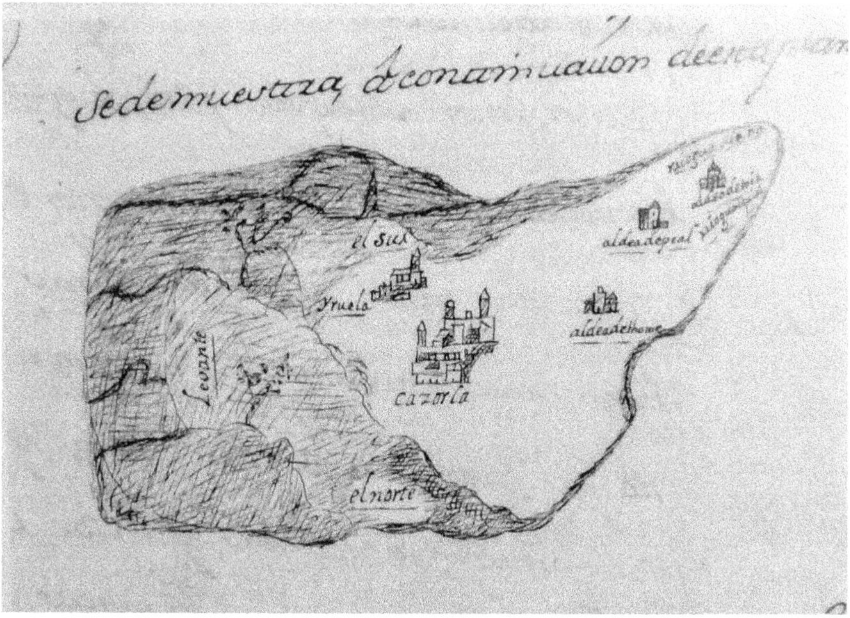

FIGURE 7.3 Map of the limits of Cazorla and La Iruela included in the *Respuestas generales* from Cazorla
SOURCE: ARCHIVO HISTÓRICO PROVINCIAL DE JAÉN, CE, LEG. 7.705

> vellota anualmente y que, por lo que haze a utilidad de leña, no la pueden hazer tampoco, por estar prohivida su corta por el Común y demás Ministros de Marina, que lo han reconocido.

After establishing that the price per *fanega* of acorns amounts to two *reales*, they state that the community obtains 600 *reales* annually from the sale of acorns produced in these woodlands.

We will end with a final example of a cadastral map: the one included in the *Respuestas generales* provided by Cazorla (see Figure 7.3) and the related textual information. When Cazorla (Jaén) was recorded in the cadastre in 1751, it shared the same limits as La Iruela, meaning that both villages enjoyed its resources, grazing and woodlands for their livestock. Curiously, the people who resided outside both population centres and in the hamlets had the option of becoming residents of either locality, with no need to move house, as a result of which residents and farmsteads belonging to both towns are "interpolated" in the same space. The village limits encompass a large area that stretches for

6.5 leagues from east to west, for 3.5 leagues from north to south, and for 16 leagues all around,[48]

> which, as the terrain was for the most part mountainous, rough and with large depressions, could be walked around in 44 hours.
>
> las que, por ser la mayor parte de dicho territorio montuoso, fragoso y de grandes profundidades, se podrán andar en 44 horas.

Obviously with such a terrain the wooded area must have been important to this village. Indeed, without going into detail about the composition of the flora in the wooded areas, the respondents state in their reply to question 4, after referring to those given over to crop cultivation, that there are rainfed areas which are planted with vines, a few olive groves, forests, pastures, scrub and woodlands. It is interesting to note that in this locality they draw a distinction between forests (*bosques*) and woodlands (*montes*), probably because it was among the wooded areas that were designated to supply timber to the Cartagena arsenal, together with the Sierra de Segura area.

It is not until the reply to question 6 that the experts and council of Cazorla refer to the composition of the flora of the wooded areas: "in the Sierra and other arable areas there are mostly holm oaks, oaks and pines with others that are fruitless, which are only good for the consumption of firewood" ("en la Sierra y otros sitios de labor están con mayor número los árboles de encinas, robles y pinos con otros infructíferos, que solo sirven para consumo de la leña").[49]

The reply to the following questions supplements this information by providing details about the location of these and a few other species: "the holm oaks, oaks, pines, terebinths, strawberry trees, junipers and other different ones are found in the Sierra of these villages and other rainfed places such as Alcoray, Burunchel, Cañamares and Tramalla" ("las encinas, robles, pinos, cornetas, madroños, henebros y otros diferentes se hallan en la Sierra de estas villas y otros parajes de secano, como son Alcoray, Burunchel, Cañamares y Tramalla").

The answer to question 10 provides key information about the characteristics, importance and size of the high forest and low woodland within this

48 AGS, DGR, 1ª remesa, lib. 324 y AHPJA, CE, lib. 7705. All information regarding Cazorla corresponds to this documentation.

49 AGS, DGR, 1ª remesa, lib. 279 y AHPGR, CE, lib. 955.

extensive area, which encompasses the villages of Cazorla and La Iruela and the hamlets of Peal, Toya and Santo Tomé. The respondents estimate "cautiously" that it could span some 100,000 *fanegas*. They state that they take a *fanega* of land to be 500 *estadales*, each consisting of 12 *celemines*: "and each *celemín* of 41 *estadales* and two thirds and a half, which makes 611 Castilian square *varas*, 7,333 *estadales* and one-third of a Castilian *vara* of Ávila, which is customary in this village" ("y cada celemín de 41 estadales y dos tercios y medio, que componen 611 varas castellanas en cuadro, 7.333 estadales y una tercia de vara castellana del marco de Hávila, que se estila en esta villa").

Of this area, they reckon that hills, uncultivated land, wood-pasture, woodlands, scrub, stony ground, gullies, paths and rocky land account for some 60,000 *fanegas* (60 percent of the total area within the limits), including the sources of the rivers, streams, rocky hills and "the sites of those population centres, churches and farmhouses" ("los asientos de dichas poblaciones, iglesias y casas cortijos").

They go on to explain the composition of this large area, stressing its possible agricultural value if rotation and sowing were practised at some point:

- 4,000 *fanegas* would be of second-class quality, could be sown and would produce the same quantities yielded by farmland of the same quality. They are not sown due to a shortage of residents of farms able to cultivate these areas or the poverty of their settlers. The respondents explain that the difficult terrain and large expanse of land within the limits call for this scattered interspersion to be able to put it into cultivation.
- 18,000 *fanegas* are populated with holm oaks, oaks, pines, strawberry trees and other fruitless trees "which are impossible to count due to their abundance and to the crags and crevices in which many of them are found" ("los que es imposible sujetar a cuenta por su muchedumbre, riscos y quiebras en que se hallan muchos de ellos").
- 30,000 *fanegas* which are also considered mountainous and other places within the village limits, most of which are only of use as grazing for livestock, although they are rough and uneven, populated with low woodland and high forest and of no use for sowing.
- The remaining 8,000 of the total of approximately 60,000 fanegas within the limits are rocky hills, with crevices and crags, part of which are impenetrable not only to people but also to livestock.

They end this account by adding that the remainder of the land within the confines up to the 100,000 *fanegas* given for the whole area within the limits of the two villages, Cazorla and Iruela, is arable, some parts irrigated and others rainfed.

After providing a detailed description of the wood-pastureland within the confines[50] and of how the pastures of the mountainous land are used, and stating that they are leased to the colegio de San Felipe Neri in the town of Baza – which currently possesses them as a judicial pledge for loans to this village – they report on the use of the holm oak and oak groves in the woodland:

> As for acorns, they are neither leased nor sold because it is common for the residents of both villages to consume them with their livestock. Were they sold, they could fetch 1,500 *reales* yearly for a five-year period, considering that not all the holm oaks and oaks bear this fruit, and the roughness of the mountains inhabited by wolves, for which reason more people and dogs are needed to guard the livestock than would be necessary were there not these disadvantages.
>
> por lo que mira al fruto de bellota, este no se arrienda ni vende porque asimismo es propio de los vecinos de ambas villas para consumirlos con sus ganados y, de haberse de vender, se pudiera dar por un quinquenio 1.500 reales cada un año, en atención a no ser todos los que llevan dicho fruto las encinas y robres, y lo fragoso de dicha Sierra y perseguida de lobos por cuyo motivo se necesita para la guarda del ganado de más gente y perros que el que dejaran de tener a no haber estos inconvenientes.

The text goes on to explain the uses of the pinewood, some of which very closely related to the navy:

> As for the exploitation or product of pines for construction timber, they also belong to those residents. A timber factory for rigging having been established at the aforesaid sawmill, either by the Royal Treasury or by contractors, it is considered that 2,000 pines could be cut every year, which, paid at four *reales de vellón* each, according to what appears to be laid down in the Royal Ordinances on woodlands, will provide this

50 In the 40,000 *fanegas* of arable land, which are those that they cultivate, they include some 1,000 consisting of several areas of wood-pasture and common land, to whose grazing rights, if sold, they assign a value: "today each *fanega* is worth in terms of grazing one *real de vellón* due to the many there are within the limits of this village and the aforesaid wood-pasture and common land are considered to have 1,000 *fanegas* of land, whose annual yield amounts to 1,000 *reales*" ("hoy valiera cada fanega por razón de pastos un real de vellón por los muchos que hay en este término y se consideran tener dichas dehesas y ejidos 1.000 fanegas de tierra, que su producto anual ha dicho precio importa 1.000 reales").

community with 8,000 *reales* yearly. As with the other products and common resources which are sold or can be sold, one quarter of this quantity belongs to the village of Iruela and the remaining three quarters of all of them to this village. The aforesaid factory for rigging and main timbers is expected to come into operation shortly because in this village there is a person entrusted with this.

por lo que hace al aprovechamiento o producto de los pinos para piezas de construcción, también son de dichos vecinos y, consiguiendo establecer en dicha Sierra fábrica de madera para arboladura o bien de cuenta de la Real Hacienda o de asentistas, se considera podrán cortarse cada un año 2.000 pinos que, pagados cada uno de ellos por cuatro reales de vellón, según parece se previene por las Reales Ordenanzas de montes, tendrá de utilidad este común anualmente 8.000 reales, de cuya cantidad, como de los demás productos y aprovechamientos comunes que se venden o pueden venderse, pertenece a la villa de la Yruela la cuarta parte y las tres restantes de todos ellos a esta villa, la cual dicha fábrica de madera de arboladura y piezas madres se espera en breve poner en egecución a causa de haver en esta población sugeto comisionado para ella.

Others uses are linked to the traditional provision of firewood, though naval interests have now come into the picture:

As for the use of firewood for the consumption of these two villages, both branches and charcoal and the other aforementioned uses are free for their residents and none is sold to other outsiders. And today, with the novelty of the new Ordinances issued by His Majesty and the person who has registered the aforesaid sawmill in his royal name, it is forbidden to cut green firewood, holm oak and oak, except at times when it is permitted to prune these trees, so that it is only possible to use dry firewood, and it is reckoned that this, and that used for the consumption of charcoal, if sold, could fetch 100 *ducados de vellón* every year.

en cuanto al aprovechamiento de leña para el consumo de estas dos villas así en rama como en carbón ésta [sic] como los demás referidos aprovechamientos son libres a sus vecinos y no hay egemplar se bendan a otros forasteros y oy con la novedad de las nuevas Ordenanzas expedidas por Su Majestad y persona que en su real nombre ha registrado dicha Sierra está prohibido se corte leña verde, encina y robre, a excepción de los tiempos en que se permite la limpia de estos árboles, con que solo se puede gastar la leña seca y, regulada esta y la del consumo de carbón si se hubiera de vender, en cada año parece pudiera importar 100 ducados de vellón.

Further information about woodlands is provided in the answer to question 13, where it is stated that "each *fanega* of land populated with holm oaks or oaks, which will grow to 35 feet, will produce 26 *fanegas* of acorns each year, which are produced in alternate years" ("cada fanega de tierra poblada de encinas o robres, que cogerá 35 pies, producirá cada un año 26 fanegas de vellota, las cuales producen un año sí y otro no"). The respondents also reaffirm the statement made in the answer to question 1: "and it is noted that although there is an abundance of holm oaks and oaks in the mountainous parts of these villages, their use is shared by the residents of both, as stated in the previous question ten" ("y se adbierte de que, aunque en la Sierra de estas villas hay muchedumbre de encinas y robres, es común aprobechamiento de unos y otros vecinos, como va expresado en la pregunta diez antecedente").

In the answer to the following question, number 14, they state that the price per *fanega* of holm oak or oak acorns is four *reales*, exactly double the amount paid for this fruit in Albolodu.

We will end the information on Cazorla by pointing out that the village claims to have some 900 lay heads of household, 700 of whom live in the centre and the rest in country houses, farmhouses and hamlets. There are many ecclesiastical *vecinos*: 29 priests, six recipients of benefices for their services and eleven chaplains. They all live in some 800 habitable houses located in the village centre, and 300 in country houses in the countryside and mountains. In addition, in the hamlet of Peal there are about 70 dwellings, some with broom roofs, a further 70 in the hamlet of Santo Tomé, and 40 in the hamlet of Thoya y Hornos. The respondents stress that these houses in the village centre, as well as the country houses, are inhabited by *vecinos* of both villages.[51]

Before ending the section on the *Respuestas generales*, it should be borne in mind that in a few specific cases information about woodlands also appears in the answer to question 40,[52] generally in connection with royally owned wooded areas, some of which are used to produce timber for shipbuilding.

51 The *Respuestas generales* provided by Cazorla and the *Libro de cabezas de casa* can be consulted in Norman Ball, introduction, *Cazorla, 1751, según las Respuestas Generales del Catastro de Ensenada* (Madrid: Centro de Gestión Catastral y Cooperación Tributaria, 1993).

52 "40. If, within the locality or village, the king has any estate or revenue that does not come under the general or provincial [tax] revenues, which are to be repealed; which ones they are, how they are administered and how much they yield" ("40ª. Si el rey tiene en el

For instance, several parishioners of the Ferrol area, such as the parish of San Salvador de Serantes, report that there are two areas of wood-pasture owned by His Majesty: one measuring eight *ferrados*, and another of second-class quality measuring five, both planted with oaks for shipbuilding. They state that Ferrol itself also has two areas of wood-pasture belonging to the king, planted with trees for building ocean-going ships and other vessels, one measuring six *ferrados* and the other four.

2 The *Libros de lo Real* and *Memoriales*: the Accuracy of the Cadastral Information

Whereas the *Respuestas generales* provide an overview of each and every one of the Castilian localities, for accurate data on assets, income and encumbrances it is necessary to turn to other cadastral documents: the *Memoriales* and the *Libros de lo real*.

The *Instrucción* annexed to the Royal Decree of 10 October 1749 established that the cadastral territorial unit was "the village" ("el pueblo")[53] and that all taxable persons, natural or legal, were to submit a declaration (*Memorial*) or list, signed by them or by a witness, of all the assets, income and encumbrances they possess in it; if they were a head of household residing in it (*vecino*), they should also include full information about their family, in the sense of extended family. Everything they declared was to be checked by the king's experts and those appointed by the village: lands were to be assayed ("paseadas") to check their size ("cabida"), use, quality, yields, location and boundaries, etc.; buildings measured by master builders (*alarifes*); people and livestock counted; annuities from leases (*censos* and *foros*) checked against receipts; and with tax revenues alienated from the Crown, their alienation proved by records of their purchase from the Royal Treasury or of their generous donation by the king, etc. Everything had to be reviewed. If a mistake or concealment was detected, the corrections were annotated in the *Memoriales* and if the declarant agreed to the correction with no further ado it was left at that. We find various corrections in the margins of the declarations and added at the end of documents, with the note "due to oversight" ("por olvido") when an asset or income had been omitted. The word "verified" ("verificado") or an alternative manner of

---

término o pueblo alguna finca o renta, que no corresponda a las generales ni a las provinciales, que deben extinguirse; cuáles son, cómo se administran y cuánto producen").

53 The criteria for defining "the village," the territorial unit, were tithed land (*tazmía*) and independent land subject to sales tax (*alcabalatorio independiente*).

expressing conformity with the data was added beside the assets that needed no correction. Only when a declarant stubbornly persisted in the concealment was a stricter process set in motion that could end in a fine.

When everything had been reviewed and checked, the information from the *Memoriales* was compiled and divided into two sections. Demographic information was recorded in the *Libro de cabezas de casa* (Book of heads of household), entered in two separate books: one for the families of laymen and another for those of clergymen. The assets, income and encumbrances were registered in the *Libro de lo real* (Book of real estate) which was also double, one for laymen and another for clergy. When all the information had been recorded, it was compulsory to convene a plenary meeting of the council by tolling the bell ("a campa tañida") and to read out in public the contents of both books and of the *Respuestas generales*, so that if anyone detected a mistake or concealment they could – and should – point it out. Once everything had been accepted, a record was issued stating that all those present agreed with the contents of both books. This was done to avoid possible future appeals, as had occurred with the cadastre drawn up years earlier in the Duchy of Milan.

Therefore, it is in the *Libro de lo real* that we can find an accurate record of all the information provided in the declarations and, with respect to the subject of this essay, about woodlands. It is also necessary to bear in mind that the information gathered in these documents is that needed to establish the tax and to glean the essential data for determining the ownership, location, composition of flora, uses and value of woodlands. As a result, they sometimes leave out certain data garnered in the declarations (*Memoriales*) that was considered "superfluous" for the purpose of the enquiries but is of interest to scholars. This process of compiling the information was employed for all the data gathered about the different branches of wealth and about the population, and therefore, where the *Memoriales* survive, it is very interesting to be able to work them as well as with the *Libros de lo real*. A couple of examples illustrate this point well: La Acebeda (now in Madrid, then part of Guadalajara) and Santo Domingo de Silos (Burgos). The council of La Acebeda states in its *Memorial*[54] that it has an area of wood-pasture:

> called Boyal, which is in the place of La Platera, at a distance of one hundred and fifty paces from this village, and it is populated by oak forest without undergrowth and low woodland of Pyrenean oak, and two thirds are of this species and one is of the aforesaid forest without undergrowth;

---

54 Archivo Histórico Nacional (AHN), Fondo Histórico del Ministerio de Hacienda (HMH), Catastro de Ensenada (CE), libro 132.

> the working cattle of the residents graze in it, and after Saint Michael's day until March the other herds belonging to those residents commonly enter. The grazing land of the aforesaid wood-pasture is of lower quality, is more or less sixty *fanegas* in size, and the boundaries are not expressed because it is the only one and very well known. It is enclosed all around and for the aforesaid grazing land nothing is paid or contributed to the Council, and from the wood it has, that needed to build residents' houses is cut, it being given to whoever requests it, for which they pay the aforesaid Council [...]. The aforesaid areas of low woodland are cut for charcoal about every ten years, and the income from this and from timber for construction is paid to the Council, which, taking into account years that produce income and those that do not, is reckoned to amount to eighty *reales* annually.

> llamada Boyal, que está al sitio de la Platera, distante de esta población ciento y cincuenta pasos y está poblada de monte hueco de roble y monte bajo de matorrales de rebollo, y de esta especie tendrá dos partes y una de dicho monte hueco, y en ella pastan los ganados bacunos de la labor de los vezinos y después de San Miguel y hasta marzo entran comúnmente los demás ganados de dichos vezinos y los pastos de la citada dehesa son de inferior calidad, tendrá de cavida como sesenta fanegas de tierra poco más o menos y los linderos no se expresan por ser única y muy conocida. Está cercada de pared alrededor, y por dichos pastos ni se paga ni contribuye cosa alguna a el Conzejo, y de las maderas que tiene se cortan las necesarias para los edificios de las casas de los vezinos, dando a el que se le ofrece, por las quales paga a dicho Concejo (...), y las citadas matas de monte bajo se cortan para carbón como de diez en diez años, cuio producto y lo de las maderas de edificios entran en propios del Conzejo, lo cual nos parece importara un año con otro, cotejado el que produze con el que no produze como ochenta reales anualmente.

In the margin of this description is a note, made by the experts who checked the statements: "In the field survey it was found to have one hundred and eleven *peonadas* and in other respects it was verified, the drawing in the margin" ("En el reconocimiento de campo se halló tener ciento y once peonadas y en lo demás comprovose, su figura al margen").

Beside this description is a drawing of the land "as it appears to the eye" ("como se aparece a la vista"). The value of its produce is also stated: this would be its taxable base, established by the experts in this inspection. Therefore the expression "eighty reales annually" ("ochenta reales anualmente") included at the end of the description of the plot of land was also written by a clerk of the court, not by the declarant.

This information is summed up in the *Libro de lo real* in the entry on the council's assets and income:[55]

> a wood-pasture called Boyal, which is within the limits of the aforesaid place of la Platera, one hundred and fifty paces from the population centre; it measures one hundred and eleven *peonadas*, and is of lower quality. It is populated with low woodland of scrub and Pyrenean oak and tall oaks, and is used for grazing the working cattle of the residents, it is enclosed on all sides. And these oaks are given freely to the residents to rebuild their houses and the low woodlands are cut every ten years for charcoal and their use is established at eighty *reales*. The tax class – in this case the twelfth of those established for the land within the limits of this village – is added in the margin.
>
> una dehesa llamada boial, que está en el término de dicho lugar a la Platera, distante a la población ciento cincuenta pasos, cave ciento y once peonadas, es de inferior calidad. Está poblada de monte bajo de matorrales y rebollo y robles altos, la que sirve para pastar los ganados bacunos de la lavor de los vecinos, está cercada de pared a todos aires. Y dichos robles se dan graciosamente a los vecinos para la rehedificación de sus casas y lo de monte bajo se corta de diez en diez años para carvón y su usufructo está regulado en ochenta reales. En el margen, se añade la clase fiscal, en este caso, la 12ª de las tierras del término.[56]

The council of Santo Domingo de Silos and its hamlets declare various woodlands, several shared with neighbouring villages. In their *Memorial*,[57] the respondents describe as follows what they call Campo la Raposa,

> A juniper woodland of rocky and craggy land which has never been cultivated and cannot be, called Campo la Raposa, Dornajo, Espeluca and Cabeza el Aliagar. It is one quarter of a league wide and three quarters of a league long, it borders to the north with the limits of the place of Contreras, to the east with the road to the aforesaid place, to the southeast

55 AHN, FHMH, CE, libro 130.

56 For how the note on the value is calculated and structured, see Camarero Bullón, "Vasallos y pueblos," 354–55.

57 Archivo de la Diputación de Burgos (ADBU), Catastro de Ensenada (CE), libro 6768.

> with the holm oak woodland of this village centre, and to the northeast with the limits of the common land of the place of Santiváñez and the said village and all of this amounts to fifty *fanegas*.
>
> Un monte enebral peñascoso y riscoso que jamás se a cultivado ni puede, yntitulado Campo la raposa, dornajo, espeluca y cabeza el aliagar, el que tiene de ancho un quarto de legua, y de largo tres quartos de legua, surca zierzo término del lugar de Contreras, al solano, camino de dicho lugar, al abregp, monte enzinar propio del casco de esta villa y a regañón, término comunero del lugar de Santiváñez y esta dicha villa y todo él haze zincuenta fanegas.

A note added in the margin of the document by the experts who "assay the farmland" ("pasean el campo") states: "the sown area has been established at fifty *fanegas*" ("Se ha regulado su sembradura en cincuenta fanegas"). The information about the area, fifty *fanegas*, is also an addition made by the experts. In the *Libro de lo real*, this area of woodland appears as:

> juniper woodland fifty *fanegas* of rocky land which they call Campo Raposa, Dornajo, Espeluca and Cabeza de Aliagar, which is only good for shared grazing for livestock. It borders to the north with the limits of Contreras, to the east with the road to that place, to the southeast with the woodland of the village centre, to the northeast with the common land of the place of Santiváñez.
>
> Un monte de enebro de cincuenta fanegas de tierra peñascosa que llaman campo raposa, dornajo, espeluca y caveza de aliagar que solo sirve de pasto común para los ganados, confronta por zierzo con término propio de Contreras por solano con camino que a dicho lugar, por ábrego con monte propio del casco de villa, por regañón con término comunero del lugar de Santiváñez.[58]

Much more could be added to the information provided by the Cadastre's *Memoriales* and *Libros de lo real* on each and every one of the woodlands and wood-pasture lands of the Crown of Castile, but the tyranny of space leaves

58 ADBU, CE, libro 6759. The *Respuestas generales* provided by Aranda are published in the Colección Alcabala del Viento: Pierre Amalric, introduction, *Aranda de Duero 1752, según las respuestas generales del Catastro de Ensenada* (Madrid: Centro de Gestión Catastral y Cooperación Tributaria y Tabapress, 1990).

room for no more than two brief examples. In some cases the designation "woodland" (*monte*) for a plot of land does not necessarily mean that this land was actually wooded at the time the Cadastre was compiled. An example of one of these areas that was no longer woodland but continued to be considered as such by the residents is found in Aranda de Duero. The entry by the council of Aranda in the town's *Libro de lo real* states that in it there is:[59]

> another woodland, which they call Monte Arandilla, or Cantaborricos, which is uncultivated and with no trees at all, one hundred *fanegas* in size, with no use whatsoever: it borders to the north with the Canta Burras road, to the east with the plot of la Presa, to the southeast with the river Duero and to the northeast with the Picocho plot, at a distance of 500 paces.
>
> Otro monte, que llaman monte Arandilla, o Cantaborricos, que se alla echo herial y sin árbol alguno, de cavida de cien fanegas, sin aprovechamiento alguno; confronta por cierzo camino de Canta Burras, solano, pago de la Presa, ábrego el río Duero y regañón, el pago de el Picocho, dista 500 pasos.

The declaration by the council of Nava de Roa (Burgos) highlights the process of deforestation the village was undergoing, as a result of which the existing wooded area did not cover residents' needs for firewood. The situation of two of its woodlands illustrates this process:[60]

> A woodland, which is called El Viejo, which is of Iberian holm oak, with a few tall crowns, which is half a quarter of a league long and wide; and another, which is named Los Montecillos, is of Iberian holm oaks or thickets of holm oaks, and is the same length and width, and inside and outside them are as many as sixty *fanegas* of sown treeless land and two meadows [...] and the grass they produce is only used for the grazing of this village's livestock; and the firewood of the aforesaid woodlands is cut by the residents [...] with great limitations, and for this reason it does not supply them with what they need, and they buy it elsewhere.
>
> Un monte, que se llama El Viejo, que es de carrasca de encina, con algunos pies altos, que tiene de largo y ancho medio quarto de legua; y otro, que

59 ADBU, CE, libro 80.
60 ADBU, CE, libro 1174.

> se intitula Los Montecillos, es de carrascas o matorrales de encinas, y tiene el mismo largo y ancho, y dentro y fuera de ellos se comprehenden hasta sesenta fanegas de sembradura de tierra blanca y dos prados [...] y la yerba que producen únicamente sirve para pastar los ganados de este pueblo; y la leña de dichos montes, para dar corta a los vecinos [...] con mucha limitación, y por lo mismo no les alcanza para la que necesitan, y la compran fuera.

The points discussed above and illustrated with examples underline the type of information the cadastral documents provide about woodlands and their ownership, size, flora, administrative situation and uses. They also show that the establishment of the taxable base – that is, their value with a view to levying the tax – depended on whether their yields (from grazing, firewood, produce, timber, etc.) were a necessary part of the locals' economy or whether any of them were sold. In the case of the former, their value was considered to be included in that of the productive livestock and arable land, and therefore their valuation for tax purposes was nil. However, if the council and residents obtained revenues of any kind from their woodlands (from cutting and selling wood, leasing pastures, etc.), this income would be subject to the single tax.

This brings us to another issue that is important to stress with a view to comparative studies: the wide range of land measurement units employed in the different parts of the kingdom. The Cadastre recorded the areas in the specific measurement units used in each place, asking the respondents to give the equivalent in square Castilian *varas*, though not all places managed to provide such information. It was not until the following century that the decimal metric system was adopted. Converting measurements into it is a task that is complex at times but doable, even if only approximately. Furthermore, the fact that it reflects this diversity also makes it a rich source of information, because it enables us to study the territorial distribution of these units and their size.

3 Conclusions

In conclusion, it should be noted that the Ensenada Cadastre (*Catastro de Ensenada*) is a unique huge collection of information on population, material resources, economy, administrative organization and geography from Castile in the middle of the eighteenth century. This allows researchers to study the forests themselves and their role in the economic activity of the localities in which they are located. However, it must be kept in mind that information on forests and natural vegetation, in general, is distributed in different documents and even within the same document in different spaces and that, to study this subject, all documents are complementary. It is essential to know what can be found in each of them and where and, consequently, to know what to ask for each cadastral document and what can be expected from it.

Webography

Websites that provide access to digitized documentation from the Ensenada Cadastre:

Andalusian Archives Web (Portal de Archivos de Andalucía) https://ws096.juntadeandalucia.es/archivaWeb/

Archivo Histórico Provincial de La Rioja: https://catastrodeensenada.larioja.org/

Familysearch: https://www.familysearch.org/es/home/portal/

PARES: https://pares.mcu.es/Catastro/servlets/ServletController?ini=0&accion=0&mapas=0&tipo=0

Abbreviations

ADBU: Archivo de la Diputación de Burgos
  CE: Catastro de Ensenada
AGS: Archivo General de Simancas
  CE: Catastro de Ensenada
  DGR: Dirección General de Rentas, 1º remesa
AHN: Archivo Histórico Nacional
  CE: Catastro de Ensenada
  FMH: Fondo Histórico del Ministerio de Hacienda
AHPGR: Archivo Histórico Provincial de Granada
  CE: Catastro de Ensenada
AHPJA: Archivo Histórico Provincial de Jaén
  CE: Catastro de Ensenada

Acknowledgements

This study has been conducted as part of the Research Projects I+D+i of the Ministerio de Ciencia e Innovación, entitled: *Avanzando en el conocimiento del*

Catastro de Ensenada y otras fuentes catastrales: nuevas perspectivas basadas en la complementariedad, la modelización y la innovación, subproyecto del proyecto coordinado *Las fuentes geohistóricas, elemento para el conocimiento continuo del territorio: retos y posibilidades de futuro a través de su complementariedad* (*FGECCT*) (PID2019-106735GB-C21) and he proyect: *Profundizando en el conocimiento de los catastros españoles de los siglos XVIII y XIX en el marco de los catastros europeos, de otras fuentes geohistóricas y de las TIC (CEyFG)* (PID2023-150367NB-100/1003080047) and as part of the Convenio FUAM-138250 de Transferencia del Conocimiento de la Fundación de la Universidad Autónoma de Madrid and the Dirección General del Catastro de España. The result was achieved under the postdoctoral contract of Ángel Ignacio Aguilar Cuesta at the Universidad Autónoma de Madrid: Ayudas "Margarita Salas" del Plan de Recuperación, Transformación y Resiliencia, funded by the European Union-Next Generation EU.

Bibliography

Aguilar Cuesta, Ángel Ignacio. *Catastrar las Castillas: racionalidad frente a despilfarro. El coste de la realización del catastro de ensenada en el Reino de Jaén*. PhD dissertation, Universidad Autónoma de Madrid, 2021.

Ball, Norman, introduction. *Cazorla, 1751, según las Respuestas Generales del Catastro de Ensenada*. Madrid: Centro de Gestión Catastral y Cooperación Tributaria, 1993.

Bravo Lozano, Jesús. *Montes para Madrid. El abastecimiento de carbón vegetal a la villa y corte en los siglos XVII y XVIII*. Madrid: Caja de Madrid, 1993.

Cadiñanos Bardeci, Inocencio. "Los Montes de Sanabria a fines del siglo XVIII y comienzos del XIX", *Anuario del Instituto de Estudios Zamoranos Florián de Ocampo* 27 (2010), pp. 237–254.

Camarero Bullón, Concepción. "La cartografía en el Catastro de Ensenada, 1750–56." *Estudios Geográficos* 231 (1998), 245–83.

Camarero Bullón, Concepción. "El Catastro de Ensenada, 1749–1759: diez años de intenso trabajo y 80.000 volúmenes manuscritos." *CT Catastro*, 46 (2000), 61–88 (Spanish version), 141–53 (English version).

Camarero Bullón, Concepción. "Vasallos y pueblos castellanos ante una averiguación más allá de lo fiscal: el Catastro de Ensenada, 1749–1756." In *El Catastro de Ensenada. Magna averiguación fiscal para alivio de los vasallos y mejor conocimiento de los reinos*, directed by Ignacio Durán Boo and Concepción Camarero Bullón, 113–388. Madrid: Ministerio de Hacienda, 2002.

Camarero Bullón, Concepción. *Madrid y su provincia en el Catastro de Ensenada. I. La Villa y Corte, II. Los pueblos*. Madrid: Ediciones del Umbral, 2001–2005, 2 vols.

Camarero Bullón, Concepción and Ángel Ignacio Aguilar Cuesta. "La cartografía, instrumento para conocer el territorio, planificar y gestionar las reformas en la España del siglo XVIII." *Manuscrits. Revista d'Història Moderna* 42 (2020), 157–78, https://doi.org/10.5565/rev/manuscrits.303.

Camarero Bullón, Concepción and Ángel Ignacio Aguilar Cuesta. "Sitios Reales menores y Sitios del Rey en el Catastro de Ensenada del Reino de Granada." In *De reinos a naciones: espacios, territorios y mentalidades*, edited by Juan Jiménez Castillo and Manuel Rivero Rodríguez, 155–92. Madrid: Polifemo, 2021.

Camarero Bullón, Concepción and Laura García Juan. "Geografía histórica de los espacios reales: Alóndiga, Aceca y Barciles, despoblados del rey en la vega del Tajo." *Estudios Geográficos* 284 (2018), 209–35, https://doi.org/10.3989/estgeogr.201809.

Camarero Bullón, and Miguel C. Vivancos: "Con 'letras antiguas y en latín': la copia de los privilegios antiguos en el Catastro de Ensenada." In *La Corte de los Borbones: crisis del modelo cortesano*, coordinated by José Martínez Millán, Concepción Camarero Bullón and Marcelo Luzzi Traficante, 237–54. Madrid: Polifemo, 2015, vol. 1.

Casals Costa, Vicente. "Conocimiento científico, innovación técnica y fomento de los montes durante el siglo XVIII." In *El Siglo de las luces: de la industria al ámbito agroforestal*, coordinated by Manuel Silva Fernández, 453–500. Zaragoza: Diputación Provincial, 2005.

Coronas Vida, Luis Javier. "Montes y arbolado en los pueblos de la jurisdicción de Burgos durante el siglo XVIII." *Boletín de la Institución Fernán González* 232 (2006), 179–222.

Diego Liaño, Concepción and Juan Carlos García Codrón. "La Corona y los pueblos en la explotación de los montes de Cantabria: deforestación y gestión del bosque en la segunda mitad del siglo XVIII." *Cuadernos de la Sociedad Española de Ciencias Forestales* 16 (2003), 215–20.

Fernández Aldana, Rafael. "Evolución de los hayedos en las cuencas de los ríos Leza, Jubera y Cidacos entre los siglos XVIII y XX, a partir del Catastro de Ensenada, de las Relaciones de Tomás López, del Diccionario de Madoz y de la clasificación y el catálogo de los montes públicos." *Zubía* 13 (2001), 113–38.

Fidalgo Hijano, Concepción and Juan Antonio González. "El entorno de las Lagunas de Ruidera en el siglo XVIII a la luz del catastro de Ensenada y la cartografía de la época." *CT Catastro* 77 (2013), 43–66.

Gil Abad, Pedro. *Junta y Hermandad de la Cabaña Real de Carreteros Burgos-Soria.* Burgos: Diputación Provincial de Burgos, 1983.

Gil Abad, Pedro. *Quintanar de la Sierra 1753, según las Respuestas generales del Catastro de Ensenada.* Madrid: Centro de Gestión Catastral y Cooperación Tributaria y Tabapress, 1992.

Konyushikhina, Nadezda. "Los cuestionarios para las Relaciones topográficas de Felipe II y las Relaciones geográficas de Indias de los años 1570." *CT Catastro* 89 (2017), 9–30.

Labrador Arroyo, Félix and Koldo Trápaga-Monchet. "La configuración del espacio y la explotación forestal de un enclave singular: el Real Sitio del Soto de Roma durante la dinastía Habsburgo." *Studia historica. Historia moderna* 39, no. 2 (2017), 293–327, https://doi.org/10.14201/shhmo2017392293327.

Labrador Arroyo, Félix. "Protection and production: Soto de Roma in the seventeenth century." In *Árvores, barcos e homens na Península Ibérica (séculos XVI–XVIII)*, edited by Rosa Varela Gomes and Koldo Trápaga-Monchet, 1–12. Zaragoza: Pórtico Librerías, 2017.

López Gómez, Antonio and Carmen Manso Porto. *Cartografía del siglo XVIII. Tomás López en la Academia.* Madrid: Real Academia de la Historia, 2006.

Luna San Eugenio, Ana. "Una propuesta para la sistematización y la difusión de la cartografía del catastro de Ensenada." In Félix Labrador Arroyo y Pablo Osma Rodríguez: *Presentar, divulgar, conocer y valorar el patrimonio: Propuestas de trabajo transdisciplinares*, directed by Félix Labrador Arroyo and Pablo Osma Rodríguez, 79–94. Madrid: Dykinson, 2023.

Matilla Tascón, Antonio. *La Única Contribución y el Catastro de la Ensenada.* Madrid: Ministerio de Hacienda, 1957.

Mongil Manso, Jorge and Javier Álvarez Martínez. "El Catastro de Ensenada y sus aplicaciones en trabajos sobre el medio natural." *Medio Ambiente en Castilla y León*, 17 (2002), 43–8.

Mongil Manso, Jorge and Javier Álvarez Martínez. "Análisis de algunos factores determinantes de la superficie forestal de las provincias de Segovia y Soria en el siglo XVIII." *Cuadernos de la Sociedad Española de Ciencias Forestales* 16 (2003), 221–25.

Moreno Fernández, José Ramón. *El monte público en La Rioja durante los siglos XVIII y XIX: aproximación a la desarticulación del régimen comunal.* Logroño: Gobierno de La Rioja, 1994.

Ortega Chinchilla, Mª José. "Cartografía del espacio vivido: los croquis del Catastro de Ensenada y del Diccionario Geográfico de Tomás López desde el enfoque de la Geografía de la Percepción." *CT Catastro* 95 (2019), 9–44.

Pérez Romero, Emilio. *Patrimonios comunales, ganadería trashumante y sociedad en la Tierra de Soria. Siglos XVIII–XIX.* Salamanca: Junta de Castilla y León, 1995.

Rey Castelao, Ofelia. *Montes y política forestal en la Galicia del antiguo régimen.* Santiago de Compostela: Universidad de Santiago de Compostela, 1995.

Rey Castelao, Ofelia. "Montes, bosques y zonas comunales aprovechamientos agrícola-ganaderos, forestales y cinegéticos." In *VII Reunión Científica de la Fundación Española de Historia Moderna*, coordinated by Francisco José Aranda Pérez, 907–996. Ciudad Real: Universidad de Castilla-La Mancha (2004), vol. 2, pp. 907–66.

Rodríguez Cancho, Miguel. "Interrogatorios del siglo XVIII. Estudio comparativo." *Norba* 2 (1981), 221–32.

Rodríguez Cancho, Miguel and Gonzalo Barrientos Alfageme, eds. *Interrogatorio de la Real Audiencia Extremadura a finales de los tiempos modernos.* Mérida: Asamblea de Extremadura, 1993–1996, 11 vols.

Rodríguez Espinosa, Eduardo and Mª Ángeles Rodríguez Doménech. *Mapas mentales y realidad en la Intendencia de la Mancha a mediados del XVIII. Superficie, población y croquis municipales del Catastro de Ensenada.* Valencia: Tirant humanidades, 2023.

Rodríguez Espinosa, Eduardo, Mª Ángeles Rodríguez Doménech and Concepción Camarero Bullón. "La representación cartográfica de los municipios manchegos en el XVIII. El Catastro de Ensenada." *Anales de Geografía de la Universidad Complutense* 40, no. 2 (2020), 499–540, https://doi.org/10.5209/aguc.72984.

Ruiz Álvarez, Raúl, "Economía de estado: La gestión de los montes del sur peninsular en el siglo XVIII." In *Raíces profundas: un viaje por las fuentes geohistóricas hasta la Ilustración,* coordinated by Concepción Camarero Bullón, María S. Gómez Navarro, Alfonso Fernández-Arroyo López-Manzanares, Miguel B. Bernabé Crespo. Cantabria: Editorial Universidad de Cantabria (2024), pp. 419–440.

Sanz Sanjosé, Gloria: "Dinámica de las masas forestales en el territorio de Riofrío (Segovia)." *Cuadernos de la Sociedad Española de Ciencias Forestales* 16 (2003), 273–78.

Sobrado Correa, Hortensio. "Rozas, estivadas y pan de vedro: el cultivo temporal del monte en la Galicia de la Edad Moderna." *Historia agraria, Revista de agricultura e historia rural* 89 (2023), 61–94, DOI: 10.26882/histagrar.089e01s.

Vallina Rodríguez, Alejandro and Nadezda Konyushikhina. "Los interrogatorios de los Catastros españoles de la Edad Moderna: fuentes geohistóricas para conocer los paisajes y las sociedades." CT *Catastro* 91 (2017), 39–63.

CHAPTER 8

Empire, Timber and Wine: the Interconnected Dimensions Landscaping Forests in the 1700s

Cristina Joanaz de Melo

1 Introduction

Through the Lisbon earthquake of 1st November 1755, nature shook up the political order of Europe. As is well known, the Portuguese capital was reduced to ruins by the seismic shocks, a huge tsunami, and a major fire. Within these events, both the naval fleet anchored in the Tagus estuary and all the artillery and timber stored in the Lisbon harbour were dragged out by the ocean waters, sank or appeared floating in fragments out in the Atlantic. Such outcomes not only weakened Portuguese naval communications and access to the oversea States but also the national defence system of the Portuguese metropolis.[1]

The aforementioned scenario, dreadfully catastrophic for Portugal, naturally favoured the other naval powers which had been competing for control over corridors in the Atlantic and Indian oceans since the late 1600s and early 1700s: the Dutch, British and French fleets.[2] The cargos of gold shipped from Brazil to Portugal alongside exotic timbers exported from India or Brazil had already become targets for piracy and smuggling in the late 17th and early 18th centuries.[3] Furthermore, following the end of the War of Spanish Succession (1714), Portugal had never recovered its dominance over the Atlantic sailing corridors.[4] By 1755, the aftermath of the earthquake weakened the national position on international sailing routes still further.

---

1 Judite Nozes, *O Terramoto de Lisboa de 1755. Testemunhos Britânicos/The Lisbon Earthquake of 1755. British Accounts* (Lisbon: The British Historical Society of Portugal-Lisóptima editions, 1990); Arnaldo Cardoso, *O Terrível Terramoto da Cidade que Foi Lisboa. Correspondencia do Nuncio Filippo Acciaiuolli* (Lisbon: Aletheia, 2013).

2 Jorge Pedreira, "From Growth to Collapse. The Breakdown of the Old Colonial System. Portugal and Brazil (1750–1830)," *The Hispanic American Historical Review* 80, no. 4 (2000): 839–65.

3 Jeremy Land and Rodrigo da Costa Dominguez, "Illicit Affairs: Philadelphia's Trade with Lisbon before Independence, 1700–1775," *Ler História* 75 (2019): 179–204.

4 Ângela Domingues and Carmen Ortiz García, "Viagens Científicas e Coleccionismo no Mundo Ibérico (séculos XVIII–XX)", *Asclepio. Revista de Historia de la Medicina y de la Ciencia* 71, no. 2 (2019): 1–4; Ângela Domingues, "Museus, Colecionismo e Viagens Científicas em

 | DOI:10.1163/9789004745698_010

Given their predicament, it would have seemed likely that the material chaos arising from the seismic and other shocks would have lowered Portuguese political aspirations towards leading the European imperial trade. However, it was precisely during this context of appalling access to the colonies, coupled with the desire to defend and rebuild the Portuguese territory, that plans to regain its position among the leading empires were drafted by the Portuguese intelligentsia.

The economic and political recovery of the Metropolis would envision the landscaping of specific geographies. Beginning with the prerogatives of land management within the scope of strengthening an exquisite product of value on international markets, fortified port wine, this was to be developed in parallel and also interweaving with other sectors: forest plantations, fruit orchards, among other crops.

The difficulties encountered in reaching the colonies in the first months of 1756 demonstrated how Portugal was a naval empire with sailing and shipbuilding remaining crucial activities for the Crown.[5] Consequently, without immediate access to forestry resources from the colonies, ensuring the regular delivery of timber for shipbuilding and defence of the realm, would swiftly become a matter of the greatest importance.

The internal reserve stocks of forestry products for supplying the royal shipyards were to be constantly assured. However, importing forest resources implied negotiating with Portuguese competitors in the timber and wood markets of northern Europe.[6] Furthermore, post-1755,[7] Portugal became still more dependent on external assistance for securing the delivery of timber and wood but, as demonstrated by new empirical data for 1756 and 1757, this was significantly met by internal supplies from nationally located royal and seigniorial forests. Thus, as colonial forests now seemed far-distant and far more dangerous to reach than when Portugal controlled the sailing routes, national forests

Portugal de Finais de Setecentos," *Asclepio. Revista de Historia de la Medicina y de la Ciencia* 71, no. 2 (2019): 1–19.

5 José Manuel Malhão Pereira (coord.) a *História da Marinha Portuguesa. Navios, Marinheiros e Arte De Navegar (1669–1823)* (Lisbon: Academia de Marinha, 2012).

6 K. Jan Oosthoek and Richard Hölzl (eds.), *Managing Northern Europe's Forests Histories from the Age of Improvement to the Age of Ecology* (NewYork-Oxford: Berghahn Books, 2018).

7 Monteiro, "A Monarquia," 331–78; Avelino de Freitas de Meneses, "A rutura e a reconciliação (1640–1750)," in *História de Portugal e Espanha: Amores e Desamores*, coord. Artur Teodoro de Matos, João Paulo de Oliveira e Costa and Roberto Carneiro (Lisbon: Círculo de Leitores, 2015), 7–152.

came to support the extraction of resources, mostly pinewood, and ensuring a significant proportion of the timber supply required by the royal shipyards.[8]

As recent research has verified, the first half of the 1700s experienced the renewal and expansion of royal woodlands across the Iberian Peninsula. To reinforce that policy, the Spanish Crown implemented a program for royal forests from 1748 onwards. Portugal implemented a similar plan after 1751, investing in planting trees along the coastal sandbanks and hilly valleys with tree-species appropriate for shipbuilding.[9]

While this forest expansion plan may have been designed prior to the occurrence of the earthquake, carrying it out from 1756 onwards would widely become an almost natural reaction of both the royal administration and the ecclesial and noble landlords. However, there are no studies available on these afforested areas and any eventual increase in the tree park in Portugal as a consequence and in response to the increased consumption of forest products by the royal service, particularly those supplied by private estates owned by noblemen or the Church.

Nonetheless, historiography has clearly portrayed the efforts endorsed by King José I (1750–1777) and the Marquis of Pombal to support the recovery of the population, the rebuilding of Lisbon, as well as some areas of the Algarve, the restoration of food supplies while also preventing epidemics in the wake of the 1755 seismic catastrophe.[10] Accordingly, a substantial body of solid essays was produced on risk assessment, engineering and architecture.[11] However, the strategic landscaping detailing the parallel projects for forests and farming, spanning the entire scope of the national territory, were never delivered.

---

8 Diogo de Carvalho Cabral, *Na Presença da Floresta. Mata Atlântica e História Colonial* (Rio de Janeiro: Garamond, 2014).

9 Francisco Rego, *Florestas Públicas* (Lisbon: Direcção-Geral Das Florestas, 2001); Nicole Devy-Vareta, "Os Avanços e Recuos Da Floresta Em Portugal – Da Idade Média Ao Liberalismo", in *Floresta e Sociedade: Uma História Em Comum*, ed. Joaquim Silva (Lisbon: Fundação Luso-Americana para o Desenvolvimento, 2007), 55–75; Alfredo José Martínez González, *Las Superintendencias de Montes y Plantios (1574–1748): Derecho y Politicas Forestales para las Armadas en Edad Moderna* (Valencia: Tirant lo Blanch, 2015); Cristina Joanaz-de-Melo, "Guerra, Impérios e Corte Joanina nas Coutadas de Caça: Alavancas de Regeneração Florestal em Portugal, em Meados do século XVIII," *Manuscrits: Revista d'història moderna,* 42 (2020): 199–220; Cristina Joanaz-de-Melo, "Floresta em Movimento: usar, regenerar, cuidar (seculos XVI–XIX)", in *Como a Fénix Renascida, Matas Bosques e Arvoredos (séculos XVI–XX): Representações, Gestão, Fruição,* ed., by Cristina Joanaz de Melo, (Lisbon: Edições Colibri, 2020), 79–130.

10 Nuno Monteiro, "A Monarquia Barroca (1668–1750): O tempo de Pombal," in *História de Portugal,* coord. Rui Ramos, Bernardo Vasconcelos Sousa and Nuno Monteiro (Lisbon: A Esfera dos Livros, 2010), 331–78.

11 Luiz A. Mendes, Victor Carlos Sousa Oliveira, João Azevedo and António Ribeiro (eds.), *The 1755 Lisbon Earthquake Revisited* (Netherlands: Springer, 2009).

Nevertheless, a simple reading of the legal documents observes how, from the 1760s onwards (and explored in greater detail below), swift changes took place in several agrarian regions across Portugal. This legislation, which reinforced the privileges attributed to the Port wine demarcated region, constituted part of a wider program of territorial landscaping, which has remained rather blurred in historical analyses.

The vineyards in the Douro valley produced such a strategic commercial commodity that, to protect the quality of port, its production became defended by draconian laws enacted in 1765. Indeed, in the settlements built in the demarcated production area, cork trees could and should be felled. This took place despite how, and throughout centuries, both the Portuguese and Spanish monarchies had imposed exclusive rights of ownership and prohibited the removal of such trees as they provided a fundamental shipbuilding resource. Why different acts for *a tree*?

Nonetheless, this would make every sense if there were alternative plantings of this species ongoing post-1765 with the prospect of growing larger cork woodlands. This correspondingly proposes that such oddness, the removal of cork trees to be replaced by vineyards, makes sense within the framework of planning for different areas of agricultural specialization throughout the kingdom.

The same legislation, which remained in effect between 1765 and 1779, prohibited vineyards in many water basins beyond the Douro valley. Allegedly, wines produced in other geographies were being sold as fortified port. Not only was the planting of vineyards forbidden but they were also actually removed in central and southern regions of Portugal, across the water basins of the Mondego and Tagus rivers and throughout the Algarve province with the legislation covering approximately two-thirds of the territory.[12] Consequently, regions where grapes were entirely forbidden in 1765 began growing other cultures.[13] In some areas, apart from cereals and vegetables, either orchards or

12 Conceição Andrade Martins, "Vinha, Vinho e Política Vinícola em Portugal: do Pombalismo à Regeneração," (PhD diss., Universidade de Évora, 1998).

Alvará, 17 de Agosto De 1765, Concedendo Jurisdicção Ao Provedor da Companhia das Vinhas Para os Casos, Que Occorrem Nos Varejos, e Visitas; Alvará, 26 De Outubro De 1765 Mandando Arrancar As Vinhas Dos Campos Do Téjo, Mondego, e Vouga, e com Providencias Sobre Vinhos; Alvará, 18 de Fevereiro De 1766 Declarando as Vinhas de Torres, e Outras Terras Comprehendidas no Alvará De 26 de Outubro de 1765; Edital, 03 de Abril de 1771 da Companhia Geral Das Vinhas Do Douro Ácerca Da Venda Das Apólices da Mesma Companhia Alvará, 28 de Agosto de 1776 Prorrogando por Mais Vinte Annos a Companhia das Vinhas do Douro, e com Novos Privilégios; Decreto, 05 de Agosto de 1779 Decreto Declarando os Sitios Próprios Para Plantação de Vinhas Depois da Ley De 26 de Outubro de 1765.

13 Joaquim Romero de Magalhães, O Algarve Económico *1600–1773* (Lisbon: Estampa, 1988).

groves of fruit-trees were also planted, generally juice-fruit or thorned trees (citrus and nut producers).

The relevance of such detail is to portray the scope for intentional investment in trees by private economic interests. Fruit tree trunks would very likely escape seizure by the Crown as their wood was not good for shipbuilding. However, such trees would be able to simultaneously provide firewood, fodder, and fruit with their timber remaining valuable for furniture making as well as for other carpentry products. Such an option was also eligible for development under the same legal framework that had imposed the removal of vineyards and their replacement by "others".

This hypothesis gains further support by taking into consideration how the royal forests underwent planting and renewal after 1751 and were indeed for shipbuilding. Portugal may thus have very smoothly gone down two different paths of forestation and renewal of its tree park, one promoted by the Crown and the other by landowners and peasants. This certainly does not mean that pine trees simply disappeared from private woods but rather that another forestation niche was able to gain ground. In 1779, the 1765 legislation was partially revoked and while this meant vineyards could again be planted, fruit trees would also remain throughout these landscapes. The combinations of these products very likely increased the availability of energy sources and food both for the livestock and the people, beyond cereals, wine and vegetables.

The core topic for reflection in the items above derives from how, in the fourteen years between 1765 and 1779, there were significant changes to the customary agrarian landscapes and patterns of production, implementing new patterns of soil occupation, which have not yet been fully studied. Indeed, what emerges encapsulates a geographical compartmentation of the Metropolis into productive areas, with specific plantations in keeping with the strategic locations of the surrounding settlements. These tended to be difficult to access by any external military force but from where transport to deliver crops was fairly easy, sailing down rivers that were not otherwise navigable in the opposite direction.

In a broader vision, the conceptual and designed approach of the Portuguese crown towards a functional rationalization of the metropolis as of territories under its tutelage in Africa, Asia and America engaged deeply in the Enlightenment background thought about rationalizing areas of politics and economy. As it is well known, Lisbon's rebuilt after the Earthquake of 1755 became the first geometrical urban space as dreamed by then ongoing debate on rationalization of daily life. The importance of this detail for the present reflexion is to highlight that the Portuguese elite that created the conditions for the undertaking of such specific urban project, was the same that developed a matrix for rational-geographic territorial landscaping design in the Portuguese Metropolis, as well across the State of India, the State of Brazil and in different States in Africa.[14]

The political Portuguese *inteligentia* -which was engaged in a common cultural cradle of the Enlightenment that quested for efficiency as well- used the most updated tools for a fast recovery both in Portugal and in colonial spaces to overcome not only the material but also the tremendous political crises that derived from natural hazards.

Indeed, while Britain and France were taking over the Atlantic routes, far from that area of the Globe, the Portuguese State of India would know a territorial expansion and increasing presence during the 1760s, contributing to create space and to keep a position in the top ranking of colonial trade agency. As R. A. Disney states "it is likely few would have predicted in 1799 that Portugal would retain its Indian possessions for another 150 years – and outlast the British themselves."[15]

Such conclusions imply not only that Portugal and native kingdoms of Africa and Asia did not surrender passively to the new military power of other European forces across the Indic, as the Portuguese capacity of reaction in far lands from the Atlantic were effective. The success of enlarging territorial dimension of the State of India, might have meant that Portugues intelligentsia would be well equipped with updated mechanisms to accomplish the projected goals of holding a position in international trade. This meant that Portugal exercised some sort of control over communications and commercial corridors, or gained new ones however from another strategic place distant from the Atlantic: Indian Ocean.

This perspective challenges a narrative produced by coaeval thinkers about the Iberian backwardness which calls for more attentive studies.[16] If Portugal provoked a political, military and economic turn over in Eastern Africa and in India in its favour this is a result that must receive proper scholar attention. It reflects how the Portuguese ruling elite of 1750s–1760s was enough engaged with coeval military and diplomatic strategies not only leading territorial management occupation of natural sites, in the Indian ocean, but decentralizing areas of interests to assure a circulation of its colonial trade. It did not necessarily have to be focused in the Atlantic disputed by too many naval competitors

In other words, both the territory of the Metropolis as the territories of other geographies under Portuguese mastering would have been subdued to

---

14 Anthony R. Disney, *A History of Portugal and the Portuguese Empire. From the Beginnings to 1807, volume 2: The Portuguese Empire* (Cambridge: Cambridge University Press, 2009).

15 Disney, "A History", 331.

16 Pedro Cardim and Nuno Gonçalo Monteiro (eds.), *Political thought in Portugal and its Empire c. 1500–1800* (Cambridge: Cambridge University Press, 2021).

a function, following the most updated economic rational of ongoing political economic thought in Europe. In Portugal, as it will be further explored, the Crown – in Ancient Regime –, was able to intervene across royal and seigniorial lands dividing the realm in areas of management for specific purpose with a fair amount of success, showing how a degree of the reinforcement of central power was unequivocally achieved.

Effectively, as it will be further developed, a profound alteration of landscapes, endured to nowadays in some regions of Portugal, from the effective implementation of royal acts proclaimed within 1765 and 1779 against vineyards plantation as of forest management. Hence, under this lent, a territorial management *from above*, altered landscapes across Portugal during the second half of the 18th century regardless of seigniorial land-rights. In specific areas (nevertheless across a wide range of the territory), there were circumstances under which intermediation seems inexistant or very fragile.

The former sequence of arguments proposes different approach and chronology for the debate on centralization, considering that in different kingdoms, chronologies given circumstances, the leading powers achieved the practices of centralization as theorized for later periods than the 18th century by Scott.[17]

Considering the impressive degree of royal intervention across several geographic regions of Portugal, forcing seigniorial powers in faraway peripherical lands to obey to the king, it appears to challenge the argument of central state inefficiency, furthermore as a crucial tool to create structural economic and commercial defences as a way to confront the informal subjugation of external political entities.

As can be observed, the approach to rural landscapes following the 1755 earthquake opens up an endless plethora of questions and opportunities for analysis This chapter seeks to explain some of these and, after the present introduction (Section 1) is correspondingly organised into five broadly independent but also interconnected sections before setting out the conclusions.

Thus, the second section, 'From landscaping to problems', discusses how landscapes can be considered a historical source for longue durée observations. As mosaics containing the multiple fragments of the different facets to soil usage, they reveal layers of landscapes across time. Thus, features of land occupation are considered *landscape strata* revealing the *archaeological conjunctural layers* of different landscapes in the exact same places. In other words, this proposes that landscapes convey the historical contexts of their same setting and, just as with any document, are susceptible to the heuristics

17 James C. Scott, *Seeing like a State. How Certain Schemes to Improve the Human Condition Have Failed* (New Haven: Yale University press, 1999).

and hermeneutics which enable us to better understand what to be looking for in the archives dealing with the respective territory.

The third section then deals with 'Conceptual affiliation'. This brief discourse presents the outlines of the theoretical framework underpinning the approach to landscapes and evolving through taking into consideration the prospects for environmental solutions.

After this reflection, there follows the fourth section: 'Literature review, methodologies and sources'. This sets out the (almost non-existent) state of the art, how we undertake the analysis of sources that are themselves the methodology, cross-referencing landscapes and written sources to inform on the primary and secondary empirical data supporting this research and its conclusions. This purpose involves the in-detail presentation of the coeval non-written, handwritten and printed primary sources just as much as the bibliography.

Subsequently, the fifth and sixth sections deliver the article's core discussions. The first, 'Naval mastery, the empire and natural hazards: a crucial turn in the colonial and internal forest supply' puts forward the geographical framework and the international political scenario driving these landscaping plans. This substantiates the arguments on how the earthquake triggered a chain of new political programs that landscaped Portugal. These interlinked profoundly with the interests of the colonial empire and frame the major questions around levels of forest resource consumption, delivery and ensuring the replenishment of sources on this side of the Atlantic.

The closing section, 'An Unconventional Methodology: the heuristics and hermeneutics of a (wine) landscape', discusses how, from a conjunctural detail, the removal of cork trees to plant vineyards by royal decree, provides an indicator of how rulers perceived territorial administration in the second half of 18th century, and the consequent landscaping, quite differently to previously and bringing about powerful changes in such processes. Thus, this part approaches the preannounced compartmentalization of the national territory for different economic purposes. This highlights how, from little details, specifically the contradictory coeval law stipulating the production of the same natural resource in different regions of the territory – cork – unveil how micro and macro politics intertwine at the local level.

Finally, the conclusions point out how natural and human factors contributed to launching and framing this landscaping of the Portuguese Metropolis, stating how apparently geographically, politically and economically disconnected sectors were profoundly intertwined in this process of attempting to better manage several areas of production and otherwise uncultivated wilderness.

2 From Landscaping to Problems

The departure point for this article is simple: how did the Portuguese landscape change following the earthquake that especially devastated Lisbon in 1755, and what improved in the forestry sector? Answering these questions became rather difficult as they ended up opening many other unstudied topics.

This seismic event inspired a myriad of cultural reactions and studies in philosophy, literature, religion, politics, and slavery, with a substantial literature on the subjects around the histories of destruction and their scientific replies even if without the equivalent efforts to understand just how landscapes evolved under such circumstances.[18] Despite the paramount survey produced in 1758 covering every parish in the kingdom, there remain no significant studies on the evolution and planning of the landscapes. Apart from Lisbon, the Algarve and the Douro Valley, studies on the territory remain noticeable by their absence.

Given this lack, we needed to find and select an object to begin dialoguing and obtain some answers. Economic studies would provide numbers but not the physical territory. Seeking another story to tell, I began asking what a major panorama might tell us before the landscape itself emerged as a source of information.

According to the post-1755 legal framework for wine production, the division of the territory to establish clusters of precise cultures projected the different territorial intentions then held by the Crown and varying substantially on that defined by former ruling monarchs. How and why was such change implemented? And what were the implications for the forestry related affairs that account for my main interest.

Setting out the political scenario surrounding territorial administration before and after 1755 became crucial to advancing. To this end, this had to incorporate imperial affairs as they constantly interweave with issues around forest management and communications. As mentioned above, this period started out with the loss of capacity to ensure imperial communications simultaneously coupled with the supreme desire to continue ruling a colonial empire.

Thus, one of the key ideas underlying this chapter is that the 1700s represent a completely different European reality to that prevailing in the former

18 Isabel Maria Barreira de Campos, *O Grade Terramoto 1755* (Lisbon: Parceria, 1998); VV.AA., *O Grande Terramoto*, *passim*. VV.AA., *Providências do Marquês*, 63–288; Rui Tavares, *O Pequeno Livro do Grande Terramoto* (Lisbon: Tinta da China, 2009); Monteiro, "A Monarquia," 331–78.

modern period. This has hitherto been clearly disregarded in the environmental historiography that induces similar conclusions about forestry management on the Iberian Peninsula between the 1500s and 1700s, a situation that is fortunately nowadays at a turning point and hence the reason contextualizing the 18th century is so important.[19]

Throughout this century, the European naval and political contexts differed profoundly from those prevailing over the 1500s–1600s.[20] From roughly 1500 to the 1650s, control over the Atlantic in terms of both navigation and military power belonged to Portugal and Spain. From the mid-1600s, the legal fleets of Britain, France and the Dutch Republic, just as much as their pirate peers, became especially interested in the cargos of gold shipped from Brazil to Portugal. Furthermore, the War of the Spanish Succession was hugely responsible both for draining Iberian economic resources and providing a pretext for the respective political allies, the Dutch, British and French, to expand their ocean sailing activities. Those crowns reinforced their own ambitions over the Atlantic American colonies for which military control of the Atlantic was necessary. Thus, economically strong, they invested in developing their military naval power, a leverage that catapulted France and Britain to navigation and military control over oceanic navigation routes.[21]

19 *Roots of Sustainability in the Iberian Empires: Shipbuilding and Forestry (14th–19th centuries)*, ed. by Koldo Trapaga-Monchet, Álvaro Aragón-Ruano and Cristina Joanaz-de-Melo (New York-Oxford: Routledge, 2023).

20 Ana Fernandes Pinto, *Uma imagem do Japão: A Aristocracia Guerreira Nipónica nas Cartas Jesuítas de Évora* (Lisbon: Fundação Oriente, 2004); Nuno Monteiro, *D. José* (Lisbon: Círculo de Leitores, 2006); Monteiro, "A Monarquia", 331–78; Fernando Dores Costa, *Insubmissão. Aversão ao Serviço Militar no Portugal do século XVIIII* (Lisbon: ICS, 2010); Cristina Osswald, *Written in Stone: Jesuitic Buildings and their Artistic and Architectural Features* (Goa: Goa 1556, 2013); Avelino de Freitas de Meneses, "A rutura e a reconciliação (1640–1750)," in *História de Portugal e Espanha: Amores e Desamores*, coords. Artur Teodoro de Matos, João Paulo de Oliveira e Costa and Roberto Carneiro (Lisbon: Circulo de Leitores, 2015), 7–152; José Vicente Serrão, Bárbara Direito, Eugénia Rodriges and Susana Münch Miranda (eds.), *Property Rights, Land and Territory in the European Overseas Empires* (Lisbon: CEHC-IUL, 2014); Diogo Ramada Curto, *Cultura Imperial y proyectos coloniales de los portugueses (siglos XV–XVIII)* (Madrid: Marcial Pons, 2019); Mafalda Soares da Cunha (coord.), *Resistências: Insubmissão e Revolta no Império Português* (Lisbon: Leya, 2021); José Miguel Moura Ferreira, "As Novas Conquistas: Florestas, Agricultura e Colonialismo em Goa (*c.* 1763–1912)," (PhD diss., Universidade de Lisboa, 2021).

21 Peter C. Mancall and Carole Shammas (eds.), *Governing the Sea in the Early Modern Era: Essays in Honor of Robert C. Ritchie* (Huntington Library, Art Collections, and Botanical Gardens, 2015); J. Land and R. Dominguez «Illicit Affairs: Philadelphia's Trade with Lisbon before Independence, 1700–1775», *Ler história*, 75: 179–204.

After the signing of the Peace of Utrecht in 1714, a profound change in the context of control over the Atlantic took place. Losing control over naval corridors, impacted on the delivery of cargo shipped from the colonies to Portugal.[22] The role of Portugal as leader and controlling the trade in colonial goods had already been heavily undermined in the run up to the 1755 earthquake.[23] Worsening this difficult material-social-political context, the economic situation also deteriorated. Port, the only Portuguese product capable of competing in the European imperial trading system, was being adulterated by commercial agency.

As Conceição Martins demonstrates, in the 1750s, Port wine consumers and international merchants, mostly in the British sphere, were interested both in lowering the prices of Portuguese merchandise and in buying other wines from Spain without jeopardising the commercial treaty of Methuen signed with Portugal in 1703. For the sake of the Portuguese economy, the respective landowners and traders wanted and needed to stay in business. The 'spirit wine from Douro' product from the 'demarcated Country' (*Pais Vinhateiro do Douro*) could not afford to have its quality diminished, altered and, least of all, falsified.[24] Thus, other crops ought to replace the wine production ongoing in other regions. However, the question this then raises is why would legislation approved in 1765, and apparently returning good results, be revoked in as short a period as by 1779?

Under the *Ancien Régime*, previously donated property rights for wine production and for properties rented for vineyards would not be easy to override. Taking back such property rights without any crime against a monarch was not easy to bring about, not even for the Marquis of Pombal. Indeed, the privileged area for wine production would fall upon a region where vineyards had first been planted under special conditions in the 1640s. In addition, although Queen Maria I (1777–1815) had a very strong personage, the 1765 Decree forbidding vineyards outside of the demarcated Douro region was soon revoked.[25]

For the purposes of this chapter, this is particularly important as, on the one hand, while annual crops could be replaced easily every single year, on the

---

22 João Daniel, *Quinta Parte do Thesouro Descoberto no Rio Maximo Amazonas: Contém hum Novo Methodo para a Sua Agricultura, Utilissima Praxe Para a Sua Povoação, Navegação, Augmento, e Commercio, Assim dos Indios como dos Europêos* (Rio de Janeiro: Imprensa Régia, 2007); José Augusto Pádua, *Um sopro de destruição: Pensamento político e crítica ambiental no Brasil escravista, 1786–1888* (Rio de Janeiro: Zahar, 2002).

23 Jorge Pedreira, *Estrutura Industrial e Mercado Colonial. Portugal e Brasil, 1780–1830* (Lisbon: Difel, 1995); Pedreira, "From Growth," 839–65.

24 Martins, "Vinha, vinho", 62–93.

25 Martins, "Vinha, vinho", 62–93.

other hand, trees planted to grow over a long-term scale (twenty to fifty years later) could not simply be uprooted.[26] Indeed, marks of this soon abolished legislation were left across several regions of Portugal, thus, the area of already cultivated forest expanded markedly.[27] In the Algarve, wine production has never since recovered, and fruit trees became a major economic driver of the province with a similar phenomenon happening in the hilly valleys of Serra da Estrela and Beira Baixa where fruit trees became an important supply of resources.[28]

Furthermore, the introduction of new species into the 1700s Portuguese landscape does not represent a disruptive kind of intervention due to long-standing overseas contacts and efforts to naturalize botanical species across the Iberian Peninsula. As part of the Mediterranean world, where hybrid landscapes had been cultivated since classical times, the 1750s Portuguese territory was a heritage not only of raiding invaders but also of settlement by many different cultures and populations down the course of centuries. This reflects in how the co-evolution and hybridization of sites reflected a broadly traditional way of managing landscapes reaching back to the Roman Empire.[29] This at least partially explains why so little relevance seems to have been attributed

26 Pinho Leal, *Portugal Antigo e Moderno: Diccionario Geographico, Estatistico, Chorographico, Heraldico, Archeologico, Historico, Biográphico E Etymologico de Todas as Cidades, Villas e Freguezias de Portugal e de Grande Numero de Aldeias* (Lisbon: Livraria Editora de Mattos Moreira, 1873–1890). José Mattoso, Suzanne Daveau and Duarte Bello, *Portugal – o Sabor da Terra. Um retrato histórico e geográfico por regiões* (Lisbon: Temas e Debates, 2010).

27 José Monteiro de Carvalho, *Diccionario Portuguez das Plantas, Arbustos, Matas, Árvores, Animais Quadrupedes, e Repteis, Aves, Peixes, Mariscos, Insectos, Gomas, Metaes, Pedras, Terras, Mineraes* (Lisbon: Officina Miguel Manescal da Costa, 1765); Francisco Soares Franco, *Diccionário de Agricultura* (Coimbra: Real Imprensa da Universidade, 1804–1806), 5 vols;, João Baptista da Silva Lopes *Corografia ou Memoria Economica, Estadistica, e Topografica Do Reino o Algarve* (Lisbon: Typografia da Academia, 1841).

28 *Anais do Conselho de Saúde Pública do Reino* (Lisbon: Typographia Lisbonense, 1838–1848); Isabel Braga, *Portugal à Mesa: Alimentação, Etiqueta e Sociabilidade 1800–1850* (Lisbon: Hugin, 2000); Jean-Baptiste Barrés, "Memórias de um oficial do grande exército," in *Linhas de Torres Vedras. Memórias Francesas Sobre a III Invasão* (Lisbon: Livros Horizonte: 2010), 25–48; Cristina Clímaco, *As Linhas de Torres Vedras. Invasão e Resistência 1810–1811* (Lisbon: Câmara Municipal de Torres Vedras-Edições Colibri, 2016); B. Péruse, "Campanhas de Portugal (1810–11)," in *Linhas de Torres Vedras. Memórias Francesas Sobre a III Invasão*, (Lisbon: Livros Horizonte, 2010), 158–89; Francisco de Sousa Lobo, *A Defesa de Lisboa – Linhas de Torres Vedras, Lisboa, Oeiras e sul do Tejo (1808–1814)* (Cascais: Tribuna da História, 2015).

29 Ana Duarte Rodrigues "The Role of Portuguese Gardens in the Development of Horticultural and Botanical Expertise on Oranges", *Journal of Early Modern Studies*, 6:1 (2017): 69–89.

to the modern planning of the territory – as long as it did not collide with any seigniorial or municipal property rights.

In fact, the Iberian Peninsula, on the European geographical scale, represents a vast platform located between the Mediterranean Sea and the Atlantic Ocean and also between Africa and Europe. This reflects in a territory with climatic and botanical regions of *transition* where there is a rather abundant diversity in botanical species.[30] Thus, in the 1760s, the idea of relocating cultures and forest species according to the areas where they were known to thrive would seem both feasible and acceptable.

Given the above, the novelty would stem from the pioneering design for rationalizing the geographies of production, widening the expanse of the dominant cultures – in lands and locations submitted to the Crown's tutelage –, whether or not the actual property of the royal family. This design resulted in a 'division of Portugal' into three major regions for wine, forests and other products for agricultural cultivation. The program or the model of such regional investment would be the same for the three sectors: defining and organizing specific sectors in those economic regions favoured by accessible transport to channel production to markets. This landscaping program perhaps also benefitted from the specific climatic, pedologic and hydrologic characteristics of the Iberian Peninsula.[31]

The major unknown factors requiring further analysis incorporate just how the rural landscapes evolved in the areas not mentioned in the legislation. Furthermore, even across the Douro, Mondego and Tagus valleys, we thus far do not know to what extent the lands developed or simply remained unchanged, ignoring the legislation in areas with lower levels of inspection by royal officers, between 1765 and 1779 when wine was once again restored to its prior taxation status.

Finally, irrespective of all the efforts undertaken by the royal administration to foster efficiency in forest and wine production to deal with sudden crises in international trade and ensure the supply of the forest resources needed for shipbuilding, defence and housing, these were not enough for the global needs required in the wake of the Lisbon Earthquake. Thus, beyond the supply demanded of the royal forests, especially for shipbuilding and defence but also for housing, how did the territory and different social actors respond

30 Orlando Ribeiro, *Portugal, o Mediterrâneo e o Atlântico: Esboço de Relações Geográficas* (Lisbon: Sá da Costa, 1987).

31 Ana Duarte Rodrigues, "The Role of Portuguese Gardens in the Development of Horticultural and Botanical Expertise on Oranges," *Journal of Early Modern Studies* 6, no. 1 (2017): 69–89.

and manage landscapes with multiple tree-species to meet such needs? Which opportunities did such a conjuncture of calamities generate in favour of improving strategies for the renewal of the natural resources produced by afforested areas? Would the disastrous scenario drive plans for rescuing orchards and woodlands, that is replanting or planting anew?

3 Conceptual Affiliation

A reflection on "cultural landscapes" could be almost obvious to address this analysis,[32] namely considering the starting point of demarcated settlements for Porto wine production, which started a business that has turned into a cultural landscape. However, that legacy would not cover other processes of landscapes evolving across the territory in the same long run interval. Indeed, the Tagus water basin as the Region of the Algarve, among others, knew layers of land cover which did not necessarily endure by human or natural care.

Perhaps, a proxy of how culture in a broad sense influenced landscaping would be more adequate for this approach. However, it remains uncomplete for the lent of analyses through cultural landscapes, because they do not cover considerable invisible strata of landscapes that once existed and were replaced in the exact same space. Furthermore, in this study, those landscapes – now invisible-, in given historical contexts, were fundamental to understand orientation, reorientation and invention of territorial management policies. along 1700s. Thus, a new approach was required in order to reach such replaced areas by new landscapes in the exact same locations, regardless of dealing with areas of the metropolis or with overseas territories of the Portugguese Empire.

As my major research interest encapsulates how landscapes evolve over the long term, this case study intertwines an approach to the concepts of co-evolution and landscape hybridization processes that perceives human agency as responsible not only for profound changes in landscapes but also for restoring and renewing natural resources.[33] A debate to be in environmental History.

---

32 Andrew C. Isenberg, *The oxford Handbook of Environmental History* (Oxford: Oxford University Press, 2014); Mauro Agnoletti, "Environmental Thinking and Cultural Values: a Reflection on Environmental Globalisation and the Mediterranean Culture", *Global Environment* 7, no. 2 (2014): 257–90.

33 Verena Winiewarter, Martin Schmid and Gert Dressel, "Looking at half a millennium of co-existence: the Danube in Vienna as a socio-natural site," *Water History* V (2013): 101–19; Ulrike Kiercheberger, "Horizons of Ecological Change: Stories of Transfer across the Indian Ocean in the Age of Empire," *Global Environment* XIII, 1 (2020): 8–29.

Therefore, my interests extend to ascertaining whether or not human agency was committed to avoiding the irrational depletion of natural resources and promoting their replenishing during the Modern Ages. We now know that certain replacement strategies were implemented to avoid the exhaustion of forests and woodlands in 18th century Portugal.

Such concerns come to light in this chapter, as in other publications published mainly in the third millennium, within reflections focused more on the human destructive agency than on the scope for human intent to produce environmental solutions. Within the framework of debates launched around the Sustainable Development Goals (2011), the discussions on integral ecology and the United Nations sustainability agenda for 2030 seem particularly important as they embody how the range of sustainability has expanded enormously since 1987, evolving from an economic-natural paradigm to one of human and natural dignity in 2015.[34]

Within this predominant idea, that landscapes evolve due to a plethora of natural and human factors, originating different strata of landscapes in the exact same area, we are able to reason beyond the ecological and environmental determinism of doom on humanity managing nature to depletion. Instead of looking for wrongdoing, I sought to consider the composition and management of landscapes as the sources that provide the contexts of their evolution, thereby developing systemic heuristics and hermeneutics for landscape frameworks in a methodological approach to how landscapes, as documents, provide an 'archive' of different historical inputs and as such susceptible to the heuristics and hermeneutics proposed under the name of the 'Landscaping Reading Methodology' as detailed below.

4 Literature Review, Methodologies and Sources

Landscapes incorporate their surroundings and can thereby serve as the main insight into the contexts shaping the environment. Over the long run, they reveal how human and natural factors interact with each other and disclose the contexts of production across centuries, explaining, counterbalancing and influencing the political discourses on territorial management. The actual

34 Pope Francis, *Laudato Si. Sobre o Cuidado da Casa Comum* (Lisbon: Paulinas, 2015); United Nations, 2015 *Transforming our world: the 2030 Agenda for Sustainable Development, United Nations A/RES/70/1 General Assembly, 21 October 2015 Seventieth session Agenda items 15 and 116 15–16301 (E) *1516301*, Resolution adopted by the General Assembly on 25 September 2015.*

landscapes may confirm or deny narratives directly or indirectly fabricated in the documents written about them. Furthermore, they also render visible the landscape leftovers that the political and economic powers disregarded in their written contents and thus raising questions about that lack of interest and the reasons over different time scales.

Thus, behind any landscape, there are natural and human factors implied in their shaping, perseverance and change. Hence, when a territory-scenario, just like a theatre stage, becomes a framed image-document, exactly like a handwritten document in my opinion, it may be treated as an object susceptible to a heuristic and hermeneutical approach. Such considerations are important because they might confirm, dismantle, or produce new results by interpreting historical documents about landscape design and management.

Territories, across their visual features, display many strata of layered physical elements. The 'remains' of former landscapes, in very heterogenous mosaics, provide a visual chronology of the evolution of the site over different timescales. In other words, we can grasp many components of the different layers of landscapes that have occupied that exact area over historical time. This amounts to a way of looking at landscape from above as if those different components represent different archaeological strata located over each other. At the surface, they are displayed, whether organized or chaotically. The observer is responsible for encountering a logic in the remaining botanical, rocky or material evidence. Those clues unveil not only successive different landscapes in the sites under analysis, which is new to our present, but also what might have disappeared; thus, the visual surveillance of any landscape may turn out highly informative about the past in its own right, generating suggestions on how and where to seek content likely available in written documents. Thus, landscapes and written information sources may interact and complement each other.

In some way, this compares to archaeological methodologies for analysing strata in depth but over a wider surface area in this case. Furthermore, rather than digging beneath the earth and into soil, this method analyses temporal strata through the fragments of landscape visible in a broader landscaped perspective and including high angled views. By considering these dimensions, over the long term, geography, with appropriate historical heuristics and hermeneutics, might provide the confirmation or the counter-evidence on how environmental or political discourses, set out in written forms in the past, might not correspond to human and natural actions but rather to coeval interests, interpretations and even the boundaries of knowledge. In other words, landscapes might be able to corroborate written information or propose further analytical approaches to the meanings and intentions of such documents.

In keeping with the considerations above and the contemporary concerns over environmental degradation, historical analysis has hitherto interpreted forest destruction literally from the coeval documentation, specifically reproducing royal accusations of local populations making undue use of the crown's woods and woodlands, described in legislation as facts for the entire Portuguese realm and throughout the Modern Ages. However, beyond their historical contexts of production and reception, those messages hide meanings. The criminalization of forest destruction throughout the absolute monarchy led us to believe that forests, especially all the royal lands, were experiencing regular destruction when this primarily addressed three main tree species: pine (*Pinus pinea* and *Pinus pinaster*), oak (*Quercus faginea*) and cork trees (*Quercus suber*). What the legislation does not mention is all the other botanic species that were growing abundantly throughout the territory in contrast with this narrative of 'forest resource scarcity'.

Thus, while the Iberian Peninsula led the way in felling trees for shipbuilding, which is undeniable fact, there is also the need to analyse, on the one hand, what followed in those sites following the felling and removal of the trunks: would new trees grow from the stumps? On the other hand, if the trees did not recover, what kind of natural growth took over in those same plots?

A line of historiography has been defending quite a significant destruction of Portuguese forests right throughout the Modern Ages. Thus, it became unexpected to find data on the early 1800s testifying to a different reality. Indeed, Portugal displayed an abundance of forestry resources with trees across all of the central mainland and coastal zones, which necessarily had to have been either planted or regenerated naturally at least thirty to fifty years earlier in both royal and private forests.[35] This forest availability recorded at and from – at least – the region centre of Portugal coincides with the timeline both for planting royal pinewoods post-1751 in coastal lands, and hypothetically with the eventual replacing of landscapes following the removal of vineyards in accordance with the decree of 26th October 1765.

35 Joanaz-de-Melo, "Forests", 264–68; *Anais do Conselho de Saúde Pública do Reino* (Lisbon: Typographia Lisbonense, 1838–1848); Isabel Braga, *Portugal à Mesa: Alimentação, Etiqueta e Sociabilidade 1800–1850* (Lisbon: Hugin, 2000); Jean-Baptiste Barrés, "Memórias de um oficial do grande exército," in *Linhas de Torres Vedras. Memórias Francesas Sobre a III Invasão* (Lisbon: Livros Horizonte, 2010), 25–48; Cristina Clímaco, *As Linhas de Torres Vedras. Invasão e Resistência 1810–1811* (Lisbon: Câmara Municipal de Torres Vedras-Edições Colibri, 2016); B. Péruse, "Campanhas de Portugal (1810–11)," in *Linhas de Torres Vedras. Memórias Francesas Sobre a III Invasão*, edited by António Araújo (Lisbon: Livros Horizonte, 2010), 158–89; Francisco de Sousa Lobo, *A Defesa de Lisboa – Linhas de Torres Vedras, Lisboa, Oeiras e sul do Tejo (1808–1814)* (Cascais: Tribuna da História, 2015).

The abovementioned legislation was intended to prevent any adulteration of port by poor quality red wines from other regions, especially the Tagus river basin where, on the one hand, areas of cereals and vegetables had recently been replaced by vineyards not only increasing the shortage of those products in Portugal but also producing bad wine which, easily transported by river and sea, was contributing towards the corruption of good wine.[36]

Furthermore, this expansion in vineyards would allegedly require the importing of 'bread'. Thus, this decree forced landlords and peasants to uproot vineyards in the most strategic fertile lands of the Vouga, Mondego and Tagus river basins; those lands were then to be cultivated with traditional crops. Nonetheless, vineyards could be maintained in those areas where wine had long been traditionally produced.[37]

However, the uprooting of these woody shrubs was to cause unexpected consequences: removing barriers of protection against flooding and inundating fertile lands with sands, dragged there by torrential floodwaters. Such a measure may have helped diminish wine forgery and the market alternatives to the Douro valley region but then became a problem for containing devastating flooding in the lowlands. In the end, the removal of vineyards to grant a monopoly to the exquisite fortified wine of northern Portugal turned out to be disastrous for the seedings and crops planted in the Tagus lowlands in the 1770s, the largest region of cereal production. Following repeated scenarios of devastated crops and people, Queen Maria I promulgated on 5th August 1779 a law that not only imposed the re-settling of vineyards in places crucial to preventing river flooding but also commissioned the plantation of two lines of trees alongside the watercourse to establish its banks within efforts to regulate torrential flows.[38]

Pinewoods and other tree coverage were identified along the Mondego water basin and the Lis Valley as far as Óbidos and from this lagoon to Torres Vedras in 1810 and the 1830s in addition to along the right bank of Tagus between the Spanish border and Lisbon according to coeval sources and recent analysis.[39]

This picture substantially contradicts the idea of an absence of forests in Portugal, at least across the Central region which covers the aforementioned areas of the Vouga, Mondego and Tagus water basins and also, for the purposes

36 Decree of 26th October 1765, "vinhas", in https://legislaçãoregia.parlamento.pt/1/V/10/73/p246, consulted on 04-09-2023.

37 *Ibidem.*

38 "Vinhas" law of 5th August 1779, in https://legislaçãoregia.parlamento.pt/1/V/68/137/p262, consulted on 04-09-2023.

39 Barrés, "Memórias," 25–48; Lobo, 20–150; *A defesa*, Péruse, "Campanhas de Portugal," 158–89.

of trees and a park of timber and wood production, the Sado water basin, in the Setubal region. This reports the steady replenishment of trees in royal grounds, a priority from 1751 onwards that the royal legislation of the 1790s and 1800s again contradicts. However, this legislation refers to a quite different area for its complaint about 'destruction' – Alenquer. This village nearby Lisbon was a possession of the House of the Queens and Princesses. Such areas was located in a different region from the woods and woodlands of the royal grounds in the upper Tagus plains and valleys, somewhere between 70km and 150km from Lisbon, formerly subject to legislation promulgated since the 15th century on account of tree depletion. Now, the woods addressed as devastated and not replanted as they should have been accordingly to the orders of the House of the Queens and even from the monarchs, correspond to an area on the right bank of the Tagus, near to its estuary, around 30 km from Lisbon. The geography of complaints about forest crimes may have switched from the upper to the lower region but there was no change in the status or ethos of the complainers: members of the royal family. In the end, combining these complaints with others reported to the king in 1796 regarding the left bank of the Tagus served as the pretext for reforming the royal properties with regulation in the 1800s removing hunting privileges in areas of primary wilderness, maintained as such for game breeding, in order to convert a proportion into cultivated and afforested landscapes.[40]

Indeed, the new royal preserve regulation, issued on March 21st 1800, opted to adjust the crops for planting to the prevailing soils as well as setting aside specific areas for forestry to produce timber and other areas for agriculture in keeping with a landscaping rational adopted by the Crown since at least the 1760s. Indeed, throughout the second half of 1700s, the Crown invested intentionally in nurturing different and more intensive production in different regions, including: developing the conditions favourable for the north region to produce an internationally traded product, a fortified wine; replanting forests, expanding and intensifying the rational and professionalized management to supply the needs of the Crown without resorting to private estates in the centre of the country as well as nurturing the conditions, directly and indirectly, for other kinds of resource intensification and specialization. The unexpected result, due to the prohibition on planting vineyards for a period of time, between 1765 and 1779, as mentioned in the introduction above, would be the further development of areas of fruit production in inland regions, from

40 Joanaz-de-Melo, "Floresta", 125–27.

Trás-os-Montes to the Algarve and as well as in coastal areas of the Alentejo and Algarve.[41]

Indeed, environmental history has hitherto overlooked two factors on the topic of *forest destruction* supra mentioned: firstly, the obviously forgotten, the specific characteristics of the Iberian geography; secondly, the role of property rights under the Ancien Régime. These remained quite significant throughout the 18th century as an obstacle to universal royal intervention across the territory. Seigniorial rights could not simply be ignored and overruled by the Crown. Such impositions required justification in a territory that was in any case not lacking in alternatives for managing its resources.

As the geographer Orlando Ribeiro points out, the Iberian Peninsula is situated between the Mediterranean Sea and the Atlantic Ocean but also between Africa and the north of Europe; botanic species from the Middle East, North Africa, Scotland and the North Sea can be found in Portugal. Such diversity enabled the capacity to develop a complete process for the enduring management of resources in a territory replete with microclimatic regions and as well as major differences in the prevailing soils.[42]

The Iberian geographies span different climates, geologies and botanic species and produced a distinctive way of sustainably managing many different ecosystems – in parallel – in the same small political territory as indeed happened in other Mediterranean regions.[43] For instance, on a speculative basis, Portugal could hardly have built caravels and other such vessels without this amazing botanical diversity that meant there was no need to search for resources outside its own territory in a first period.

Thus, Portugal would not need to make recourse to timber from elsewhere for its initial transoceanic navigation take-off. Timber and other resources were present in enough variety up to the point that such a small scale of many resources was no longer able to keep up with the pace of consumption broadly from the late 1400s onwards.

Over the course of the 1600s, as a wide range of studies demonstrate, the development of shipbuilding with naval empires and naval warpower complexified the entire extent of industry and the scale of resource extraction in Europe with clear evidence that many components of such units,

41 BAHMOP, Direção Geral do Comércio, Agricultura e Manufaturas (DGCAM) – RA-3S, núcleo 2, years 1852–54; núcleo 3, years 1852–54; núcleo 4, years 1852–54; núcleo 6, 1852–57; núcleo 8, years 1852–57; núcleo 11, years 1852–57.

42 Ribeiro, *Portugal, o Mediterrâneo*, 1–6, 40–41, 101–102.

43 Mattoso, Daveau and Bello, *Portugal*, 98–684; Agnoletti, "Environmental" 257–90.

including wood, were obtained from other areas of Europe as well as from other continents, especially South America and Asia.[44]

However, unanswered questions remain. It would seem quite logical that, as forest resources were being obtained from overseas regions, there was a recovery across the Iberian Peninsula woodlands over the 1500s and 1600s. Otherwise, it would have been impossible to continue destroying the forests in the same areas as the royal acts clearly state forest crimes were perpetrated against the royal woods in the exact same royal forests spanning intervals of approximately fifty (50) and one hundred (100) years. Thus, the same sources, hence, the legislation promulgated for the royal woods testifying to their illicit destruction at far distant intervals, constitute the same official sources that demonstrate the existence of forest in the exact same spots across different chronologies. However, this replenishment process would necessarily be slower and at a rhythm that would not even easily compare with the shorter time scale required for clearing forests. These three tree species (pine, cork and oak) were addressed in the legislation handed down by the Portuguese rulers as exclusive to the monarchs and with this royal preserve remaining in effect from the 1400s into the 1700s.[45] However, irrespective of these specific trees, the bushes, woods, and woodlands held a far broader range and quantity of forest resources and, secondly, those same landscapes kept on undergoing renewal, fruited and again removed in *cycles* down the course of history. They simply did not remain *destroyed* throughout the centuries ahead as the Crown kept on using them. To destroy forests, they had to be there.

Thus, the problem of considering the replenishing of resources as scarcity may commonly stem from a period of waiting while the process of tree planting and renewal was ongoing. Furthermore, events in Portugal and Spain

---

44 Jan Glete, *Warfare at sea, 1500–1650: Maritime Conflicts and the Tranformation of Europe* (Routledge: Nueva York, 1999); Lazaro Benedito da Silva, Andreia Moraes Ferreira, Sara Santos Araújo and Marta Catarino Lourenço, "Transporte de madeiras brasileiras para Portugal nos séculos XVIII e XIX," *Brazilian Journal of Development* 6, no. 7 (2020): 728–45, Ana Crespo Solana, "The Global Timber Trade and Shipbuilding in the 16th–18th centuries: Interdisciplinarity, research problem and the ForSeadiscovery project," in *Roots of Sustainability in the Iberian Empires: Shipbuilding and Forestry (14th–19th centuries)*, ed. by Koldo Trapaga-Monchet, Álvaro Aragón-Ruano and Cristina Joanaz-de-Melo (New York-Oxford: Routledge, 2023), 54–74.

45 Koldo Trapaga-Monchet, Álvaro Aragón-Ruano and Cristiana Joanaz-de-Melo, "The Game of the Demiurge in the Gardens of Chronos: Woods play hide-and seek in the long run through sustainable management," In *Roots of Sustainability in the Iberian Empires: Shipbuilding and Forestry (14th–19th centuries)*, ed. by Koldo Trapaga-Monchet, Álvaro Aragón-Ruano and Cristina Joanaz-de-Melo (New York-Oxford: Routledge, 2023), 1–8.

served to justify increasing investment in expanding wood production in 1700s, especially in concentrated areas.[46]

As already referenced, some landscapes were transformed literally by decree (1765); and accordingly, the following legislation would also have impacted on the same sites after 1779. However, while vegetables, cereals and shrubs can easily be replaced, trees planted to produce fruits and timber, long term investments, could not be so easily removed. Thus, some features of these landscapes help by adding information to the written documents and better interpreting their meaning.

The analyses and perspectives of this study incorporate territorial observations of the landscape features that reveal the past designs of the Portuguese physical territory that lie somehow hidden in the official discourses produced by the most powerful actors. The landscape, in conjunction with the coeval written documents from past periods, allows us to relate and better understand the evolution of the national terrain over the course of time. In other words, applying a very simple expression, we must understand how the territory *speaks to us*. This becomes especially important as reading landscapes via sightseeing may provide crucial clues of the past periods still surviving in the territories, and uncovering a broader reality only partially detailed in the written sources. Such is clearly the case with the royal forestry acts of the 1700s that mention the ongoing destruction of resources while omitting other realities concerning the local management of resources, namely examples of the intentional replenishment of woodlands undertaken deliberately by human action. The Portuguese royal decrees on protecting forests due to their scarcity throughout the 1700s, do not reflect the same reality as the decrees in the 1500s and 1600s and require contextualizing within the management of the territory in articulation with other aspects of the farming and agricultural markets.

Thus, the laws addressed for obeying 'by the others' can be misleading[47] as the discourse on the scarcity of timber – important for achieving royal

46 *Regimento para o Guarda Mor dos Pinhais de Leiria e Superintendente da Fábrica da Madeira* of 18 October 1751, available at Antonio Delgado da Silva, *Collecção da Legislação Portugueza desde a ultima compilação das ordenações, legislação de 1750 a 1762* (Lisbon: Typografia Maigrense, 1830), 68–90; Rosa Varela Gomes and Koldo Trapaga-Monchet (coords), *Árvores, Barcos e Homens na Península Ibérica (Séculos XVI–XVIII)* (Lisbon-Zaragoza: IAP/IHC-Pórtico, 2017); Martínez González, *Las Superintendencias*, 333–57; Cristina Joanaz-de-Melo, "Forests in Portugal 1750s–1820s: a History of Forest Compensation," in *Roots of sustainability in the Iberian Empires: Shipbuilding and Forestry (14th–19th centuries)*, ed. Koldo Trapaga-Monchet, Álvaro Aragón-Ruano, and Cristina Joanaz de Melo (New York- Oxford: Routledge, 2023), 251–77.

47 Diogo Ramada Curto, *Cultura Escrita, Séculos XV a XVIII* (Lisbon: ICS, 2007).

interests – never mentions, for instance, those areas of royal forests in recovery and growth.[48] They also fail to address all the wide diversity and density of other tree species available across the territory that the licenses to extract resources and the fines and penalties imposed for poaching in royal preserves provide.[49] They almost never address the focus underlying this discourse and which is susceptible to contradiction by the data on the daily management of royal forest resources for all the Iberian Peninsula already produced on the 16th century.[50]

Thus, to avoid rushing into anachronic conclusions when analysing the environmental policies of past periods which imply transformations in the geography, we also need to take into account the factors that limited or encouraged the human shaping of landscapes to return a thorough comprehension of the ongoing historical contexts of the binding rules and the daily life of the populations in the terrain and actually making the landscapes out of forests, woodlands, parks, gardens and orchards.

Therefore, the availability and delivery of forest resources to royal shipyards, especially after the 1755 earthquake, need explaining. From which regions where they extracted and, more importantly, which regions with available 'scarce species of trees' were not sourced in the wake of the earthquake for the rebuilding efforts not only of Lisbon but also the national naval fleet, defence structures and for so many other purposes.

In the first months after the natural disaster, mainland European Portugal was able to deliver the timber and wood necessary for military logistics and shipbuilding. This clearly stems from Portugal having available the different kinds of timber and species of wood necessary for the various crafts. Thus, the Portuguese landscapes provided forest goods produced not only from the tree species listed in the legislation but also from many others for the reconstruction works.

---

48 Biblioteca e Arquivo Histórico do Ministerio de Obras Púbricas (BAHMOP), Montaria-mor do reino (MMR), núcleo 8: years 1605–1833, núcleo 9: years 1596–1721; núcleo 10: years 1738–1833.

49 BAHMOP, MMR núcleo 1, years 1583–1833; núcleo 2, years 1724–1833; núcleo 16 years 1583–1833; núcleo 17 years 1722–1883; núcleo 37, years 1771–1883.

50 Cristina Joanaz de Melo, *An Analysis of the Royal Preserves in Portugal. Issues of Privilege, Power, Management and Conflicts* (Sheffield: Wildtrack, 2015); Rosa Varela Gomes and Koldo Trapaga-Monchet (coords), *Árvores, Barcos e Homens na Península Ibérica (Séculos XVI–XVIII)* (Lisbon-Zaragoza: IAP/IHC-Pórtico, 2017); Félix Labrador Arroyo and Koldo Trapaga-Monchet, "Recursos Naturales en la Península Ibérica: los aprovechamientos forestales e hídricos (siglo XV–XIX)." Special issue in *Tiempos Modernos. Revista Electrónica de Historia Moderna* 9, no. 39 (2019): 279–543; Trapaga-Monchet, Aragón-Ruano and Joanaz-de-Melo, "The game," 1–28.

In accordance with the above, the Portuguese landscapes would hold a quite variety and abundance of tree species. Those in short supply mentioned in the royal legislation were pines, oaks and cork oaks trees. The Crown's discourse on the scarcity of forest resources simply did not address every Portuguese tree species but rather those required for military purposes and with some of these being fruit trees otherwise considered as part of the farming world. Hence, the landscaping of 1700s Portugal evolved swiftly and included many human and natural influences across several swiftly changing conjunctures that demanded the updating of territorial management in the metropolis and as well as across the empire.

Thus, framing such intricate environmental processes requires contextualizing the internal and the external political and economic affairs. This approach seeks to interlink port production, the imperial political background, the geographical and climatic characteristics of the Iberian Peninsula and the surveying of colonial forests.

To explain such a complex framework, I drew mostly on secondary sources. As regards the impacts of both political and natural changes on managing the terrain, I consulted both primary and a few secondary sources.

To grasp the landscape evolution process in keeping with the political changes taking place in the first half of the 1700s in the Atlantic, Brazil and Portugal, I collated both political data and information on natural resource availability post-1751 and 1755 and analysed mostly primary sources in the Portuguese National Library: the Correspondence of Brazilian viceroys with Ministers of State in the Lisbon Court from the 1750s to 1807, interweaving these findings with the legislation.

As the foundations of this chapter depend on the availability of forest resources this requires fully demonstrating. For this purpose, I explore handwritten sources on the delivery of forest goods to the harbour of Lisbon, specifically charcoal, firewood, timber and logs, as registered by the 'Tagus Customs Office': the books of Lisbon Port and customs records registering the internal delivery of such resources. This documentation is deposited in Torre do Tombo National Archive, in Lisbon, in the 'Sugar Customs' (Alfandega do Açúcar) section.

The Customs records information on wood deliveries from 'Tagus and Brazil' begin in January 1756. The first record dates prior to any arrival of Brazilian or other colonial timber for the purpose of reconstructing housing in the wake of the earthquake. Especially in 1756 and 1757, there are almost daily and abundant cargos of forest resources, wood, timber/logs, firewood as well as other products, disembarking and being stored in the port of the capital of Portugal and delivered from sites in the Metropolis and not from the colonies. These record books, while registering the delivery and payments for such goods, do not necessarily specify the region of extraction in Portugal. However, they do sometimes, and also on occasion, inform on the seigniorial ownership of the providers of such goods, identifying the name of the respective seigniorial house or religious congregation. Thus, although the sources under analysis for this chapter enable the sequential identification of internal deliveries for decades, they do not provide the full identity of the suppliers. However, the origins of the deliveries, when not corresponding to royal forests or properties, had to have been dispatched from seigniorial estates.

As this collection contains no equivalent records for harbours in the north, on the delivery of equivalent resources to the harbours of Oporto (on the mouth of the Douro), Caminha (on the mouth of the Minho) or Viana do Castelo (on the mouth of the Lima), where oak for cooperage would be in demand, we may only observe that goods from those regions also arrived in Lisbon. Although impressive in extent, these records of the internal supply of such resources might not cover all the regions where resource extraction took place after 1755. However, resources from across all the metropolis were clearly funnelled into Lisbon.

Furthermore, we are able to complete the information on the centre region by recourse to the records produced in administrating the royal forests, confirming whether such grounds alternated extraction with replanting to deliver resources also for shipbuilding and housing. Achieving these required inputs from several documental sources. The Bureaux of Royal Forests Corpus and that of the Administration of Royal Forges were consulted to check on suppliers of timber and wood for the monarch's shipyards and charcoal and firewood for the royal glass factory, forges as well as the social consumption of forest goods by the Court. In order to gain a better insight into the delivery of logs for shipbuilding, I also quote records from the Portuguese Navy Archive (Arquivo da Marinha) but indirectly through published bibliographies. Some monographs cite original documents and I deploy this information to reinforce how royal forests underwent replenishing and kept up supplies to royal navy arsenals throughout 1700s via different geographies and scales of both renewal and extraction.

I also consulted primary coeval printed sources, field trip information featuring geographical descriptions of the territory, memoirs and diaries written by foreign soldiers while crossing Portuguese lands, which detailed the territory across different chronologies, providing descriptions of landscapes conveying the natural features of soil occupation and which enable some degree of comparison with the information produced under the auspices of the monarchy and legislation.

Finally, along the text, I also apply the preamble of the published legislation to collect data on landscape descriptions, especially humanized and cultivated areas where there were significant anthropic interventions both weakening and restoring resources and landscapes.

5 Naval Mastery, the Empire and Natural Hazards: a Crucial Turn in the Colonial and Internal Forest Supply

Post-1755, there was a paramount and entirely unpredictable need for forest goods in the wake of the Lisbon earthquake of 1st November. Following the earthquake itself, a tsunami destroyed and dragged away entire houses, people, and wood stocks. The impacts on the naval sector were devastating as the war-commercial fleet stationed in the Tagus estuary was globally destroyed and the timber stored in *Paço da Madeira* (the king's warehouse for the wood and forestry resources received via Lisbon harbour) or elsewhere in the port was pulled into the waters of the estuary and out along the maritime coast of Lisbon. The seismic shocks caused candles to fall in churches (and probably homes on All Saints Day) triggering fires which ripped through wooden houses all over the city leaving it and the port in utter ruins.[51]

In such a conjuncture, it might be understandable that the idea of sourcing colonial supplies of timber for reconstructing Lisbon would prevail. However, forest history has not yet studied the internal delivery and availability of Portuguese forests in this process. This observation does not seek to claim that timber and wood from Brazil did not arrive in Portugal. However, historiography has hitherto not really considered that Portuguese colonial forests in the 1700s would not provide the strategic reserve of timber and the other wood products necessary for the Portuguese Metropolis, and still less so in the wake of the 1755 earthquake, as a plausible hypothesis. Albeit such a feasibility might effectively be demonstrated.

In recent analysis, Lazaro Silva et al. present an expansion in the diversity of tree species leaving Bahia for Portugal and arriving in Lisbon for the supply of nurseries and botanical gardens and as well as for delivery to the royal shipyards.[52] Such data seems to reinforce the idea of how Brazilian forestry became

51 Carlos Oliveira, *O Grande Terramoto de Lisboa Volume I, Descrições*, 23–86; VV. AA, *Providências do Marquês*, 28–288; Araújo, *O Terramoto*, 28–40; Tavares, *O pequeno*, 1–20; Nicholas Shrady, *O dia do Fim. Ira, ruína e razão no Grande Terramoto de Lisboa* (Alfragide: Texto editores, 2014).

52 Silva, Ferreira, Araújo and Lourenço, "Transporte," 728–45.

a major supplier meeting the monarch's imperial interests in maintaining timber reserves in the shipyards coupled with exploiting exotic tree species and gaining a leverage in international trade in Europe over the 1700s.

Such information would confirm other reliable records on the shipping of cargos of exotic trees from Brazil. Furthermore, the coeval Portuguese surveys of Brazilian forests, reported to the Crown after 1756–57, and also of the State-of-India (Goa), following the conquest of the State of Ponda, in 1763 (Ferreira 2021), and of Brazil again in the 1790s, report a huge diversity and wide expanses of such resources that would also sustain this indisputable fact: timber and exotic woods were loaded in Brazil and left its ports with the official intention of delivering their entire cargos to Lisbon. What we might be lacking here is a comparison between the cargos loaded and those that arrived at their end point.

Up to the present, the historiographical literature on the wood and forest resources shipped from the Portuguese colonies to Portugal has not yet been compared with details on their effective delivery to their final destination, Lisbon. Such records on the cargos unloaded from Brazil might not coincide with the records of the goods sent from the original harbours. This would corroborate the Diogo Cabral thesis that dismantles the idea of Brazilian timber being a supplier of Portugal shipyards as most of the cargos would be smuggled to intermediate resupply posts in the Atlantic to avoiding being robbed by pirates on the high seas.[53]

Taking into consideration his former observations on the legal and illegal trade in forest resources around the Atlantic, Cabral proposes another contribution for the timber trade and imports to Portugal. From his point of view, the importing business in Portugal would have incurred less risk buying timber from Canada, via Great Britain. Sailing across the north of the Atlantic would be shorter (30 days) than the trip to Brazil (60 days) and the return journey (90 days). Considering the huge risk of cargo loss and the lack of income from the naval corridor, companies and traders in Portugal would rather have imported timber from Canada. Nonetheless, Lazaro Silva et al. analyses the exotic woods sent to Portugal from Brazil but not necessarily applied for housing and shipbuilding[54] with a wide selection of exquisite wood products applied to more sophisticated carpentry, confirming the production of goods mentioned in the trade accounts and balances studied by Jorge Pedreira.[55]

53 Diogo de Carvalho Cabral, *Na Presença da Floresta. Mata Atlântica e História Colonial* (Rio de Janeiro: Garamond, 2014).

54 Silva, Ferreira, Araujo and Lourenço, "Transporte," 728–45.

55 Pedreira, *Estrutura Industrial*, 260–375; Pedreira, "From Growth", 839–65.

Thus, the reconstruction of the urban settlements destroyed by the earthquake, military means and facilities and even shipbuilding, in a conjuncture of emergency, would have been supplied by wood from national forests. Indeed, considering the distance of travel from Lisbon to the ports of South America easily took five to six months,[56] following the 1755 earthquake, Portugal could have received timber from Brazil to help with reconstruction only within 150 and 180 days, by May 1756 in the best of all hypotheses. As kindly pointed out by Trapaga Monchet in the course of editing this text, even in the most positive scenario, orders to prepare timber for shipping to Portugal could only have been sent immediately after 1st November, and the desired wood would then have to be felled, cut, worked, stacked, and stored in the port and only ready to leave three months after the natural disaster.

This does not state that timber and forest goods were not sent from the colonies but certainly makes us wonder about the extent. Whether information flew by pigeon mail, cargos still had to be prepared for shipping and it took time to fell trees, saw and cut them into adequate pieces, before packing and finally shipping. On beholding the deeper data, they all seem complementary with each adding something to a still incomplete story. In this polyhedric dimension of different contributions to better understanding the pushing for afforestation in Portugal in the 1700s, study of the Sugar Customs documental archive informs us that, despite the daily unloading of wood, firewood, timber, wooden boards and charcoal produced in national woods, parks and forests, Portugal did not become forest free between 1756 and 1779.[57]

This does not contradict colonial exotic woods getting sent to Portugal and unloaded in the Lisbon harbour. More than discussing the origins of forest goods extracted in the colonies, such data unlocked the quite extensive and hitherto invisible afforested landscapes across Portugal, from the north to the south, apparently maintaining quite a major delivery capacity without turning Portugal into a treeless landscape despite periods of intensive extraction during the 1750s and 1760s, after the aforementioned earthquake of 1755 and again after another earthquake in 1761, which was less destructive and therefore demanded less replacement timber.[58]

Accordingly, this documentation importantly registers timber deliveries only to the international port of Lisbon and no other national seaports. Nonetheless, Lisbon was the leading port in the Portuguese Empire. Thus, at this

56 Cabral, *Na Presença*, 377–96.

57 Arquivo Nacional Torre do Tombo (ANTT), Paço da Madeira (PM) 026, years 1756–1764; ANTT, PM 027, years 1757–1779.

58 ANTT, PM 026, years 1756–1764.

significant destination, the forest goods received in the first year after the 1755 earthquake, hence between 1756 and 1757, were sent to Lisbon's harbour in the majority from Portuguese woods. The wood, timber, charcoal and other forest products stored in the royal warehouses in Lisbon port, before dispatch to the royal arsenals, shipyards, and sawing engine, were sourced from Portuguese estates.[59] Therefore, we understand that significant volumes of forest products were shipped into Lisbon. However, we are not able to provide any exact quantitative data or even approximative figures on tons of branches, trunks and charcoal from national woods unloaded in the Tagus estuary given the mostly qualitative nature of the information available. For instance, cargos and piles get designated as important, significant, huge or simply a particular amount of a specific good. Sometimes the records detail quantities but in measurements that we can no longer convert into contemporary standardized systems. One hypothesis for development in future studies would involve calculating such cargos in accordance with the scale of the vessels operating in national ports. Nevertheless, even this endeavour would still require many reservations. What this source thus far demonstrates is that there were daily cargos of forest products undergoing delivery to Lisbon, thereby supplying the royal arsenals and the population from national woods and woodlands.

Furthermore, we should recall that the most precious cargos shipped from Brazil were gold, and not timber even if the latter was highly important for specific activities. However, rendering Portugal incapable of sailing would serve quite perfectly British and French intentions of taking over that commerce.

In addition, the historiographic literature has already addressed how building vessels overseas was a strategy already under development both in Brazil and in India since the 17th century.[60] This correspondingly attenuates the issues around colonial timber applied for shipbuilding in the metropolis. Furthermore, such a position returns us to the matter of just which territories across Portugal supplied all the national shipyards and other needs for wood.

As mentioned earlier in this chapter, Portugal spans a myriad of ecosystems in a zone of transition both between the Mediterranean and the Atlantic and Africa and the bulk of Europe. In my view, one of the major issues underemphasised in the forest history of Portugal derives from its geography. Climatic and soil diversity, alongside the availability of water, play very important roles regarding this issue of forest species diversity with the Iberian Peninsula

59 ANTT, PM 026, years 1756–1764.

60 Lopes, 1820; Pádua, *Um Sopro*, 60–62; Cabral, *Na Presença*, 159–175; Ferreira, "As novas conquistas", 1–51, 117–150.

hosting both Mediterranean and Atlantic habitats. Across the Iberian platform, nature is truly diverse.

The diversity of species, the social consumption of forest resources and the property regimes beyond the royal grounds do account for a significantly larger quota of the kingdom than in other European Modern Age monarchies. Hence, the narrow perspective gained from reducing Portuguese forests, whether belonging to the crown or others, to *Pinus pinea, Pinus pinaster*, *Quercus suber* and *Quercus faginea* or any other species of oak is unacceptable as a sample of a territory so deeply rich in botanical diversity. Therefore, we need to step outside and read the landscape as it presents itself to us. Indeed, such an understanding makes it mandatory to seek out far more information about landscape design than the contents provided by royal legislation on the tree species required for shipbuilding. Beyond those, Portugal holds a quite wider variety of forest trees.

6 An Unconventional Methodology: from Landscapes to Problems and Archives – The Heuristics and Hermeneutics of a (Wine) Landscape

Port attains an exquisite flavour. The fortified wine is produced in the vineyards along the Douro valley in Portugal where their terraces enchant and visually impact through all their splendour. Nevertheless, the contemporary beauty of this monoculture spans a very significant geographic range where, from the 15th to the 18th centuries, other trees also occupied this terrain, including holm, cork, chestnut, oak, and other *Quercus* species. Thus, let us now pose an odd question: should shipbuilding be such a strategic activity, why were the cork tree savannas of the Mediterranean section of the Douro valley not recruited for the supply of shipbuilding products?

It is important to remember that the Portuguese Crown had been complaining about a shortage of such tree-plants ever since the 15th century. If they were available after 1755, why did the Crown not impose exclusivity over those trees in the Douro valley, designating them 'royal sticks' (*paus reais*), thus, tree species exclusively reserved for the royal household? This juridical regime had been imposed on royal estates and grounds as well as across some private seigniorial lands in the Tagus and Sado river basins with this imposition remaining in effect until at least 1824 for unfenced hunting grounds and timber woods.[61] As the forest resources were the same in this northern region, why did

61 Joanaz-De-Melo, *An Analysis*, 21–50.

landscapes containing the same forest species receive different treatment in the respective legal and territorial management policies?

One reason would derive from the timber produced there being needed for cooperage, combined with the low cost of transport. Both reasons might provide justification or the wood may have been delivered to the small-scale shipyards of the king in the Leiria region, via the small maritime ports of S. Pedro de Muel and S. Martinho do Porto,[62] both just a short distance (by sea) from the royal sawing engine of Marinha Grande. Another justification would arise from the difficulties in rescinding property rights in wide areas of seigniorial lands distant from the Court. The Crown would certainly have had less implementation capacity than in the central region, where the monarchs, the royal family and much of the aristocracy lived.

Thus, given such availability, is the role of Brazilian timber in Portugal not overrated? From the history of trees point of view, the availability, especially after the 1755 earthquake, the adventure of removing cork trees for good, and thus ending wood supplies to the royal shipyards from the Douro, became a fantastic and obsessive mystery to me. I was utterly astounded to grasp how the cork tree forests of northern Portugal escaped the royal narrative of the scarcity of such an irreplaceable shipbuilding resource. As Portugal was a naval empire in decline with other European empires competing for the same timber resources, what motivations might have justified excluding such forests?

The Crown could not ignore the overriding need to defend the Metropolis and deal with the many scenarios of public calamities, indigenous uprisings in the colonies and with the naval war fleet sunk or floating in pieces along the Tagus estuary and Portuguese coast reflecting a period of enormous fragility. Communications had to be ensured, and the oft-mentioned cork trees were still relevant for shipbuilding and with oak needed for cooperage. The economy had to somehow recover a relevant position in international trade and commerce and exploiting an already existing and successful product in external markets might account for part of this strategy.

In fact, a project for the upper Douro valley region was already in motion when the 1st November earthquake struck and its socioeconomic and political implications across Portugal delayed the Crown asserting strict control over that production.[63] Regardless of the national recovery priorities, supporting a viable and established product also required due consideration. Paradoxically,

62 As examples, Frigate *Nossa Senhora das Brotas* built at S. Martinho (1697–1721), holding 52 pieces of artillery, Small *Frigate Bom Jesus de Mazagão* built in S. Martinho, (1698–1712) arming 36 pieces of artillery, Pereira, *História da Marinha,* 68 and 81.

63 Martins, "Vinha, vinho," 19–61.

the Portuguese loss of mastery over the Atlantic trade might well have resulted in significant investment in port production which drove other territorial changes and the redesign of several landscapes across the Portuguese territory.

At this point, I wish to make a special note. Considering the vast bibliography on port wine, I would clearly state that I am not producing either a literature review or critical analysis of the state of the art on such wine history. Regardless of its amazing and exquisite flavour, what only matters to this text is its role in attaining a better understanding of the wider politics of landscaping forests and other tree policies in Portugal during the 1700s.

Although this might at first appear a little farfetched, port wine would bring about investments in forests and orchards as an alternative to vineyards (but not to cereals as demonstrated by the 1779 legislation). Thus, the indirect meaning of not removing cork trees for shipbuilding tells us how in the Douro valley region, the cork tree was not a resource for extraction and export to the royal shipyards and neither were oak or pine trees.

Thus, like any written document, landscapes are moulded by their contexts; the intention of production may lie behind calling for heuristic and hermeneutical analyses. Hence, landscape reading becomes a methodology for interpreting political contexts leading us from landscapes to researching archives and again from the archives to better understanding how the different pieces of the coeval landscape patchwork fit into the puzzle of the territory as a whole.

In my view, as landscaping pre-defines the political context this becomes a third fundamental tool for historical environmental analyses, alongside time and space: landscapes also encapsulate historical context in each time and space, driving us from its constituent elements to demonstration in documentation. Landscapes thereby become an expression of all the political, cultural and natural contexts prevailing; would they have become time-contexts by themselves?

In the end, the ground, the physical territory merges into the intertwined political and natural dynamics; they evolve in parallel with the contradictory royal narratives on resources and goods that told of a scarcity and shortage of trees for shipbuilding in one region where the same crown invested in restoring and expanding trees for shipbuilding, while another narrative defended the increase of an export product and provided for mature cork trees, adequate for shipbuilding to be felled, such as those in the inland stretch of the Douro valley.

Throughout all this process of protecting precious trees for shipbuilding, as the Douro valley was designated to produce wine, the cork tree forests that were abundant in that region with a similar climate to the Portuguese Alentejo province (extending from the river Tagus to the Algarve), remained untouched by royal legislation throughout the 1700s. In other words, the Douro valley did not become demarcated with particular tree species reserved solely for the royal arsenals and shipyards. If there was such shortage, why would royal forests in the south not coping with the supply of cork trees, which are not a tropical forest species and could therefore hardly be imported from Brazil. Had the southern forests replenished naturally or were they effectively managed to supply forest resources from those existing? Why did the Crown release the Douro from such an obligation? The answer may stem from the proximity or distance and the respective transport costs incurred.

If port wine travelled directly from Oporto to European ports or to Brazil and the barrels were shipped to Lisbon for transport back to the north, this would be unnecessarily expensive, and the Douro valley timber imports would be unattractive due to the cost to the crown. The sawing engines, alongside the royal shipyards and arsenals were centrally located. The transport of timber and forest products would be cheaper from any region south of the river Mondego. That would probably justify the need to mark out specific trees in the centre region for the royal arsenals.

Indeed, the legislation on changing, extending and improving royal forests always addresses landscapes in the central and south regions: the royal estates. However, the House of Bragança owned lands and controlled municipalities across all of Portugal. Furthermore, in 1751, the administration's option for planting a future source of wood for royal shipbuilding was located contiguously to a previous forest of royal pinewoods; this expansion took place on sandbanks within a distance that would ensure minimal costs for transporting these crucial shipbuilding resources.[64]

Thus, the expansion of planted areas specifically to produce wood for shipbuilding would intentionally concentrate either on lowlands and sandbanks near the sea or on slopes and perimeters located close to rivers, to provide fast, secure and cheap transportation and storage. Although apparently contradictory, the expansion of woodlands was carefully handled. They attended only to royal landscapes under the Crown's tutelage, within the perimeter of the royal grounds which thus did not include lands belonging to other seigniorial landowners but only royal lands not subject to any formerly donated property rights, whether for farming or any other purpose.[65] In parallel, such endeavours experimented in areas under the ownership of the royal family,

64 *Regimento para o Guarda Mor dos Pinhais de Leiria e Superintendente da Fábrica da Madeira* of 18 October 1751, available at Silva, *Collecção. Legislação de 1750 a 1762*, 68–90; Joanaz-de-Melo, "Guerra, Impérios", 210–14.

65 Edital do Almirantado de 25 maio de 1799; Edict from the Admiralty of 25th May of 1799.

that is properties of the House of Bragança. Finally, the major forest plantings took place on properties under the tutelage of the crown and belonged to the kingdom, such as coastal lands borders and sandbanks, riverbeds and banks, and highland slopes and summits. After 1751, such tree planting then focused on areas where the costs of operation were least expensive: those nearby naval means of transport.

Indeed, the central region experienced many different strata of elements occupying the soil in the same royal estates, sometimes forested, sometimes left to grow wild but also placed under cultivation, especially on the left bank of the Tagus and near the Sado river estuary as explained below.[66]

Different strata of forests, alongside other agriculture products, occupy the same territorial spots over the long run. Landscapes change in the exact same places. This was the case in some of the Portuguese royal grounds, at least since the 1650s, when the same spots like Coruche, Salvaterra de Magos, Muge, Benavente, Golegã, Lamarrosa, Santarém, Alpiarça Almeirim, Benavente, Alfeite, Almada, Alhos Vedros, Alenquer, Xira, Setúbal, Comporta, Alcacér do Sal, Santa Margarida do Sado, Zambujeira, Azeitão, Sintra, Queluz, Óbidos, Vieira, S. Martinho, Muel, were addressed differently by royal narratives and even legislation in keeping with the ongoing changes in the ways human handled the different natural resources.

In this case, authors looking through the forestry regulations in effect during the 1600s and 1700s failed to notice that the devastated pinewoods were then provided for cultivation. This privilege, granted to the population, occurred under the Habsburg dynasty, in 1605, also as a narrative of grace. In the former regulation, exclusivity over tree felling had been imposed on both cork and pine trees. The same lands without trees appropriate to producing wood for shipbuilding, would nevertheless provide the crown with taxes while the woodland was resting and renewing.[67] Indeed, in the following period after the separation of the dual monarchy and the coronation of João IV in 1640, the restoration of the original demarcation of the perimeters of the royal forest-hunting preserves to their 1605 boundaries took place only ten years later, in 1650,[68] when those same lands had already undergone substantial replenishment of their trees in the period since 1605. In this year, according to documents issued under Philip II, some of the royal grounds along the Tagus had

66 BAHMOP, MMR, núcleo 1, years 1721–1883; núcleo 2, years 1777–1821; núcleo 16, years 1700–1833; núcleo 17, years 1722–1833.

67 *Regimento do Monteyro Mor do Reyno de Portugal* (*1605*), in Carlos Maria Baeta Neves, "Dos Monteiros-mores aos Engenheiros Silvicultores," *Anais do Instituto Superior de Agronomia* XXVIII (1965): 119–61.

68 José Justino Silva, *Collecção Chronologica da Legislação Portugueza. Segunda serie, 1648–1656*, (Lisbon: Imprensa de F. X. de Souza, 1856), 68–71.

become treeless. Forty years later, the pine trees, at least, would have renewed naturally and become a forest of trees fit for the supply of trunks, timber and wood to the royal shipyards and arsenals.

Following a similar logic, the laws of 1790 and 1797,[69] which complained about forest resource scarcity (stipulating a new administration of the woodlands in the Queens and Princesses estates), cover a region in close proximity to Lisbon, on the right bank of the Tagus River. Furthermore, the new regime for royal hunting grounds in the 1800s, which fully separated the administration of woods for professional timber production from wildernesses for hunting, also addresses only the royal grounds and pine woods in the central region. This reflects a clear intention to continue investing in this sector where it was least expensive for the coffers of the royal household: in the centre Region, closer to the royal Shipyards.

Indeed, it seems unthinkable that, after the sinking of the fleet, the extraction of cork trees in the Douro valley would be ignored were there was not an alternative reserve of timber for shipbuilding. Even after the conquest of Pondá (1763) and further inland exploration in Brazil, these supply options would simply have been too risky. They were oceans far, too distant under emergency supply.

Thus, out of the desired shaping of wine landscapes by decree in 1765 other landscapes would indirectly emerge in reaction to this legislation, especially forests, orchards and farming beyond the Douro region. This protected area for fortified port wine production implied destroying trees suitable for shipbuilding and other carpentry services. However, while this action was specifically implemented for the production of this wine, the Crown marked out another region to become the supplier of forest resources, at least to meet imperial needs. Areas where royal forests, woods and woodlands already existed received greater investment, often expanding into contiguous lands (for pinewoods) as well as improving on their management. This region would stretch across the centre of the kingdom, covering the lands of the upper Mondego and Tagus river basins as well as the sandbanks along Leiria's coastline. Consequently, other lands, where forests and wine were not to be exploited as the main products for the export trade and imperial communications, received other crops. Therefore, the process of shaping a wine region indirectly resulted in a specific region of a functional forestry for the Crown, and, elsewhere in the national territory, the vineyards were replaced by introducing new plants or

69 Alvará March 17th 1790 [Act abolishing the Superintendence of the Pine Nuts of Leiria, establishing internal Administration, and the respective Regulation]; Alvará of 9th December 1797 (Act from Secretary-General of the Navy of 9th December 1799); Edital do Almirantado of 25th May 1799 (Edict from the Admiralty of 25th May 1799).

simply expanding the areas of production of already proven and robust crops, such as figs in the Algarve.

The Crown's investments in forests from 1751 onwards was directed towards new plantations along costal sandbanks. Even considering the economic aspects, there were certainly other lands more appropriate for planting with forests than these coastal lands located around 100km to the north of Lisbon. However, according to natural law, these lands were available, free of any cultivation and duly under the tutelage of the Crown for the strategic protection of the realm and its subjects: a region that needed the monarch's protection against predictable factors that would endanger human lives, such as coastal flooding or military incursions.

The pinewoods seeded post-1751 along the Portuguese coastal sandbanks met economic aims as well as provided protection. They were planned both as barriers to defend the lands and the populations against factors of erosion and destruction while delivering the forestry products needed for military affairs, for the navy and beyond.[70]

As a matter of fact, it would seem logical that a geography of wine production could be deepened and widened across an already tested area. Thus, the legislation to protect and incentivise port wine production, promulgated in 1765, prohibited the planting of trees in certain regions of vineyards and ordered the removal of the latter from other lands where the resulting wines might be easily transported elsewhere by river or sea within the scope of guaranteeing the product's quality for international trading.

This demonstrates how the national landscaping planned by the Crown impacted differently on different regions in keeping with the variations in their respective geographies and positions: from ensuring a product that enabled the Portuguese Empire to maintain a position in the international trading system to afforestation for strategic communications. Furthermore, following the enactment of such legislation, the resulting landscape changes also convey how private initiative adjusted to those interests and the new agricultural framework to ensure their own resourceful production in the short and long term.

---

70 *Regimento para o Guarda Mor dos Pinhais de Leiria e Superintendente da Fábrica da Madeira* of 18 October 1751, available at Silva, *Collecção da Legislação 1750 a 1762*, 68–90; Nicole Devy-Vareta, "Fomento e ordenamento florestal nas regioes litorais durante a Época Moderna," in *O Litoral em Perspectiva Histórica* (*Séc. XVI a XVIII*), ed. Inês Amorim, Amélia Polónia and Maria Helena Osswald (Oporto: Instituto de História Moderna, 2002), 165–176; Ana Isabel Alves Lopes, "Governar a Natureza": o assoreamento da foz do rio Cávado, em Fão – causas, impactos e respostas sociais (1750–1870)" (Master's Dissertation: Universidade de Porto, 2019).

The aforementioned *division of the territory into regions of resources* appears quite rational: vineyards in the Douro far from the sea- free from piracy radings-, but from where the transport of wine -trhough river flows-, could easily take place; managing forests for the production of timber and wood far from arable lands and also along the sea coast or in valleys where the logs could be transported down the tributaries of the main rivers connecting with Lisbon and the royal arsenals. Such improvement also seems to have been driven by achieving lower transport costs leaving other regions to deploy the raw materials available on private lands for private business interests, which favoured the generation of income that boosted the taxation income of the royal household.

Likewise, it would have been impossible to develop a major area of fortified wine production in the lowlands of the Tagus and Sado rivers for two key reasons: the very poor existing techniques for wine production and conservation and these regions hosting some of the most consolidated seigniorial estates. Imposing the planting of vineyards on private lands would threaten the existing donated property rights as aristocratic privileges were not easily removable, not even by the almost mythological figure of Pombal.

In parallel to these aspects, the expansion of a culture into remote areas (the arid parts) of such territories might also mean the intensification rather than merely amplifying a secular way of managing the same territory: extensive agriculture on poor lands. Whereas in other parts of Europe, the Americas, the lands of the Mediterranean and Maghreb, resources are abundant in density but quite uniform over huge distances, Portugal experiences an amazing diversity of botanical species over shorter distances and with waste lands still susceptible to the cultivating of vines.

7 Conclusions

The Lisbon Earthquake of 1755 generated huge impacts that included the redesign of landscapes across Portugal. The major driver was the need to economically recover while producing a stock of resources spanning the commodities necessary to running a naval empire. Within this framework, the Crown intervened through defining areas for specialisation whether for products destined for trading on international markets or for the resources and raw materials capable of fully autonomously supplying the royal shipyards.

For a short period, some farming landscapes were changed and 'designed by Decree' and leaving a footprint on the landscape. Over the 1760s and 1770s, the Portuguese Crown planned for the Douro valley region in the north of Portugal to raise its output of an already valuable commodity; the central region would continue producing forest goods, establishing new plantations of woodlands on summits, slopes and sandbanks; the central inland upper mountains around the Mondego valley, as well as the highlands and lowlands of the Algarve region, would experience the removal of vineyards, replaced both by annual agricultural crops and fruit and nut orchards.

The colonial timber supply did not contribute to the early efforts either to rebuild Lisbon or the national naval fleet in the 1750s. The urgent deliveries of timber and other forest goods were sourced from within Portugal. The bulk of the historiographic literature maintains that Portugal was then 'forestless'. However surprising, the reality was quite another and this availability is the fascinating and intriguing issue that requires deepening.

This argues that the lack of alternative sources for supplying forest goods in contexts of emergency, whether due to conflicts in the Atlantic colonies, naval limitations on reaching the far distant colonies due to the French, British and Dutch fleets taking control of the sailing routes or the natural hazards (such as the rebuilding tasks following the Lisbon earthquake on 1st November 1755), drove the Portuguese forest product market – sourced from royal, noble and church lands across the realm, thus the Metropolis – delivered to royal arsenals and shipyards. Despite such a supply arriving in the port of Lisbon throughout the 1700s, Portugal did not become stripped of woodlands or wilderness and even less so at the expense of colonial forests.

Regardless of the Portuguese loss of control over sailing corridors in the Indian and Atlantic oceans to the French, British, and Dutch fleets in the second half of 1700s, the Portuguese Crown wished to retain its status as a player in international trade. A unique and exquisite product was required to fulfil such an aim and with production and delivery fully controlled by Portugal to avoid piracy, storms and cutting the risk of lost cargos.

Port wine was chosen as the new Portuguese red gold. However, such an option would require a wide geography of production to compete with any eventual colonial plantations and a 1765 decree marked out a perimeter in the Douro for the exclusive planting of vineyards. Cork trees were removed or felled; a tree species identified as in perpetual short supply for the imperial navy ever since the 16th century. Thus, the proposition here is that the felling of cork trees was only possible due to the existence of alternative areas of forest supply from both royal and private woods and also only after there was the awareness of the new dense, varied and wide expanses of forests growing in the colonies. This would become insurance for coping with the relocating of afforested areas and new monocultural farming experiments on a large scale. Indirectly, the information provided in the surveys of Brazil

(1750s–1790s)[71] and the details returned by the conquest of the Region of Ponda in the State-of-India by Portuguese troops (1763) would inform about the abundant and diverse forests in Portuguese areas of influence and control. Such a new awareness might have impacted on the building of vessels in Portuguese shipyards. However, once the resistance of these woods to the corrosion of ocean waters had been tested, it would nevertheless be safer to build ships both in the State of Goa and in Brazil rather than shipping cargos of woods and running the risk of smuggling or being preyed upon by the other, now competitive, fleets eager for the same exotic materials. Thus, the colonial forests would have been insurance with the scope for transferring a proportion of Portuguese shipbuilding to the colonies and releasing the pressure on the Portuguese royal forests for such purposes.

Furthermore, the increase in demand for wood products in the royal internal market encouraged improvements and better forest management both in royal and seigniorial estates for the former to gain income and the latter to supply a strategic reserve of woods in the Metropolis, with the Crown as the most likely eventual buyer.

Summing up, by the 1800s, Mediterranean territories were developing a dual system of managing the land as a legacy from the past: replenishing through seeding and extracting resources in order to continue renewing the wood supply but never fully exhausting the available supply of trees. This thus broadens the literature on forests in Portugal and the Empire in the 1700s through analysis incorporating a highly complex set of players: the enlightenment, physiocracy, piracy and ocean trade within the Indian and Atlantic oceans. Furthermore, the former interlinked factors influenced how political and economic thinking adjusted military fragility to economic development, specifically in justifying the transfers and naturalization of species on a large scale with the aim of improving the general living conditions and not only of better achieving the economic interests of the Crown and social elite. Indeed, Portuguese rulers and thinkers were interested in trying to reposition the national forest park. Correspondingly, new forestry species were planted in European soils to observe whether the forests might recover faster with species from the colonies and beyond the royal interest in producing woods for shipbuilding. In this sense, Portugal attempted to introduce exotic species in order to faster compensate for the removal of resources needed by all of society.

---

71 Juliana Gines Bortoletto, *Os desenhos Botânicos da Viagem Filosofica ao Brasil. Uma Rede Política, Científica e Criativa* (PhD Dissertation, Universidade de Coimbra, 2018), 25–206; Domingues, "Museus, Colecionismo", 1–19.

This hypothesis, seeding new species for the first time to ascertain whether they would prosper, constituted a natural step within the framework of both the Portuguese and Spanish cultures. Such practices would derive from a geographic approach as a patchwork of many different ecosystems enrichened with botanical species was a millennium old characteristic due to the layers of occupation left by many peoples and since at least the Roman Empire and the resulting acculturation.

Such a subconscious geographic model might have shaped a different way of dealing with the Portuguese territory, that is, planting trees in areas not traditionally hosting forests prior to the 1750s. It is important to recall that even after the 1755 and 1761 earthquakes, and again following the floods of the 1770s-1780s, the Portuguese territory still managed to supply forest goods for the royal shipyards and arsenals as well as for social purposes, against all the historiographical odds. This outcome would have resulted from previously embedded practices of replenishing known tree species, introducing new trees while also expanding the areas planted for social purposes in a territory, the Iberian Peninsula, that displayed an amazing abundance of different ecosystems capable of hosting a wide range of many different tree species.

Throughout the Iberian Peninsula, a patchwork of so many different shapes and areas of cultures emerged, reflected in the forests, that this constituted a pre-existing model of thinking the territory.

The consequences for establishing new landscapes of forests and farming – in parallel – in Portugal between 1755 and the 1870s, would then represent the *normal* (rural farming) set up. Diversity and micro slots of farmed lands alongside large areas of forest were totally compatible and such a pattern was anything but new. Indeed, small and large crop plantations coexisted side by side with woods and cereals or other plantings, such as the fruit orchards that would remain in the landscape even after permission was once again granted for the planting of vineyards in 1779. Hence, whether cultivating many crops simultaneously or monocultures of vines along the valley slopes or of trees on sandbanks or expanding areas of orchard, such measures would be new but still part of an already existing pattern. The adjustment of new forests planted outside traditional farming areas, rather than some rupture, would embody a development of the pre-existing ways of managing Iberian territories.

Thus, geographies producing diversity but low amounts of each product compelled the management of limited quantities of resources to ensure they lasted over the long run, coupled with the external factors preventing the supply of forest goods from the colonies, demanded solutions for coping with sudden increases in demand. Such difficulties in Portugal incentivized and enabled a redesign of landscapes to establish reserves of resources – for exporting and maintaining the international trading position, to supply arsenals with wood and to feed the population after the overwhelming natural events that caused massive and widespread damage to the Portuguese metropolis but nurturing the opportunities and the conditions for reasoning how best to deal with and transform nature and, ultimately, to cope with it.

Some of the options resulted in better control of the watercourses while others drove increased levels of flooding. Such a conjuncture created an optimal opportunity for the expansion and implementation of new forests and farming settlements and despite the known abundance of such resources in the colonies that were nevertheless unreachable on the scale and in the time required for the needs prevailing in Portugal. Indeed, the great distance and the lack of military naval power across the Portuguese imperial routes throughout the 18th century fostered self-sufficiency in the metropolis and not dependency on the colonies.

On the other hand, such landscape evolution in Portugal did not put an end to the extraction and consumption of natural resources in the colonies. However, they were transformed where they would not necessarily be stolen or looted, such as in the shipyards in Brazil and Goa that began and continued building vessels for the Portuguese commercial fleet.

The new scientific-economic developments, as responses to the political and natural setbacks, underwent implementation by the Crown on lands belonging to the royal household or in areas where security depended on the Crown and thus not on the estates of every landowner.

Abbreviations

ANTT: Arquivo Nacional Torre do Tombo
  PM: Paço da Madeira
BAHMOP: Biblioteca e Arquivo Histórico do Ministério de Obras Públicas
  DGCAM: Direcção Geral do Comércio, Agricultura e Manufacturas
  MMR: Montaria-Mor do Reino

Bibliography

Agnoletti, Mauro. "Environmental Thinking and Cultural Values: A Reflection on Environmental Globalisation and the Mediterranean Culture". *Global Environment* VII, no. 2 (2014): 257–90, https://doi.org/10.3197/ge.2014.070202.

Anais do Conselho de Saúde Pública do Reino. Lisbon: Typograhia Lisbonense, 1838–1848.

Araújo, Ana Cristina. *O Terramoto de 1755. Lisboa e a Europa*. Lisbon: CTT, 2005.

Barreira de Campos, Isabel Maria. *O Grande Terramoto 1755*. Lisbon: Parceria, 1998.

Barrés, Jean-Baptiste. "Memórias de um oficial do grande exército." In: *Linhas de Torres Vedras. Memórias Francesas Sobre a III Invasão*, 25–48. Lisbon: Livros Horizonte, 2010.

Bortoletto, Juliana Gines. *Os desenhos Botânicos da Viagem Filosofica ao Brasil. Uma Rede Política, Científica e Criativa*, History of Art Doctoral Degree Thesis supervised by Professor Maria de Lurdes dos Anos Craveiro and awarded by the Faculty of Letters, the University of Coimbra. Coimbra: 2018, pp. 25–206.

Braga, Isabel. *Portugal à Mesa: Alimentação, Etiqueta e Sociabilidade 1800–1850*. Lisbon: Hugin, 2000.

Cabral, Diogo de. *Na Presença da Floresta. Mata Atlântica e História Colonial*. Rio de Janeiro: Garamond, 2014.

Cardim, Pedro and Nuno Gonçalo Monteiro (eds.), *Political Thought in Portugal and its Empire, c. 1500–1800*. Cambridge: Cambridge University Press, 2021.

Cardoso, Arnaldo. *O Terrível Terramoto da Cidade que Foi Lisboa. Correspondencia do Nuncio Filippo Acciaiuolli*. Lisbon: Alethéia, 2013.

Carvalho, José Monteiro de. *Diccionario Portuguez das Plantas, Arbustos, Matas, Árvores, Animais Quadrupedes, e Repteis, Aves, Peixes, Mariscos, Insectos, Gomas, Metaes, Pedras, Terras, Mineraes*. Lisbon: Officina Miguel Manescal da Costa, 1765.

Clímaco, Cristina. *As Linhas de Torres Vedras. Invasão e Resistência 1810–1811*. Lisbon: Câmara Municipal de Torres Vedras-Edições Colibri, 2016.

Costa, Fernando Dores. *Insubmissão. Aversão ao Serviço Militar no Portugal do século XVIIII*. Lisbon: ICS, 2010.

Crespo Solana, Ana. "The Global Timber Trade and Shipbuilding in the 16th–18th centuries: Interdisciplinarity, research problem and the ForSeadiscovery project." In *Roots of Sustainability in the Iberian Empires: Shipbuilding and Forestry (14th–19th centuries)*, edited by Koldo Trapaga-Monchet, Álvaro Aragón-Ruano and Cristina Joanaz-de-Melo, 54–74. New York-Oxford: Routledge, 2023.

Cunha, Mafalda Soares da (Coord.). *Resistências: Insubmissão e Revolta no Império Português*. Lisbon: Leya, 2021.

Curto, Diogo Ramada. *Cultura Escrita, Séculos XV a XVIII*. Lisbon: ICS, 2007.

Curto, Diogo Ramada. *Cultura Imperial y Proyectos Coloniales de los Portugueses (siglos XV–XVIII)*. Madrid: Marcial Pons, 2019.

Daniel, João. *Quinta Parte do Thesouro Descoberto no Rio Maximo Amazonas: Contém hum Novo Methodo para a Sua Agricultura, Utilissima Praxe Para a Sua Povoação, Navegação, Augmento, e Commercio, Assim dos Indios como dos Europêos*. Rio de Janeiro: Imprensa Régia, 2007.

Devy-Vareta, Nicole. "Fomento e Ordenamento Florestal nas Regiões Litorais Durante a Época Moderna." In *O Litoral em Perspectiva Histórica (Séc. XVI a XVIII)*, edited by

Inês Amorim, Amélia Polónia and Maria Helena Osswald, 165–76. Oporto: Faculdade de Letras, 2002.

Devy-Vareta, Nicole. "Os Avanços e Recuos da Floresta em Portugal – Da Idade Média ao Liberalismo". In *Floresta e Sociedade: Uma História em Comum*, edited by Joaquim Silva, 55–75. Lisbon: Fundação Luso-Americana para o Desenvolvimento, 2007.

Disney, Antonhy. *A History of Portugal and the Portuguese Empire, From the Beginnins to 1807, volume 2: The Portuguese Empire*. Cambridge: Cambridge University Press, 2009.

Domingues, Ângela and Carmen Ortiz García. "Viagens Científicas e Coleccionismo no Mundo Ibérico (séculos XVIII–XX)". *Asclepio. Revista de Historia de la Medicina y de la Ciencia* 71, no. 2 (2019): 1–4, https://doi.org/10.3989/asclepio.2019.12.

Domingues, Ângela. "Museus, Colecionismo e Viagens Científicas em Portugal de Finais de Setecentos." *Asclepio. Revista de Historia de la Medicina y de la Ciencia* 71, no. 2 (2019): 1–19, https://doi.org/10.3989/asclepio.2019.12 178.

Ferreira, José Miguel Moura. "As Novas Conquistas: Florestas, Agricultura e Colonialismo em Goa (*c.* 1763–1912)." PhD diss., Universidade de Lisboa, 2021.

Francisco, Papa. *Laudato Si. Sobre o Cuidado da Casa Comum*. Lisbon: Paulinas, 2015.

Franco, Francisco Soares. *Diccionário de Agricultura*. Coimbra: Real Imprensa da Universidade, 1804–1806, 5 vols.

Glete, Jan, *Warfare at Sea, 1500–1650: Maritime Conflicts and the Tranformation of Europe*. New York: Routledge, 1999.

Gomes, Rosa Varela and Koldo Trapaga-Monchet (coords.) *Árvores, Barcos e Homens na Península Ibérica (Séculos XVI–XVIII)*. Lisbon-Zaragoza: IAP/IHC-Pórtico, 2017.

Hanna, Mark G. (2015). "Well-Behaved Pirates Seldom Make History: A Reevaluation of English Piracy in the Golden Age." in *Governing the Sea in the Early Modern Era: Essays in Honour of Robert C. Ritchie*, edited by Mancall, Peter C. and Shammas, Carole 129–168. San Marino, California: Huntington Library, Art Collections, and Botanical Gardens, 2015.

Isenberg, Andrew C. *The oxford Handbook of Environmental History*. Oxford: Oxford University Press, 2014.

Joanaz-de-Melo, Cristina "Guerra, Impérios e Corte Joanina nas Coutadas de Caça: Alavancas de Regeneração Florestal em Portugal, em Meados do século XVIII." *Manuscrits: Revista d'Història Moderna* 42 (2020): 199–220.

Joanaz-de-Melo, Cristina. "Floresta em Movimento: usar, regenerar, cuidar (seculos XVI–XIX)". In Joanaz de Melo. In *Como a Fénix Renascida, Matas Bosques e Arvoredos (séculos XVI–XX): Representações, Gestão, Fruição*, edited by Cristina Joanaz de Melo, 79–130. Lisbon: Ediçõs Colibri, 2020.

Joanaz-de-Melo, Cristina. "Forests in Portugal 1750s–1820s: a History of Forest Compensation." In *Roots of Sustainability in the Iberian Empires: Shipbuilding and Forestry*

(*14th–19th centuries*), edited by Koldo Trapaga-Monchet, Álvaro Aragón-Ruano, and Cristina Joanaz de Melo, 251–77. New York- Oxford: Routledge, 2023.

Joanaz-de-Melo, Cristina. *An Analysis of the Royal Preserves in Portugal. Issues of Privilege, Power, Management and Conflicts*. Sheffield: Wildtrack, 2015.

Kierchberger, Ulrike. "Horizons of Ecological Change: Stories of Transfer across the Indian Ocean in the Age of Empire." *Global Environment* XIII, no. 1 (2020): 8–29, https://doi.org/10.3197/ge.2020.130102.

Labrador Arroyo, Félix and Koldo Trapaga-Monchet. "Recursos Naturales en la Península Ibérica: los aprovechamientos forestales e hídricos (siglos XV–XIX)." *Tiempos Modernos: Revista Electrónica de Historia Moderna* 39, no. 2 (2019): 279–543.

Land, Jeremy and Rodrigo da Costa Dominguez. "Illicit Affairs: Philadelphia's Trade with Lisbon before Independence, 1700–1775." *Ler História* 75 (2019): 179–204 https://doi.org/10.4000/lerhistoria.5936.

Leal, Pinho. *Portugal Antigo e Moderno: Diccionario Geographico, Estatistico, Chorographico, Heraldico, Archeologico, Historico, Biographico E Etymologico de Todas as Cidades, Villas e Freguezias de Portugal e de Grande Numero de Aldeias*. Lisbon: Livraria Editora de Mattos Moreira, 1873–1890.

Lobo, Francisco de Sousa. *A Defesa de Lisboa – Linhas de Torres Vedras, Lisboa, Oeiras e Sul do Tejo (1808–1814)*. Cascais: Tribuna da História, 2015.

Lopes, Ana Isabel Alves. "Governar a Natureza": o Assoreamento da Foz do Rio Cávado, em Fão- causas, Impactos e Respostas Sociais (1750–1870)." Master's Dissertation: Universidade de Porto, 2019.

Lopes, João Baptista da Silva. *Corografia ou Memoria Economica, Estadistica, e Topografica Do Reino o Algarve*. Lisbon: Typografia da Academia, 1841.

Magalhães, Joaquim Romero de. *O Algarve Económico 1600–1773*. Lisbon: Estampa, 1988.

Mancall, Peter C. and Shammas, Carole (eds.). *Governing the Sea in the Early Modern Era: Essays in Honor of Robert C. Ritchie*. Huntington Library, Art Collections, and Botanical Gardens, 2015.

Martínez González, Alfredo José. "La elaboración de la Ordenanza de Montes de Marina, de 31 de enero de 1748, base de la política oceánica de la monarquía española durante el siglo XVIII". *Anuario de Estudios Americanos* LXXI, no. 2 (2014): 571–602, DOI:10.3989/aemer.2014.2.07.

Martínez González, Alfredo José. *Las Superintendencias de Montes y plantíos (1574–1748: Derecho y Políticas Forestales para las Armadas en Edad Moderna*. Valencia: Tirant lo Blanch, 2015.

Martins, Conceição Andrade. "Vinha, Vinho e Política Vinícola em Portugal: do Pombalismo à Regeneração." PhD diss., Universidade de Évora, 1998.

Mattoso, José, Suzanne Daveau and Duarte Bello. *Portugal – o Sabor da Terra. Um Retrato Histórico e Geográfico por Regiões*. Lisbon: Temas e Debates, 2010.

Mendes-Victor, Luiz A., Carlos Sousa Oliveira, João Azevedo and António Ribeiro (eds.). *The 1755 Lisbon Earthquake Revisited*. Netherlands: Springer, 2009.

Meneses, Avelino de Freitas de. "A Rutura e a Reconciliação (1640–1750)." In *História de Portugal e Espanha: Amores e Desamores*, coordinated by Artur Teodoro de Matos, João Paulo de Oliveira e Costa and Roberto Carneiro, 7–152. Lisbon: Circulo de Leitores, 2015.

Mesquita, José Carlos Vilhena. "A Economia Agrária do Algarve, na Transição do Antigo Regime para o Liberalismo (1750–1836)." *Estudos III* (2009): 143–95.

Monteiro, Nuno. "A Monarquia Barroca (1668–1750): O Tempo de Pombal." In *História de Portugal*, coordinated by Rui Ramos, Bernardo Vasconcelos Sousa and Nuno Monteiro, 331–78. Lisbon: A Esfera dos Livros, 2010.

Monteiro, Nuno. *D. José*. Lisbon: Círculo de Leitores, 2006.

Neves, Carlos María Baeta. "Dos Monteiros-mores aos Engenheiros Silvicultores." *Anais do Instituto Superior de Agronomia* XXVIII (1965): 19–172.

Nozes, Judite. *O Terramoto de Lisboa de 1755. Testemunhos Britânicos/ The Lisbon Earthquake of 1755. British Accounts*. Lisbon: The British Historical Society of Portugal-Lisóptima edições, 1990.

Oosthoek, K. Jan and Richard Hölzl (eds.). *Managing Northern Europe's Forests Histories from the Age of Improvement to the Age of Ecology*. New York-Oxford: Berghahn Books, 2018.

Osswald, Cristina. *Written in Stone: Jesuitic Buildings and their Artistic and Architectural Features*. Goa: Goa 1556, 2013.

Pádua, José Augusto. *Um Sopro de Destruição: Pensamento Político e crítica ambiental no Brasil Escravista, 1786–1888*. Rio De Janeiro: Zahar, 2002.

Pedreira, Jorge. "Guerras, Afinidades e Nacionalismos (1750–1910)." In *História de Portugal e Espanha: Amores e Desamores*, edited by Artur Teodoro Matos, João Paulo Oliveira e Costa and Roberto Carneiro, 153–328. Lisbon: Círculo de Leitores, 2015.

Pedreira, Jorge. "From Growth to Collapse. The Breakdown of the Old Colonial System. Portugal and Brazil (1750–1830)." *The Hispanic American Historical Review* 80, no. 4 (2000): 839–65.

Pedreira, Jorge. *Estrutura Industrial e Mercado Colonial. Portugal e Brasil, 1780–1830*, Lisbon: Difel, 1995.

Pereira, José Manuel Malhão (coord.). *História da Marinha Portuguesa. Navios, Marinheiros e Arte De Navegar (1669–1823)*. Lisbon: Academia de Marinha, 2012.

Péruse, B. "Campanhas de Portugal (1810–11)." In *Linhas de Torres Vedras. Memórias Francesas Sobre a III Invasão*, 158–89. Lisbon: Livros Horizonte, 2010.

Pinto, Ana Fernandes. *Uma imagem do Japão: A Aristocracia Guerreira Nipónica nas Cartas Jesuítas de Évora*. Lisbon: Fundação Oriente, 2004.

Ramos, Rui, Bernardino Vasconcelo Sousa and Nuno Monteiro (coords.). *História de Portugal*. Lisbon: A Esfera dos Livros, 2009.

Rego, Francisco. *Florestas Públicas*. Lisbon: Direcção-Geral Das Florestas, 2001.

Ribeiro, Orlando. *Portugal, o Mediterrâneo e o Atlântico: Esboço de Relações Geográficas*, Lisbon: Sá da Costa, 1987.

Rodrigues, Ana Duarte. "The Role of Portuguese Gardens in the Development of Horticultural and Botanical Expertise on Oranges." *Journal of Early Modern Studies*, 6:1 (2017): 69–89. DOI: 10.5840/jems2017614.

Scott, James C. *Seeing like a State. How Certain Schemes to Improve the Human Condition Have Failed*. New Haven: Yale University Press, 1999.

Serrão, José Vicente, Bárbara Direito, Eugénia Rodriges and Susana Münch Miranda (eds.). *Property Rights, Land and Territory in the European Overseas Empires*. Lisbon: CEHC-IUL, 2014.

Shrady, Nicholas. *O Dia do Fim. Ira, Ruína e Razão no Grande Terramoto de Lisboa*. Alfragide: Texto editores, 2014.

Silva, Antonio Delgado da. *Collecção da Legislação Portugueza desde a ultima compilação das ordenações, legislação de 1750 a 1762*. Lisbon: Typografia Maigrense, 1830.

Silva, Lazaro Benedito da, Andreia Moraes Ferreira, Sara Santos Araújo, Marta Catarino Lourenço. "Transporte de Madeiras Brasileiras para Portugal nos Séculos XVIII E XIX." *Brazilian Journal of Development* 6, no. 7 (2020): 728–745, DOI:10.34117/bjdv6n7-860.

Tavares, Rui. *O Pequeno Livro do Grande Terramoto*. Lisbon: Tinta da China, 2009.

Trapaga Monchet, Koldo, Álvaro Aragón-Ruano and Cristiana Joanaz-de-Melo. "The Game of the Demiurge in the Gardens of Chronos: Woods play hide-and seek in the long run through sustainable management." In *Roots of Sustainability in the Iberian Empires: Shipbuilding and Forestry (14th–19th centuries)*, edited by Koldo Trapaga-Monchet, Álvaro Aragón-Ruano and Cristina Joanaz-de-Melo, 1–30. New York-Oxford: Routledge, 2023.

Trapaga Monchet, Koldo. "Guerra y deforestación en el Reino de Portugal (siglos XVI–XVII)." *Tiempos Modernos: Revista Electrónica de Historia Moderna* 39, no. 2 (2019): 396–425.

United Nations. *Transforming our world: the 2030 Agenda for Sustainable Development, United Nations A/RES/70/1 General Assembly, 21 October 2015 Seventieth session Agenda items 15 and 116 15-16301 (E) *1516301*, Resolution adopted by the General Assembly on 25 September 2015*. https://www.un.org/ga/search/view_doc.asp?symbol=A/RES/70/1&Lang=E.

VV.AA. *O Grande Terramoto de Lisboa. Volume II: A Protecção*. Lisbon: FLAD/Público, 2005–2006.

VV.AA. *Providências do Marquês de Pombal que se Deram no Terramoto que Padeceu a Corte de Lisboa no ano de 1755*. Vol. III, Lisbon: FLAD/Público, 2005–2006.

Winiwarter, Verena, Martin Schmid and Gert Dressel. "Looking at Half a Millennium of Co-existence: the Danube in Vienna as a Socio-natural Site." *Water History* V (2013): 101–19.

Webography

Legislation: *Ordenações Filipinas* in www.legislacaoregia.com.
Alvará de 09 de Dezembro de 1797, (Act from the Secretary-General of the Navy of 9th December 1799).
Alvará de 17 de Março de 1790, (Act abolishing the Superintendence of the Pine Nuts of Leiria, establishing an internal Administration, and respective Regulation).
Edital do Almirantado of 25th May 1999, (Edict from the Admiralty of 25th May 1799).

CHAPTER 9

Shaping and Boosting the Economic Viability of a Royal Space for Agricultural and Livestock Production: the Royal Site of La Florida (1787–1814)

Félix Labrador Arroyo

The first years of Charles IV's reign (1788–1808) saw the creation of a new royal site. Given over to recreational uses and agricultural and livestock production, this estate also ensured territorial continuity between the royal properties of the Casa de Campo and the Monte de El Pardo.[1]

The Royal Site of La Florida was the result of merging two large estates acquired by the monarch. The first was the Príncipe Pío hill (Montaña del Príncipe Pío), which was purchased from Prince Pío's heirs in 1792 using royal revenues; the second, La Moncloa, which was owned by Manuel Godoy, was exchanged for other royal land. As we shall see, they were joined by smaller adjacent properties such as the Granjilla (farm) of the Hieronymite friary and land belonging to the Duke of Alba and the Seminary of Nobles of Madrid, among others, and together spanned a total area of 526 hectares of land that was predominantly arable (just over 90%).[2] This expanse of land was located by the river Manzanares, which provided abundant water resources that were used to create vegetable gardens and orchards and cultivate produce that required irrigation, an aspect that will be discussed in greater detail in due course.

Despite the economic situation, further areas were incorporated into the royal patrimony during Charles's reign. There was an economic rationale behind all these new acquisitions: to obtain profits and revenues through agricultural, livestock and forestry exploitation rather than simply using these properties for the purposes of image and representation of the Crown or as hunting grounds. For example, Castellanos de la Dehesa, a single property (*término redondo*) belonging to the Marquis of Castellanos, and Cercado de San Sebastián with its mill were purchased on 12 February 1795 and

1 For the earlier evolution of this space see María Teresa Fernández Talaya, *El Real Sitio de La Florida y La Moncloa. Evolución histórica y artística de un lugar madrileño* (Madrid: Fundación Caja Madrid, 1999), 40–47; on the land incorporated then, 54–59.

2 Archivo General de Palacio (AGP), Reinados, José I, caja 66, exp. 11.

© FÉLIX LABRADOR ARROYO, 2026 | DOI:10.1163/9789004745698_011 subsequently incorporated into San Ildefonso on 8 October of the same year.[3] A few years later, between 1802 and 1805, the *soto* (riparian woodland) and castle of Aldovea and the *sotos* of Galapagar, Gordo, Vaciabotas and Quintana were acquired along with the depopulated areas of Daralcalde and Viveros and the *soto* of Mejorada and Baezuela. These were all incorporated into Aldovea and later became part of San Fernando, although they were often managed by the administrator of La Florida.[4]

Likewise, on 25 July 1804 the Coca pinewood, an important resin-producing area, was incorporated into the site of San Ildefonso, adding to the already significant forest mass of Riofrío, Pirón and Valsaín.[5] Important farming and pasture areas became part of Aranjuez not long afterwards: the *dehesa* (wood-pasture) of La Saceda, which belonged to the Madrid convents of La Victoria and Las Descalzas Reales; the *dehesas* of Híjar, Las Cabezadas de la Saceda and El Campillo, owned by the municipality of Yepes, were acquired on 15 June 1806; and the *soto* of Añover de Tajo, the *dehesa* of La Alhóndiga and the *quintos* (estates) of Valjuanote and Valquemado were acquired on 2 February 1808.[6] In addition, the Sacedón baths were incorporated into the royal patrimony in 1802, being used for therapeutic-medicinal purposes and farming during the following years.[7]

Although it may seem contradictory, these acquisitions were made at a time of crisis when the traditional mechanisms for managing the royal finances were proving insufficient to sustain the Crown's expenditure, leading to the consideration of confiscatory measures. These measures culminated in the decrees of 19 September 1798, which were extended to the royal patrimony by means of a royal edict (*cédula*) of 21 October 1800 whereby it was agreed to sell off assets deemed unnecessary, and by a royal edict of 17 April 1801, where it was agreed to redeem the emphyteutic leases on royal properties. This meant, de facto, freeing up patrimonial assets. Both provisions were reinforced by another royal edict of 17 January 1805, based on the royal order of 6 November 1799.[8] However, all the incorporations stemmed from economic concerns,

3 AGP, Administración General (AG), leg. 1283, exps. 12 and 13.

4 Félix Labrador Arroyo, "Entre la costumbre y la transformación: el Real Sitio de San Fernando (1790–1814)," in *Arqueología, Historia y Medio Ambiente: visiones cruzadas*, eds. Luis Alberto Polo Romero and Koldo Trápaga-Monchet (Madrid: Dykinson, 2019) 127–29.

5 AGP, AG, leg. 359.

6 AGP, AG, leg. 1274, exp. 3bis.

7 Aurelio García López, *La Isabela y Baños de Sacedón* (Guadalajara: Bornova, 2003), 38 and 83.

8 Miguel Artola, *La Hacienda del Antiguo Régimen* (Madrid: Alianza, 1982); Fernando Cos-Gayón, *Historia jurídica del Patrimonio Real* (Madrid: Imprenta de Enrique de la Riva, 1881); Marta Friera Álvarez, *La desamortización de la propiedad de la tierra en el tránsito*

as these assets had an agricultural, livestock or forestry value. This economic aspect of the royal patrimony had begun to gain importance during the reign of Charles III thanks to the influence of Enlightenment ideas.

1 Shaping the Royal Site of La Florida (1787–1807)

The process of configuring this new space began in 1787 when the 8th Duchess of Alba sold part of the land of the Principe Pio hill to the king. This area, which stretched down from the San Vicente city gate and included buildings and land in the Leganitos district, was acquired for the sovereign with the intention of building the royal monastery of San Pedro de Alcántara there. Land belonging to the Duke of Osuna and small private individuals was also incorporated around this time. In the end, however, the monastery project was not realized.[9]

The purchases continued nonetheless, now with the aim of creating a new Royal Site. On 7 July 1792 the Princess of San Gregorio sold the Crown the country estate of La Florida and the Príncipe Pío hill for 1,900,000 *reales*. The money was deposited with the Diputación de los Cinco Gremios de la Villa de Madrid (the association of the five major guilds of Madrid) for 4 years with a return of 3%. A few days later, on 26 July, the Crown acquired the horticultural land and house of La Junquera and Romanillos, which belonged to Doña María Pérez de Castro, for 370,000 *reales*. In addition, at the end of the year, on 17 December, a portion of land 8½ *celemines* in size located in the area where the chapel of San Antonio was being built was acquired from the Hieronymite friars of Madrid.[10]

As part of this purchasing process, the royal treasury allocated a monthly sum of 50,000 *reales* (600,000 *reales* yearly) to works, beginning on 1 January 1792. A general plan showing the land purchased by the king from the Marquise of Castel Rodrigo (*Plan general que muestra el terreno que compró el rey a la marquesa de Castel Rodrigo …*) was likewise drawn up by José de Merlo and Manuel de la Ballina.[11] However, due to the Crown's economic straits, a

del Antiguo Régimen al Liberalismo (*La desamortización de Carlos IV*) (Gijón: Caja Rural de Asturias-Fundación Foro Jovellanos del Principado de Asturias, 2007), and Encarna García Monerris and Carmen García Monerris, *Las cosas del rey. Historia política de una desavenencia* (*1808–1874*) (Madrid: Akal, 2015), 169.

9 Rosario Die Maculet, "Lejos de la Corte. El destierro alicantino de la princesa Pío en la segunda mitad del siglo XVIII," *Revista de Historia Moderna* 30 (2012): 85.

10 AGP, AG, leg. 1257, exp. 13, leg. 1258, exp. 1 and Fernández Talaya, *El Real Sitio*, 16, 47, 59 and 114–19.

11 Fernández Talaya, *El Real Sitio*, 59–60.

TABLE 9.1 Income and expenses of La Florida from 1794 to 1802

Period	Income	Expenses
April 1794–February 1795	275,000	275,079 *reales* and 32 *maravedíes*
May 1795–April 1796	350,000	357,926 and 29
May 1796–April 1797	300,000	418,883 and 18
May 1797–April 1798	300,000	327,705 and 26
March 1798–April 1802	1,200,000	1,509,784 ½

SOURCE: AGP, AG, CAJA 10383, N.D., AND CAJA 10,404, EXPS. 10, 19, 21 AND 23

royal decree of 1 March 1793 halved the allocation to 300,000 *reales* per year (see Table 9.1). This amount failed to cover the ordinary expenses of the wages of the people employed in the tasks of gardening, farming, caretaking, management of works and stables; the gratuities, uniforms and privileges that the king had awarded to employees; the lighting of streets belonging to the Casa del Duende; aid and bonuses granted by royal order; land leases; masses held at Nuestra Señora del Puerto; and notarial and ordinary expenses relating to stables, livestock, tack and carriages.[12]

The purchases continued during the following years as they exceeded the amounts allocated. On 22 February 1795, the *huerta* (horticultural land) of La Moncloa, located on the road to El Pardo, which Manuel de Godoy had acquired not long earlier, in 1792, in exchange for two country houses in Aranjuez (the Cortijo del Rey and the Cortijo de Canosa), was incorporated into La Florida. A few months later, on 25 April 1795, the Granjilla, on the road to El Pardo, San Bernardino and Tejares, among others (see Table 9.2), were acquired from the Hieronymites; they spanned an area of 747 *fanegas* and 140 *estadales* and were valued at 1,679,508 *reales* and 28 *maravedíes*, including the part that the monarch took on 12 February 1793.[13]

12 AGP, AG, caja 10404, exp. 5.

13 Fernández Talaya, *El Real Sitio*, 49. It was acquired in exchange for 4 *censos* at an interest rate of 3%: an annuity of 15,000 *reales* payable by the Principality of Asturias on 500,000 *reales*' capital; an annuity of 18,000 *reales* payable by Valdecorzana on 600,000 capital; an annuity of 6,300 *reales* payable by the Marquise of Hinojosa on 270,000 *reales*' capital; and an annuity of 13,500 *reales* payable by the Marquise of Guadalcázar on 450,000 *reales*' capital. They had also been offered the option of being paid in the form of buildings in Madrid or country estates in Valdemoro. AGP, AG, caja 10404, exp. 15.

TABLE 9.2 Land previously owned by the Hieronymites that was incorporated in 1795[a]

Horticultural and other land belonging to the Hieronymites	**598 *fanegas* and 10 *celemines***	**859.091 *reales* and 24 ⅔ *mrs***
Farmhouse		391,547 and 8
Pipes, mines, manholes and reservoir		48,095
Casilla de los Huertezuelos		4,144 and 8
Pipes, mines and other		9,263
Casa del Pastor		36,798
Casa Nueva, beside San Antonio		38,114 and 8
Casa de la Huerta Grande		59,499
Casa del Lavadero		12,960 and 17
Total		1,459,512 and 29 ⅔ *mrs*

a The land incorporated in 1793 is not included

SOURCE: AGP, AG, CAJA 10404, EXP. 16

Additional incorporations (see Table 9.3) were 9 *fanegas* and 6 *celemines* of land belonging to the Marquise of Castejón in the Migas Calientes botanical garden; 2 *fanegas* and 9 *celemines* of the *huerta* of Los Almendros belonging to the Count of Noblejas; the 3 *fanegas* of the *huerta* of Belén, owned by Don Fernando Ruiz de Alarcón; another 13 *fanegas* belonging to Don Juan Palas; 6 *fanegas* of land acquired from Don José Rivera; 9 *fanegas* acquired from the community of San Martín, 7 from Don Juan Facundo Caballero, as well as from the heirs of Don Bartolomé Valero, 4 *fanegas* from the abbot of Vivanco, 1 *fanega* from the nuns of Santo Domingo and 8 from Don Juan Albalá, among others (see Table 9.3). All this amounted to an outlay of almost 622,900 *reales* in the form of *censos* (annuity payments), cash or *fanegas* of the Dehesa de Amaniel, an area of wood-pasture land owned by the monarch.[14] The incorporations were constant and gave shape to an enclosed space with sizeable agricultural and forestry resources.

On 3 December 1795, 20 *fanegas*, 10 *celemines* and 4 *estadales* of land belonging to the Duchess of Alba beside the Cantarranas stream, valued at 16,758 *reales* and 8 *maravedíes*, were purchased to be incorporated into La Moncloa.[15]

14 AGP, AG, caja 10404, exp. 16.

15 AGP, AG, caja 10404, exp. 28.

TABLE 9.3 List drawn up by José de Merlo in spring 1795 showing the cost of the land, buildings, waterwheels, pipes and reservoirs

Owners	Land and other assets (value in *reales* and *maravedíes*)		Amount
Marquise of Castejón	Land	38,000 *reales*	
	Factory and mines	180,135	291,115
	Pipes	68,004	Paid in *censos*
	Trees and plants	24,976	
Count of Noblejas	Land	11,000	16,740
	Waterwheel and pond	5,515	Paid in *censos*
	Trees	225	
Don Fernando Ruiz de Alarcón	Land	12,000	61,205
	Factory and mines	48,430	Paid in *censos*
	Trees and plants	775	
Don Juan Palas	Land	28,695 and 28	28,695 and 28 Paid in *censos*
Don José Rivera	Land	4,200	4,200 Paid in plots of the wood-pasture land of Amaniel or uncultivated land in Parla
Community of San Martín	Land	5,850	5,850 Land in Amaniel
Don Juan Facundo Caballero	Land	4,900	4,900 Payment with land in Amaniel was proposed, but it was made in *censos*

TABLE 9.3 List drawn up by José de Merlo in spring 1795 showing the cost of the land, buildings, waterwheels, pipes and reservoirs (*cont.*)

Owners	**Land and other assets (value in *reales* and *maravedíes*)**		**Amount**
Heirs of Bartolomé Valero	Land	4,200	4,200 Paid in money as he had debts
Abbot of Vivanco	Land	2,000	2,000 Paid in land in Amaniel
Nuns of Santo Domingo	Land	2,916 and 22	2,916 and 22 Land in Amaniel
Don Juan Albalá	Land	5,200	5,200 Land in Amaniel
Don Pedro Bezinduaga	Land	2,750	2,750 Land in Amaniel
Don Diego Ordóñez	Land	2,100	2,100 Land in Amaniel
Don Francisco Benavides	Land	4,500	4,500 Land in Amaniel
Pedro Alcalde	Land	2,500	2,500 Land in Amaniel
Marqués de Corbera	Land	7,600	7,600 Paid in *censos*
Don Cayetano de Urbaneja	Land	3,239 and 6	3,239 and 6 Paid in land in Amaniel

Fruit/vegetable farmer and lessors			
Francisco Santos	Riverside produce and land	300	52,271 and 25
	Half of the Hieronymites' shed	650	Paid in money due to debts
	Produce and equipment of the Hieronymites' *huerta* (horticultural land)	48,766 and 25	
	Trees of the Hieronymites' *huerta*	595	
	Fruit and other produce of the Huerta de la Condesa	1,960	
Marcelo Laurel	Trees and plants of the Hieronymites' *huerta*	544	30,007 and 8
	Produce, livestock and equipment	27,513 and 8	Paid in money due to debts
	A shed, half of others, a pen and one part	1,950	

TABLE 9.3 List drawn up by José de Merlo in spring 1795 showing the cost of the land, buildings, waterwheels, pipes and reservoirs (*cont.*)

Owners	Land and other assets (value in *reales* and *maravedíes*)		Amount
Santiago Olmedilla	Produce, livestock and equipment of the Huerta de Alarcón	10,974 and 17	11,425 and 17 In land in Amaniel and part in money due to debts
	Trees	451	
Jacobo Rivas	Produce, livestock and equipment of the Huerta de Castejón	7,002 and 17	7,002 and 17 Paid in money to settle debts
María Teresa del Hoyo	Produce of the Huerta de Noblejas	6,844 and 17	26,629 and 17 Paid in money to settle debts the owner had
	Trees	114	
	House and factory of the Huerta de Noblejas	19,675	
Manuel Losada	Produce of Pedro Alcalde	75	1,597 and 17 Paid in land in Amaniel
	Produce of Bezinduaga	82 and 17	
	Produce of Juana Albalá	1,440	
Mauro López	Produce of San Martín	1,620	2,745 Paid in land in Amaniel
	Produce of Palas	990	
	Produce of Don Diego Ordóñez	135	

Fernando Sánchez	Produce of Bartolomé Valero	980	980 Paid in money
Don Francisco Garralda	Produce of Vivanco abbey	100	125 Paid in money
	Produce of the nuns of Santo Domingo	25	
Widow of Tomás Espinosa	Produce of Francisco Benavides	405	405 Paid in money
Discalced Carmelites	*Censo* taken out on the botanical garden by the Marquise of Castejón	20,000	20.000 Paid in *censos*
Payments for appraisers and deeds		20,000	20,000
Total			622,901 All in *censos* except for 51,923 *reales* and 28 *maravedíes* in plots of the king's land in Amaniel and 145,620 and 33 in cash

SOURCE: AGP, AG, CAJA 10404, EXP. 16

By the end of 1795, the Crown had thus spent more than 5,200,000 *reales* on buying land for the territorial configuration of the Royal Site of La Florida.[16]

The process of acquiring arable land continued in the following years. For example, in 1796, 6 *fanegas* of land appraised at 4,200 *reales* and belonging to José García de Rivera, a farmer resident in Parla in the area known as Las Cruces de San Bernardino, were incorporated. The following year saw the addition of some 10 *fanegas* of arable land belonging to the Seminary of Nobles beside the San Bernardino gate, acquired from the Marquis and Marquise of Carpio on 23 March 1730.[17] A particularly notable incorporation was made on 23 July 1802: horticultural land (the Huerta de Alba) purchased from the Duchess of Alba's heirs, together with the Bordador plots, for 2,830,313 *reales* and 7 *maravedíes*, half in cash and the other in public debt securities (*vales reales*).[18] Nine *fanegas* of farmland on the Pimiento hill (the site of what is now the Hospital Clínico) were acquired that December, and land stretching from the Huerta de Alba to the boundaries of El Pardo was purchased in 1803.[19] A further 30 *fanegas* and 4 *celemines* of land that was part of the Dehesa de Amaniel were likewise incorporated shortly afterwards.[20]

The new royal site accordingly took shape between 1787 and 1804. It was largely productive and, as shown in Table 9.3, predominantly agricultural, in accordance with the new rationale applied to the royal sites, in which economic criteria prevailed. When its configuration was complete, in January 1804 it was decided to fence this space at a cost of 1,749,546 *reales*, though the French occupation prevented it from being completed:[21]

> la grandeza de espíritu y delicado gusto del Sor Rey Carlos Quarto en las obras q proyectaba, se manifiestan bien claramente en las Posesiones del Rl Sitio de la Florida, pues a no haber ocurrido las novedades y vicisitudes de tiempos tan calamitosos que sobrevinieron, llevando aquellas a la debida perfección, hubiera perpetuado su memoria quedando un monumento hasta los siglos más remotos. Pero angustiado y poseído

16 AGP, AG, caja 10404, exp. 16.

17 AGP, AG, caja 10404, exp. 22. The process of incorporating this land triggered conflicts with the parish churches that were entitled to tithes on them, which were no longer paid when they became royal property. This was the case of the parish church of San Martin, in 1798. AGP, AG, caja 10404, exps. 17 and 24.

18 Fernández Talaya, *El Real Sitio*, 179. AGP, AG, caja 10405, exp. 20.

19 AGP, Reinados, José I, caja 66, exp. 11.

20 Agustín Gómez Iglesias, "La Dehesa de Amaniel o de la Villa," *Anuario del Instituto de Estudios Madrileños* 2 (1967): 33–81.

21 AGP, AG, caja 10405, exp. 31.

> de los más tiernos sentimientos, considerando a la Nación en un estado deplorable, no solo mando la suspensión del trabajo, sino que dispuso al mismo tiempo se aprovechasen muchos de los materiales que había prevenidos, sirviesen para las obras más precisas y cerrar la posesión, que no se acabó de ejecutar.[22]
>
> The greatness of spirit and exquisite taste of His Lordship King Charles the Fourth in the works he planned are clearly visible in the Properties of the Royal Site of La Florida. For had it not been for the changes and vicissitudes of the calamitous times that befell him, bringing them to completion would have perpetuated his memory in a monument that would have remained for centuries to come. But in his anguish, and possessed of the most tender feelings, considering the Nation to be in a deplorable state, he not only ordered the work to be suspended but also arranged for many of the available materials to be used for the most necessary tasks and to enclose the property, which was not completed.

The process of fencing royal spaces became common practice from the time of Ferdinand VI. Stemming from reasons of security and control, it was also a money-saving measure as it prevented animals from straying beyond the confines of the royal estate and causing damage – for which the Crown was liable – to the surrounding land. Furthermore, fencing created an enclosed and private royal space over which the Crown exercised full rights.

2 The Productive Value of the Royal Site (1807–1814)

2.1 *The Situation of the Royal Site in 1808*

The economic significance of this new royal space was perfectly clear from the beginning of the 1800s, when unirrigated land, plots of vineyards and vegetable gardens/orchards began to be put into production. This was made possible by the proximity of the Manzanares river and hydraulic structures already in place such as pipes and waterwheels (see Table 9.3), to whose maintenance various items of expenditure were allocated, as shown in Table 9.5 under the heading 'plumbing expenses' (*gastos de cañería*). For instance, in 1808, according to a report by Nicolás Thelis and Florencio Martín dated 6 August 1814, La Florida had an income of 693,877 *reales* and 14 *maravedíes* (see Table 9.4) from

22 AGP, AG, caja 10405, exp. 31.

TABLE 9.4 Income of La Florida in 1808 (in *reales* and *maravedíes*)[a]

Farm products	*Revenue*	*Total*
Harvest (8,815 *fanegas* of all kinds of grain; unspecified but probably mainly wheat, barley and oats)	330,843	
Straw (31,212 *arrobas*)	46,818	405,711
Land leased for broad beans and peas	10,300	
Sale of unusable mules and donkeys	9,850	
Produce from the Bordador vineyard	4,000	
Products of the henhouse and dovecote	3,900	
Livestock products		
Cattle	12,609	
Sheep	45,429	62,505
Pigs	4,467	
Garden produce		
From La Florida or Montaña	30,415 and 16	
From La Moncloa and Huertezuelos	64,332 and 16	224,439 and 44
From El Paso and Belén	66,136 and 20	
From Alba	63,554 and 30	
Product of the beehive	1,222	1,222
Total		693,877 *reales* and 14 *mrs*

a Another list included a few changes: for instance, the amount of straw obtained from La Florida increased by 5,000 *arrobas*, boosting income by 6,000 *reales*, and that of the Huerta de La Moncloa by 1,222 *reales*. This brought the total income up to 701,099 *reales* and 31 *maravedíes*

SOURCE: AGP, AG, CAJA 10405, EXP. 32

agricultural and garden produce and livestock commodities (major irrigation systems were installed during those years: pipes, waterwheels, etc., to ensure a sufficient water supply); this is not counting the amounts already earmarked and ordinary expenses of 551,724 *reales* and 11 *maravedíes*, plus 141,265 for wages, leaving a profit of 888 *reales* and 3 *maravedíes*.[23]

23 AGP, AG, caja 10404, exp. 4.

TABLE 9.5 Expenses of La Florida in 1808 (in *reales* and *maravedíes*)

Expenses of the royal farmhouse	
Salaries	50,005
Extraordinary days' wages	134,604
Fodder for 83 mules and 11 donkeys	41,219
Livestock expenses	
Salaries, fodder, equipment and expenses for the cows	13,837
Salaries, shearing, equipment and maintenance of dogs for the sheep	35,450
Salaries, bran and maintenance of a dog for the pigs	14,946
Expenses of the vegetable gardens/orchards	
Salaries, days' wages, piecework and equipment for La Florida	59,180 and 16
Salaries, days' wages, piecework and equipment for La Moncloa and Huertezuelas	56,131 and 17
Salaries, days' wages, piecework and equipment for El Paso and Belén	55,347 and 30
Salaries, days' wages, piecework and equipment for Alba	69,788 and 16
Plumbing expenses	
Days' wages and materials	21,215
Total	551,724 and 11
Salaries related to auditing, revenues, guards, widows, orphans and retired employees, and works	141,265
Total	692,989 and 11

SOURCE: AGP, AG, CAJA 10405, EXP. 32

The economic and productive significance of the Royal Site of La Florida grew when riparian woodlands (the *sotos* of Aldovea, Daralcalde and Viveros) and other land acquired by Charles IV in the area of the Henares and Jarama rivers were brought under the administration of La Florida. These royal spaces continued to be managed by the same administrator, Don José de Merlo, until the minister Azanza separated them in April 1808.[24]

24 Nevertheless, in June 1809 Joseph I brought them together again under the direction of Don Ramón de Alvarado, who remained in the post until he was dismissed, at which point they were separated again. On 7 June 1813, following the departure of the French, the intendent was asked if he wished to again combine them under the same management. AGP, AG, caja 10405, exp. 22.

Even so, attempts were made to further reinforce the economic nature of the Royal Site of La Florida during those years through a plan submitted by Don Juan Blesa.[25] It envisaged planting 200,000 fruit trees given the characteristics of the land and the abundance of channelled water resources (running water or *agua de pie*). This would require little spending and would make 'this property delightful, joyful and pleasant and a stimulus for HM's enjoyment and full relaxation'.[26] Accordingly,

> en el plano del Camino Nuevo y falda de la Montaña, tomando por la derecha del Tejar hasta la gruta del palacio que fue de Pío, otras 4 carreras de arbolado y lo mismo en el declive que forma el terreno hasta que descansa en la huerta. Igual operación podrá practicarse en toda la extensión que se advierte desde el Tejar y Huerta de las Minas, formando con el tiempo las parras caminos cubiertos, procurando que los sarmientos sean de buenas calidades. Estos tres plantíos tienen agua de pie para su riego hasta que prendan.[27]
>
> On the level ground of the New Road and the side of the Hill, to the right of El Tejar as far as the grotto of the palace that belonged to Pío, another 4 rows of trees [can be planted] and likewise on the slope formed by the land running down to the vegetable garden. A similar operation can be carried out throughout the entire expanse of land visible from El Tejar and Huerta de las Minas, with the vines forming covered paths over time, endeavouring that the vine shoots are of good quality. These three plantations have running water to irrigate them until they take root.

To boost the site's income, the plan stated that the San Gil brick factory, which had 4 kilns, should supply to the royal site and the proceeds from the sale of between 50,000 and 60,000 tiles produced there annually be incorporated into its revenues.[28] It also recommended stepping up milk production by giving the cows fodder of salted broad beans and oat, barley and carob straw at night. This milk could be sold at the San Vicente gate and at the fusiliers' barracks. It is important to note that while the production of this dairy cattle was upped, that of Aranjuez was then having problems.

---

25 AGP, AG, caja 10405, exp. 31.

26 'risueña, alegre y placentera esta posesión y estímulo para la diversión y puro desahogo de S.M.'

27 AGP, AG, caja 10405, exp. 31.

28 AGP, AG, caja 10405, exp. 31.

Money could also be made by bringing 600 goats to the pastures of Príncipe Pío hill, which would generate an additional 50,823 *reales* and 28 *maravedíes* in income. Blesa stated that the breeding of black pigs from Extremadura could be started up with 20 sows and could lead to an income of 70,000 *reales*. The report gives an idea of how this area was viewed from a purely economic perspective. Revenues would be obtained not only by maximising the available agricultural, horticultural and livestock resources and improving crops and produce but also by introducing animal species. These modifications would result in the consolidation of a multifunctional space and a landscape that could bring aesthetic and leisure benefits for the royal family by making it 'delightful, joyful and pleasant and a stimulus for HM's enjoyment and full relaxation.'[29]

Another means of increasing income was to lease the vegetable gardens/orchards and plots suitable for growing cereals (*tierra de pan llevar*) that were inside the fence, reserving the right of way and the buildings for the monarch. This meant that 60,000 *reales* could be obtained from leasing the horticultural and arable plots of La Florida stretching from the lower pond to the dividing wall of the Areneros road, including the Huerta del Platero and Pío gardens; 15,000 from the croplands inside the wall, from the Plazuela de San Antonio to the Huerta de Marcelo Laurel; another 60,000 from the Huerta de Marcelo Laurel and La Chispa; 60,000 from the botanical garden and Huerta de La Moncloa; 40,000 from the Huerta de Alba, excluding the Dehesa de la Villa; and another 34,000 from the hills below the Dehesa de la Villa. The arable lands of La Granjilla and the Huerta de Alba were not included in this project. It was reckoned that the total revenues would amount to 641,303 *reales* and 28 *mrs*.[30] The envisaged sums were high and somewhat unrealistic, as according to a report of 1814 income from leases of arable and horticultural plots on La Florida came to a total of 93,000 *reales* (Table 9.6).[31]

Charles IV's abdication in favour of his son Ferdinand VII early in 1808 did not lead to any changes in the economic value of the royal site. During the short time Ferdinand remained in power (3 months), an economic approach to the royal patrimony – the rationale behind the running of La Florida – was maintained. On 22 March 1808 the monarch announced a reduction in his hunting grounds and gave instructions for plots of land owned by him to be leased

29 '[siendo] risueña, alegre y placentera esta posesión y estímulo para la diversión y puro desahogo de S.M.', AGP, AG, caja 10405, exp. 31.

30 The project was never implemented on account of the French occupation. AGP, AG, caja 10405, exp. 31.

31 AGP, AG, caja 10404, exp. 4.

TABLE 9.6 Value of leased land[a]

	Lots	*Fanegas*	Price per *fanega*	Value
Príncipe Pío hill		120		25,000
	1	196 ½	50	8,580
	2	60 ½	40	2,289
	3	83	38	2,905
	4	90 ½	50	7,000
	5	105 ½	36	4,820
	6	114	36	3,762
	7	103 ½	40	4,000
	8	87 ½	38	4,260
	9	98	38	
	10	86	38	6,936
	11	101	36	
Huerta de la Moncloa		18	800	7,000
Huerta del Paso		18	800	6,000
Huerta de Alba		18	800	7,000
Huerta del Botánico		3 ½	800	1,220
Huerta de Noblejas		3	800	770
Huerta de Belén		2	800	770
Los Huertezuelos and annexed plots		3 ½	600	600
Vineyard on La Moncloa hill with 100 olive trees and 1,000 vines				688
Total				93,600

a The differences with respect to the final value in some cases are due to the fact that not all *fanegas* of land produced the same and were worth the same.

SOURCE: AGP, AG, CAJA 10404, EXP. 4

out so that they would be put to cultivation.[32] The Crown was still in need of money, given the debt incurred as a result of the alliance with Napoleonic France, which required it to pay the sum of 6 million livres a month from the signing of the Peace of Amiens in order to guarantee neutrality. Payments were maintained after Spain joined in the war in 1804.

Months later, on 14 November 1808, the secretary of state and finance in exile was entrusted on behalf of the supreme government committee with

32 AHN (Archivo Histórico Nacional), Consejos, lib. 1504, no. 28.

drawing up a set of regulations on the exploitation of game and meadows on the royal sites 'to produce a proportionate benefit for the royal treasury'[33] after Ferdinand VII had lifted the ban on hunting in the royal preserves. The committee also agreed that this secretary's office would be responsible for drawing up economic plans to curb expenditure and boost the income of the royal sites.[34] On 23 November the clerk of the counting-house (*veedor*) at Casa de Campo accordingly submitted a report to the committee on 'improving the use of the lands hitherto devoted to hunting [...] By giving each piece of land of those hitherto devoted to the leisure activity of hunting a more profitable use'.[35] La Florida was to be an example of this economic management and of the use of land for agricultural and livestock purposes.

2.2 *The Value of the Royal Site under the French Government (1808–1813)*

The advent of the French merely increased the economic value of this royal property, as Title IV of the Statute of Bayonne (articles 21 to 24) considered the royal estates to be fundamental to the maintenance of the king and his royal household.[36] The royal patrimony would be used to obtain the resources required by the Crown Treasury to cover its needs. A distinction was thus drawn for the first time between state revenue and royal revenue.

The new administration needed to be apprised of the situation of the royal patrimony in order to to implement measures to enable it to comply with the provisions of Bayonne. Accordingly, at the beginning of July 1808 the French administrators drafted a report on the situation of La Florida in which they also described the staff of officials and indicated that it provided the Crown with an annual profit of 36,000 to 37,000 *reales*.[37] This report sprang from the logic of gathering information in order to subsequently implement the measures that were deemed necessary. Shortly afterwards, a report was drawn up on the state of the 1808 harvest yielded by the unirrigated lands, which, as can be seen in Table 9.7, produced both cereals and grains for human food and straw for animal consumption. At the end of the month, on 25 July, Joseph I commissioned Stanislas Girardin d'Ermenonville and Jean-Baptiste Maximilien Villot

33 'produzca una utilidad proporcionada al real erario'.

34 AHN, Estado, leg. 22A, no. 10.

35 'el mejor aprovechamiento de los terrenos consagrados hasta aquí a la caza [...] Dando a cada terreno de los hasta aquí consagrados a la diversión de la caza un destino más provechoso', AGP, AP, Casa de Campo, leg. 15.

36 Félix Labrador Arroyo, "Desamortización o reforma. Los Sitios Reales en un contexto de cambios (1790–1814)," *Memoria y Civilización* 25 (2022): 224.

37 It stated that there were 1,000 sheep and rams, 40 pigs, 21 cows, 14 calves, 72 mules, 2 horses and 6 asses. AGP, Reinados, José I, caja 66, exp. 11.

TABLE 9.7 List drawn up by Florencio Martín and José Miranda of crops harvested from unirrigated land in 1808

	Fanegas of arable land	*Fanegas* of grain	*Fanegas* produced	Straw in *arrobas*
Durum wheat	193 and 4	198 and 1	2,425 and 6	10,915
Beardless wheat	99 and 7	98 and 8	755	3,898
Barley	229 and 6	424 and 2	4,928	14,784
Rye	19 and 6	11	227	650
Oats	22 and 6	32 and 6	279 and 6	560
Chickpeas	34 and 6	24	170	255
Broad beans	4	8 and 6	38	50
Grass peas	9	10	75	80
Peas	3	2 and 6	17	20
Total	614 and 11	809 and 5	8,915	31,212

SOURCE: AGP, AG, CAJA 10405, EXP. 12

de Fréville, in whom the new monarch fully confided, to ascertain the value of the Crown's real estate and properties.[38]

Joseph I's departure from Madrid on 31 July 1808 following his defeat at the battle of Bailén hindered any further progress in the enquiries about the situation of the royal site. The court did not return until the beginning of December 1808. During the days that ensued the Duke of Campo-Alange, as master of the horse, asked the officials in charge of the royal sites for detailed information about the products, revenues, liabilities and staff of each one.[39] The superintendent of the royal household and patrimony, the Count of Melito, also requested reports. On 22 January 1809 Florencio Martín, acting as administrator in lieu of José Merlo, who had gone to Bayonne with Charles IV, completed a detailed statement of the properties and annexed plots belonging to La Florida estate, as well as its products and annual expenses, credits, debts,

38 The seventh article of the decree they received requested them, after ascertaining the situation of the Crown patrimony, to state the value of its properties to determine whether this figure attained the annual amount of a million *pesos fuertes* stipulated by art. 21 of the Statute. AGP, Reinados, Fernando VII, caja 308, exp. 22.

39 AHN, Estado, leg. 3101.

liabilities and stocks.[40] Antonio de Mollinedo submitted information on El Pardo on 25 January 1809, and the steward of El Escorial, Francisco Carmona, did likewise on 30 January. For his part, the chief official (*alcalde*) of Aranjuez, Domingo Gaspar Pérez, stated on 5 February that the economic situation of that royal estate was disastrous, and José Manglano, the governor of San Ildefonso, reported on 21 February that the royal estate barely generated any income as the pinewoods were administered directly by the royal treasury.[41]

The situation described in the reports that continued to be sent from La Florida was not very encouraging. The events of 3 and 4 December, when the French again surrounded the capital following their victory at Guadarrama, had caused significant losses to the royal site. According to a list submitted on 18 March 1809, these amounted to 290,000 *reales*' worth of animals and 329,810 *reales* for the theft of produce from the farmhouses and granaries: 270 *fanegas* of wheat, 3,400 of barley, 150 of oats, 130 of rye, 38 of broad beans, 17 of peas, 60 of grass peas, 130 of chickpeas and 25,000 *arrobas* of straw, as well as horticultural produce (very substantial figures considering that year's harvest, as well as the income generated by the Royal Site of Florida, as can be seen in Table 9.4, as they almost equalled its annual income).[42] It was also pointed out in this report that the royal estate's cultivated land comprised 1,100 *fanegas* under the responsibility of the head gardener, with the help of 5 foremen or overseers, 32 individuals on daily wages and various officials of the vegetable gardens/orchards and hands, as required.

Shortly afterwards, on 22 March 1809, a report was submitted on the state of the products and stocks of the royal site from 1 December 1808 to 18 March 1809. This report on the first three and a half months of French administration indicated that there was 66,356 *reales* and 8 *maravedíes'* worth of agricultural produce on the royal site and 33,912 *reales* and 16 *maravedíes* in garden produce, while the expenditure on daily wages during these months totalled 67,519 *reales* and 20 *maravedíes* for the arable land and 29,491 *reales* and 33

40 AGP, Reinados, José I, caja 66, exp. 11.

41 Juan Mercader Riba, *José Bonaparte. Rey de España. 1808–1813. Estructura del estado español bonapartista* (Madrid, CSIC: 1983), 76–8.

42 The list does not mention the harm caused by the lack of two covered wagons, two carts, a great deal of equipment and farm implements or the damage to the buildings. Lettuces, curly endives, cabbages, celery, beans, aubergines, tomatoes, cucumbers, brussels sprouts, red cabbages and broccoli were grown in the vegetable gardens of La Florida. Summer and winter pears, apricots, peaches, cherries, morello cherries, strawberries and olives were also harvested. AGP, Reinados, José I, caja 65, exp. 7.

maravedíes for the horticultural land. This report did not reflect the lost income resulting from the events of 3 and 4 December.[43]

Slightly later, on 26 March 1809, the administrator Florencio Martín submitted a list of employees with their date of incorporation and their monthly and annual wages and the names of those who were on duty on 3 December 1808, when the French arrived.[44] The staff responsible for inspections at the royal property were the comptroller and 2 clerks, as well as 2 physicians and a surgeon. For gardening there were a senior gardener, 8 gardeners and 2 purveyors of fruit and flowers. Guard duties were performed by a deputy forester with the rank of captain, a corporal, 12 soldiers, a doorkeeper and 4 guards at the gates (18). The works branch comprised an architect, a works administrator, 3 clerks of works for masonry, stonework and carpentry, a plumber, an assistant, a hand, two senior overseers and two overseers (12). The royal chapel, which was in the parish church of San Antonio, had 2 deputies, a priest sacristan, an acolyte and a washerwoman and ironer. In total, 52 people were employed in the upkeep and improvement of the Royal Site of La Florida, in addition to those who were hired for specific tasks. This gives an idea of the importance of this space and the Crown's determination to improve it.

The steady flow of information from La Florida to the palace continued, always on economic matters. For example, on 6 April 1809 Florencio Martín completed a list of the horticultural and agricultural produce from April 1808 to the end of March 1809, pointing out the 329,810 *reales*' worth of produce that was lost in December, as well as the expenses incurred, which totalled 660,517 *reales* (Table 9.9).[45] Likewise, La Florida had a monthly allocation of 25,000 *reales* from the general treasury for the payment of wages, works, materials and repairs, but since April 1808 only 6,000 had been paid.

Also on 6 April Ramón Alvarado was appointed administrator of the royal site. The following day, together with Florencio Martín and the overseer Pedro Gumucio, he inspected the animals there. The visit revealed a significant reduction in their numbers. Of the just over 1,150 mentioned in the report of 1 July 1808, there were now 330: 80 mules, 11 donkeys, 215 sheep, 11 cows, 1 bull, 7 goats and 3 kids.[46] On the 9th Alvarado submitted a list of employees at La Florida whose posts could be done away with in order to streamline the administration.[47] Not long afterwards, however, following the abolition of the vegeta-

43 AGP, Reinados, José I, caja 65, exp. 7.

44 AGP, Reinados, José I, caja 66, exp. 11 and AG, caja 10405, exp. 21.

45 AGP, Reinados, José I, caja 66, exp. 11.

46 AGP, Reinados, José I, caja 66, exp. 11.

47 AGP, Reinados, José I, caja 66, exp. 11.

TABLE 9.8 Income from the produce of La Florida (April 1808–March 1809, in *reales* and *maravedíes*)[a]

Garden produce		
From La Florida	21,531 and 16	
From La Moncloa	55,505 and 16	188,208
From El Paso and Belén	54,338 and 20	and 32
From Alba	56,834 and 14	
Agricultural produce		
2,111 *fanegas* of durum wheat at 44 *reales* per *fanega*	92,884	
485 *fanegas* of beardless wheat at 28 *reales* per *fanega*	13,580	
74 *fanegas* of rye at 18 *reales* per *fanega*	1,332	126,508
17 *fanegas* of chickpeas at 110 *reales* per *fanega*	1,870	
5,352 *arrobas* of straw	12,842	
Produce of the vines	4,000	
Livestock produce		
From the sheep	33,288	
From the pigs	4,467	52,608
From the cattle	10,953	
From the dovecote and henhouse	3,900	
Losses, thefts, etc.	*329,810*	
Total		697,134 and 32

a AGP, Reinados, José I, caja 66, exp. 11.

ble gardens and orchards of the Casa de Campo estate, orders were given on 17 April for the related officials to be incorporated into the staff of La Florida.[48]

Nevertheless, Alvarado continued to call for a reduction in the workforce at La Florida. For instance, on 12 May 1809, when sending superintendent Melito a new set of regulations for Soto de Aldovea where he was interim administrator, he commented 'if only a similar economy were established at La Florida!'.[49] Alvarado's interest in improving the administration clashed with his manner of keeping the accounts. In September the counting-house warned Melito that those submitted for La Florida were not in order. As a result, he had to step

48 AGP, Reinados, José I, caja 69, exp. 8.

49 '¡ojalá que se estableciera en la Florida igual economía!', AGP, Reinados, José I, caja 65, exp. 11.

TABLE 9.9 Expenses of La Florida (1808–9, in *reales* and *maravedíes*)

Salaries paid until mid-November 1808	161,502 and 17
Bonuses and days' wages of labourers hired on a daily basis and extraordinary farm wages	51,883
Weeding, clearing of fallow land and purchased shrubs	9,487 and 17
Harvesting and fruit picking expenses	41,748
Stables and other expenses	6,230
70 *fanegas* of beardless wheat and 12 of carob purchased for sowing	3,608
Blacksmith's and cartwright's accounts paid up to December	19,714 and 11
Equipment purchased and tools	2,547
Livestock expenses	41,824 and 32
Bonuses, piecework and daily wages of officials and hands up to December in the vegetable gardens/orchards	124,669 and 17
Accounts paid to the cartwright and blacksmith for the vegetable gardens/orchards	20,812 and 33
Stables expenses for all the vegetable gardens/orchards	11,430 and 22
Total	495,458 and 23
4 months' expenses from December to March, and fodder for the livestock up to the end of July, and what was owed of it in March	165,059 and 14
Total	660,517 and 37

SOURCE: AGP, REINADOS, JOSÉ I, CAJA 66, EXP. 11

down from his post on 26 October 1809 and Messrs Ardena and Pardo were entrusted with inspecting La Florida.[50]

Before resigning, Alvarado reported that stocks in the storehouses of the royal site amounted to 80 *arrobas* of chickpeas for sowing, 60 *fanegas* of grass peas for sowing and a month's worth of straw and barley for the working animals used in horticultural and agricultural production. On 19 July he wrote to the Count of Melito in response to the request for information about the harvest of the royal site of La Florida. He reported that he expected to gather 6,000 *fanegas* of barley and 3,000 of durum and beardless wheat,[51] which

50 AGP, Reinados, José I, caja 56, exp. 15. A summary of the file in exp. 12.

51 At the time the price of barley in Madrid was 13 and a half or 14 *reales* per *arroba*, so that the harvest was worth 81,000 *reales*, whereas durum wheat was worth 38 *reales* per *arroba*, amounting to 112,000 *reales* (there was no mention of the price of beardless wheat).

could fetch about 170,000 *reales*, leaving a portion to cover the needs of the royal site; double the amount could be obtained if they held on to it and sold it in winter. However, during these months consideration was being given to the possibility of selling La Florida and other royal sites to raise money for the royal coffers. Alvarado tried to demonstrate the site's economic capacity and the advisability of keeping it in the royal patrimony. He wrote to Melito: 'I repeat to Your Excellency my sadness at perhaps seeing the glory that would surround me disappear, yielding a net profit of 200,000 *reales* even in the first year of my administration, when it was necessary to buy everything'.[52]

Over the course of 1809 an area of 296 *fanegas* of land, 1 *celemín* and 25 *estadales* was sown with 320 *fanegas* and 6 *celemines* of wheat at La Florida; a further 198 *fanegas*, 1 *celemín* and 26 *estadales* were sown with 332 *fanegas* of barley at La Florida; 25 *fanegas* were sown with 18 *fanegas* of carob; 8 *fanegas* with 4 *fanegas* and 6 *celemines* of rye; 28 *fanegas*, 1 *celemín* and 6 *estadales* with 20 *fanegas* of oats; and a further 15 *fanegas* were sown with grass peas, 2 *fanegas* and 6 *celemines* with peas, and 40 *fanegas* with chickpeas. These figures were provided by a report of 5 February 1810 drawn up by the new administrator, Don José Álvarez del Valle, who had been second in charge of the general counting house of the royal stables and interim overseer of the royal woodlands of the Casa de Campo estate.[53]

Alvarado's efforts to curb expenditure and boost revenues do not appear to have achieved the desired effects. Income from the sales of the royal site's produce diminished, as a result of which it was not possible to pay the wages for November and December 1809. To increase revenues, on 24 December the Palace issued a general instruction stating that the auditors-administrators of the royal patrimony were to ensure the 'decency, decorum and subsistence' (*decencia, decoro y subsistencia*) of their administrations. It likewise gave orders that reforms carried out in any of them should be justified by the 'splendour befitting its magnificence in order that it is not only in good repair from now on but produces substantial revenues for the Crown allowance, as provided for in the Constitution'.[54]

52 'repito a V.E. mi dolor de ver desaparecer tal vez la gloria, que me rodearía, dando un beneficio neto de 200,000 reales aun en el primer año de mi administración, en que ha sido menester comprar todo', AGP, Reinados, José I, caja 65, exp. 19.

53 AGP, Reinados, José I, caja 66, exp. 6. The expense of harvesting these grains was 56,156 *reales* and 22 *mrs* (5,500 for weeding, 22,104 for sowing, 22,500 for threshing and other related expenses, and 6,052 and 22 for separating the grain from the chaff). AGP, Reinados, José I, caja 31, exp. 2.

54 'esplendor propio de su magnificencia, para no solo esté reparado en adelante, sino que produzca rentas quantiosas para la dotación de la Corona, según lo prevenido en la

This new instruction was to be implemented by the new administrator of the royal site, Don José Álvarez del Valle, who attempted to augment its income and reduce expenses. However, production decreased. This was clearly stated in the reports of 27 July and 3 August 1810, which indicated that the sale of agricultural produce, except for the *fanegas* kept for sowing, would fetch just over 243,000 *reales*, which could be increased to 300,000 if they waited until the market supply was lower to sell; the addition sum of nearly 300,000 *reales* could be earned from the sale of the produce of the vegetable gardens and orchards, livestock and the lease of 100 *fanegas* of land. In other words, it was hoped to obtain more than 544,000 *reales*,[55] a substantial sum but well below the nearly 700,000 earned in 1808, equivalent to a fall in income of 77.71% (reference year 1810) (Table 9.10). Nevertheless, the administrator was able to send 83,346 *reales* and 4 *maravedíes* to the general treasury for the Crown's maintenance (Table 9.11).

La Florida produced some 7,915 *fanegas* of wheat, barley, rye, oats, chickpeas, broad beans, grass peas and carob beans, as well as 25,000 *arrobas* of straw, which could fetch 239,820 *reales*, not including the produce of the vegetable gardens and orchards. These amounts, despite having decreased with respect to previous years, showed that the property was still one of the most productive Crown estates. Indeed, even compared to 1808, with the exception of the livestock, the increases were significant and demonstrated that the changes implemented were having the desired effect. Indeed, they were higher than the 6,000 *fanegas* of wheat and barley from the royal site of Gózquez, which fetched 198,000 *reales*, and the 157,500 *reales* earned from the 6,550 *arrobas* of wheat, barley, oats and broad beans from San Fernando-Aldovea.[56]

In order to improve the administration of the royal site and boost its income, it was decided to appoint the administrator of El Pardo as its director. Accordingly, on 5 August 1811 a letter was written to Pedro Bouchard to inform him that 'the king having learned that it was advisable, under the current circumstances, for the royal properties of El Pardo and La Florida to be brought under the same administration',[57] he was to take charge of both, although he was instructed to keep separate accounts for each of the two sites. Bouchard was also entrusted with curbing the expenses of the officials, being asked to make staff cuts to keep expenditure down to no higher than 22,770 *reales* a

Constitución', ARAS (Archivo Real Alcázar de Sevilla), caja 405, exp. 6.

55 AGP, Reinados, José I, caja 31, exp. 2.

56 AGP, Reinados, José I, caja 31, exp. 1.

57 'haviendo conocido el rey que conviene en las circunstancias del día que estén bajo una misma administración los reales dominios del Pardo y de la Florida'.

TABLE 9.10 Calculation of the income of La Florida (1810, in *reales*)

Produce	*Fanegas*	Value at normal prices	Value at high prices
Wheat	2,960 *fanegas* at 40 *reales*	118,400	148,000
Barley	4,200 *fanegas* at 15 *reales*	63,000	84,000
Rye	40 *fanegas* at 18 *reales*	720	960
Oats	230 *fanegas* at 10 *reales*	2,300	2,760
Chickpeas	25 *fanegas* at 200 *reales*	5,000	6,000
Peas	50 *fanegas* at 60 *reales*	3,000	3,000
Grass peas	130 *fanegas* at 46 *reales*	5,980	7,800
Carob beans	280 *fanegas* at 14 *reales*	3,920	4,480
Straw	25,000 *arrobas* at 1½ *reales*	37,500	37,500
Vineyards		4,000	4,000
Total		243,820 *reales*	298,500
Vegetables from the Alba garden		54,735	
Vegetables from the Moncloa and Huertezuelos gardens		53,800	
Vegetables from the El Paso and Belén gardens		43,402	
Vegetables from the La Florida garden		38,572	
Fruit and grapes		10,000	
Milk and lard		15,000	
Goats' milk, sale of lambs, wool		15,000	
Pigs sold and for breeding		7,000	
Lease of 100 *fanegas* of arable land		12,000	
Total		544,009	

SOURCE: AGP, REINADOS, JOSÉ I, CAJA 31, EXP. 2

TABLE 9.11 Income and expenses of La Florida in 1810 (in *reales* and *maravedíes*)

Month	Income	Expenses	Balance
January	25,353	13,524 and 26	11,828 and 8
February	18,542	14,625 and 17	3,916 and 17
March	18,440 and 20	19,551 and 31	−1,111 and 11
April	19,509 and 13	13,866 and 19	5,642 and 28
May	26,951 and 20	23,218 and 32	3,732 and 22
June	40,917	24,524 and 8	16,392 and 26
July	21,945 and 12	21,843 and 23	101 and 23
August	47,589 and 18	52,390 and 20	−4,801 and 2
September	67,498 and 4	30,766 and 17	36,731 and 21
October	37,895 and 1	29,355 and 28	8,539 and 7
November	17,629 and 4	29,282 and 23	−11,653 and 19
December	32,628 and 18	17,948 and 4	14,680 and 14
Total	374,948 and 6	290,899 and 10	84,048 and 30

SOURCE: AGP, REINADOS, JOSÉ I, CAJA 56, EXP. 29

year. This involved a fairly significant reduction in the number of employees whose wages were paid out of the property's income.[58]

Throughout his term in office, although 'savings' were attempted, it was not possible to reduce staff expenditure by much. Personnel expenses went from just over 5,100 *reales* per month to slightly more than 4,900 *reales* per month.[59] Efforts were also made to boost income, for which leases were promoted as a management model. For example, the 120 *fanegas* of arable and horticultural land on the Principe Pio hill were leased to Don Manuel Tejera for 5 years for 18,400 *reales* annually. Another 196 *fanegas* and 6 *celemines* – the first of the lots into which the hill was divided – were leased to Don Francisco García for 8,580 *reales* per year; 60 and a half *fanegas* (the second lot), to Don Francisco Oliva for 2,289 *reales;* 83 *fanegas* (the third lot) to Jerónimo Arjona for 2,905 *reales*; 90 *fanegas* and 7 *celemines* (the fourth lot) to Tiburcio López for 4,348 *reales*; 105 *fanegas* and 4 *celemines* (the fifth lot) to Blas Montero for 3,476

58 AGP, Reinados, José I, caja 56, exp. 29.

59 AGP, Reinados, José I, caja 54.

reales; and 114 *fanegas* (the sixth lot) to Raimundo Montero for 3,762 *reales*. A total of 769 *fanegas* and 11 *celemines* were thus leased for 43,760 *reales*.[60]

A major dispute arose that year, 1811, over the payment of tithes. The royal estates had been exempt from payment but the Napoleonic administration revoked this right by means of a decree of 24 July 1811.[61] As the administrator pointed out, the payment of tithes was detrimental to the royal patrimony: 'we all know that the funding from the royal patrimony is very far from being complete. And it would be much more so if the incorporated estates were subjected to all general or municipal exemptions.[62] The administrator's views were backed by the advisor to the royal household, Don Domingo Agüero y Neira, who justified why tithes should not be paid in his report of 27 August 1811 to the Count of Melito:

> No puede, porque el estado no ha organizado aún su sistema de contribuciones reales, y cuantas medidas toma en el día son más bien un resultado de la exigencia de las circunstancias que de un sistema adoptado, las reclamaciones para el reintegro del real patrimonio serían muy difíciles por no decir imposibles de liquidar ... No se debe porque la dotación de la corona aún no está cubierta, ni se ve el momento en que podrá ser cumplido este deber y no será justo que sobre no haberse completado se disminuya aún con exenciones en lo poco que actualmente la constituye, cuyo reintegro no puede ser efectivo en la actualidad, pues que ni siquiera lo es la consignación mensual sobre el tesoro público.

> It cannot, because the state has not yet organized its system of royal contributions, and whatever measures it takes at present are more the result of circumstantial requirements than of an established system, and demands for payment of royal patrimony would be very difficult if not impossible to meet ... It should not because the funding of the Crown has yet to be covered, nor is it possible to envisage when this task will be fulfilled, and it is not fair that when it is not yet complete the little there is currently should be further diminished, even with exemptions, as

60 As the area of land was too large for a single lessee, it was decided to divide it up into smaller lots (*suertes*) to ensure higher revenues and be able to lease them more easily. AGP, AG, caja 10405, exp. 22.

61 AGP, Reinados, José I, caja 66, exp. 28.

62 'todos sabemos que la dotación del real patrimonio se halla muy lejos de estar completa. Y lo estaría mucho más si las fincas agregadas se sujetan a toda exención general o municipal', AGP, Reinados, José I, caja 66, exp. 28.

> payment cannot be effectively made at present, for not even the monthly allocation to the public treasury is.

In the end La Florida did not have to pay these tithes. Despite all odds, La Florida continued to be a significant agricultural estate, as demonstrated, for one thing, by the fact that its related revenues increased following the arrival of the French (see Tables 9.4, 9.8 and 9.10). On 21 July 1812 Joseph I decided that the harvests of El Pardo and La Florida should be sold in Madrid and used to supply the city (poor harvests had undoubtedly been conducive to such a measure).[63]

However, this measure was not implemented to a great degree as the French abandoned the capital only weeks later. As occurred at the rest of the royal sites, the vicissitudes of the military confect affected La Florida with respect both to the normal course of the production processes and to military needs. The arrival of English troops in Madrid in August 1812 spurred the departure of the Madrid government. The interim finance council (*Junta Interina de Hacienda*) requested information about the state of La Florida as it did for the other royal sites in Madrid.[64]

The task of drawing up the reports was entrusted to the administrator José Álvarez del Valle, who held the post from 12 August to 24 October 1812,[65] and the comptroller Don Florencio Martín, both of whom had been dismissed from their government duties by the French. The first of the reports listing the staff employed there was submitted to the Junta on 26 August. Another, describing the condition of the buildings, was sent shortly afterwards on 30 August 1812.[66] They stated that the palaces of Alba and La Moncloa were 'well cared for and decorated' (*bien tratados y adornados*), though the farmhouses were 'in middling condition' (*un estado medio*) and included an inventory of the objects in them: at La Moncloa, 5 pine tables, 8 walnut tables, 2 walnut cabinets, 4 gilt tables, 36 chairs from Vitoria, 4 fireguards, 1 screen with no cover, with bronze mounts, 10 blue taffeta curtains, 10 green taffeta curtains, 2 white marble busts,

63 AGP, Reinados, José I, caja 67, exp. 6.

64 Gloria Martínez Leiva, "La destrucción del patrimonio arquitectónico durante la Guerra de Independencia: El Buen Retiro y el Edificio Villanueva," in *La Guerra de la Independencia, Jornadas de Arte e Iconografía*, dir. José Manuel Pita Andrade (Madrid: Fundación Universitaria Española, 2009), 255–57. This interim council worked in collaboration with the Regency Council to guarantee Ferdinand VII's rights.

65 AGP, AG, caja 10386, n.d.

66 AGP, AG, caja 10405, exp. 14.

87 paintings of different sizes, 103 miniature paintings, 3 pictures without frames and 6 'cane-coloured' taffeta curtains.[67]

The report likewise mentioned the situation of the few animals that remained, as the French had taken much of the livestock with them: 366 sheep, 39 mules, 4 oxen and 5 donkeys – barely one-third of the more than a thousand animals there were before the outbreak of the Peninsular War.[68] Such a decrease not only led to a substantial reduction in the revenues from livestock but also made farm work and transportation more difficult. La Florida was reported as having little more than 1,200 *fanegas* of land, of which 1,100 were unirrigated, for growing cereals, 83 of vegetable gardens and orchards, and 72 *aranzadas* of vines and olive trees. Five or 6 pairs of mules needed to be brought: otherwise the produce of the vegetable gardens and orchards would go to waste and the land could not be worked. The contents of the storehouses were also listed (Table 9.12).[69] The authors reported that 600 *fanegas* of grain worth 17,000 ducats had been taken from the estate's granaries and supplied to the joint Anglo-Spanish army. The report also included a list and inventory of the farming equipment (among other items, irons for branding livestock, tools for pitching, cauldrons, tinplate pots, churns, copper creaming vessels, tubs, pitchers, hoes and wrenches).

TABLE 9.12 Livestock and produce remaining in the storehouses at La Florida

Sheep	366	Existing grain in La Granjilla	*Fanegas*
All kinds of cattle	22	Chickpeas	27
Oxen	2	Grass peas	32
Horses	2	Lentils	1 ¾
Colts	2	Carob beans	68
Donkey foals	2	Peas	3 ½
Pigs	13	Barley	1,100
Beehives	12	Ears of wheat	200
Carts	12	Wheat and barley straw	5,500 *arrobas*
4-wheel wagon	1	Oil	20 *arrobas*

SOURCE: AGP, AG, CAJA 10405, EXP. 14

67 AGP, AG, caja 10405, exps. 12 and 18.
68 AGP, Reinados, José I, caja 66, exp. 11.
69 AGP, AG, caja 10405, exp. 14.

TABLE 9.13 Situation of the 83 *fanegas* of vegetable gardens/orchards

Garden	Expenses	Days' wages	Produce	Total expenses	Total produce	Profit
La Florida	21,663	26,645	50,000	192,558	218,500	25,952
El Paso	22,211	17,155	40,500	*reales*		
La Moncloa	31,737	20,805	64,000			
Alba	31,537	20,805	64,000			

SOURCE: AGP, AG, CAJA 10405, EXP. 16

Soon afterwards, on 6 October 1812, Francisco Antonio Álvarez and José Álvarez del Valle submitted a plan for making the site more economically viable and restoring it to its former splendour.[70] The proposal involved leasing 400 of the 1,100 *fanegas* of arable land, from which 37,400 *reales* would be obtained at no expense to the Crown; the 83 *fanegas* of vegetable gardens/orchards, for 8,000; and the meadows and pasturelands, for 3,000 *reales* annually. An additional 8,000 *reales* could be earned from the yields from the 72 *aranzadas* of vines and olive trees (Table 9.13). Furthermore, the farmhouse of La Granjilla would be leased for a small sum, but the tenant would be obliged to repair it, as well as the 7 small buildings used to keep the livestock. This would provide a further 56,400 *reales*.

The authors of the plan advised against exploiting the livestock: although a considerable profit could be obtained – more than 9,000 *reales* – the expenses involved would be substantial (Table 9.14). Despite the profit earned, revenues fell significantly from 62,505 *reales* in 1808 (Table 9.14) to 34,134 *reales* – in other words, by 54.6% with respect to the income recorded in 1808.

As the assets would be leased, a staff of only 10 officials would be required. This would amount to a saving of 130,508 *reales* in expenses: on payments of daily wages, the salaries of the permanent staff, manual workers, harvesting, etc. (Table 9.15). The heftiest farming expenses at La Florida were fodder for the livestock, whose price had shot up during the war, followed by sowing arable land and harvesting, and the days' wages paid for these tasks.

The estate's income could be augmented by leasing a further 700 *fanegas* of lower-quality land for 14,000 *reales* and by selling the grain and livestock

70 AGP, AG, caja 10405, exp. 16.

TABLE 9.14 State of the livestock

	Days' wages	Expenses	Products	Total expenditure	Total products	Profit
Sheep	12,957	2,450	20,134			
Cattle	8,030	1,200	14,000	24,637	34,134	9,497

SOURCE: AGP, AG, CAJA 10405, EXP. 16

TABLE 9.15 Farming expenses at La Florida

Daily wages	Extraordinary wages	Manual labourers	Fodder for livestock	Harvesting expenses	Expenses of weeding vineyards and olive groves	Amount of *fanegas* sown	Total
16,790	11,400	7,284	38,820	23,344	6,000	26,870	130,508

SOURCE: AGP, AG, CAJA 10405, EXP. 16

on the royal site. It would be possible to obtain 42,224 *reales* (from 130 *fanegas* of wheat, 57 of carob beans, 16 of chickpeas, 19 of grass peas and 1¾ of lentils, as well as 13 pigs, 9 rams, 4 goats, 2 blind horses and 1 mad colt) plus a further 107,330 *reales* and 14 *maravedíes* for the wheat, barley and hay supplied to the allied armies and the city of Madrid (739 *fanegas* of wheat, 355 *arrobas* of straw and 283 *fanegas* and 9 *celemines* of wheat for Madrid).[71]

These proposals for improvements and reducing expenses were not put into practice, as the French re-entered Madrid early in December 1812. On returning to the previous situation, they appointed Juan Vidal as administrator of El Pardo and La Florida by means of a letter of 11 December 1812.[72] Vidal attempted to keep up the agricultural production, endeavouring to achieve the highest possible revenues. Given the economic straits and the situation caused by the vicissitudes of the war, he opted for reinforcing the model of leasing the

71 AGP, AG, caja 10405, exp. 16.

72 AGP, Reinados, José I, caja 67, exp. 6.

property. The 18 January 1813 edition of the *Diario de Madrid* thus announced a 5-year lease on the Príncipe Pío hill for an annual sum of 14,000 *reales* and a 6-year lease on the Fuente de la Reina vegetable garden/orchard beside the road to El Pardo for 2,200 *reales* per year. During that period the pasturelands were leased for 750 *reales* to Juan Herreros, Basilio Herreros, Juan Altares, José Basala and Jerónimo Arjona,[73] and the first lot of land of the Príncipe Pío hill was leased to Don Francisco García and Don Casimiro Sanz, residents of Madrid, on 6 February 1813.[74]

However, few further measures were adopted during those months as the French withdrew from Madrid permanently on 17 March 1813. After the senior officials of the Regency Council that had remained in Spain to safeguard Ferdinand VII's interests arrived at this royal site, a plethora of reports was produced on the situation not only of the buildings but also of the economic resources.

The revenues of La Florida, like those of other royal sites such as the Casa de Campo, Valladolid, the Buen Retiro and El Escorial, were diminished by the military occupation and looting, despite the French-implemented plans to streamline it and improve its management.[75] This was pointed out in the report that the temporary lord steward, the Marquis of Sales, and Don José González Manrique, the king's attorney, submitted to the finance committee (Comisión de Hacienda) at the Cortes of Cádiz early in 1813[76] and in various reports sent from La Florida, such as the one drawn up on 7 June by the intendent general Don Francisco Antonio de Góngora.[77]

The years of war and the changes in the administration of these places had plunged them into a difficult situation. This was stated, with regard to La Florida, in a report sent to the intendent general. The document acknowledged that this royal property had been the best farming estate during Charles IV's reign but had now almost gone to waste, as was reflected in the detailed inventory of all the buildings and outbuildings on the estate, as well as the farm equipment, crops and livestock.[78] Its income had dwindled to a mere 43,760 *reales* from leases of land.[79]

---

73 AGP, AG, caja 10405, exp. 22.

74 AGP, AG, caja 10405, exp. 36.

75 On the looting of assets during the Peninsular War, see, among others, María Dolores Antigüedad del Castillo-Olivares, *El patrimonio artístico de Madrid durante el Gobierno Intruso (1808–1813)* (Madrid: UNED, 1999). Some reports are held in AGP, AG, leg. 403.

76 AHN, Consejos, leg. 13564, exp. 1, no. 18.

77 It reported on the condition of the buildings and stated that no livestock and little equipment remained. AGP, AG, caja 10405, exp. 22.

78 AGP, AG, caja 10405, exp. 12.

79 AGP, AG, caja 10405, exp. 22.

Before Ferdinand VII's arrival work was undertaken to put the royal site in order and boost its income. One of the first measures adopted, after dismissing all those that were approved under the French government, was a plan to lease unirrigated land for 4 years. Submitted on 20 July 1813,[80] the scheme was approved by the intendent of Madrid, Don Francisco Antonio de Góngora, on the 29th of the month.[81]

The plan achieved the expected effects. In August 1813 Don Isidro Luengas and Don José Ardison took out 5-year leases on the Huerta de Alba and Casa de la Granjilla for 7,000 *reales* annually; on 9 September the eighth lot of Príncipe Pío hill (87 *fanegas* and 9 *celemines*) was leased by Benito González for 4,260 *reales*. The following day the lease of the seventh lot of Príncipe Pío hill, suitable for cereal crops (103 *fanegas* and 3 *celemines*), was granted to Isidro Luengas and José Ardison for 4,000 *reales*; and a 4-year lease of the ninth, tenth and eleventh lots (285 *fanegas*) was awarded to Don Isidro Luengas for 6,936 *reales*. The Huerta de Belén (2 *fanegas*) was leased for 770 *reales* per year for 5 years and the first lot (196 *fanegas* and 6 *celemines*) to Don Francisco García for 8,580 *reales* annually for 4 years; and the second lot (60 *fanegas* and 6 *celemines*), to Don Francisco Javier Oliva for 2,298 *reales* annually for a 4-year period. Also on 10 September the third lot (83 *fanegas*) was leased to Jerónimo Arjona at 2,905 *reales* annually for 4 years; the fourth lot (90 *fanegas* and 7 *celemines*) to Don Sebastián Cuesta for 7,000 *reales* annually for the same period; the sixth lot (114 *fanegas*) to Raimundo Montero for 3,762 *reales* annually; the Huerta de Noblejas (3 *fanegas*) to Don Francisco Oliva for 770 *reales* yearly; and the produce of all the vines and olive trees to Don Bernardo Tomé for 1,920 *reales*.

On 20 September, the 120 *fanegas* of Príncipe Pío hill were leased for 5 years to Don Sebastián Cuesta for 25,000 *reales,* and the fifth lot (105 *fanegas* and 4 *celemines*) to Don Marcos Cubillo for 4 years for 4,820 *reales*. At the beginning of December a 5-year lease of the Huerta de la Moncloa (18 *fanegas*) was awarded to Pedro Cuesta and on 10 December the pastureland was leased to Don Francisco García for 4 years for 3,700 *reales* annually and the Huerta del Botánico (3 ½ *fanegas*) to Jerónimo Arjona for 1,220 *reales* per year, for 5 years. The following year, on 4 February 1814, the vineyard located in the upper part of the Moncla hills was leased to Don Francisco García for 5 years for 688 *reales* annually, and the Huerta del Paso to Pedro Cuesta for 5 years for 6,000 *reales* a year.[82]

---

80 AGP, AG, caja 10386.

81 AGP, AG, caja 10404, exp. 1.

82 Report by Manuel de Retes, 16 April 1814. AGP, AG, caja 10403, exp. 3.

3 Conclusions

In the late 1700s and early 1800s the royal sites of the Spanish Crown underwent major changes with respect to how they were considered and the economic approach from which they were run. These changes stemmed from the financial situation caused by wars and the need to find new economic resources, as well as from the influence of the ideals of the Enlightenment, a period in which the Crown needed to take the initiative to promote the country's development. Hunting and representation of the power of the sovereign and his family had no place in the shaping of the new royal sites.

The Royal Site of La Florida, established when the Crown was in the grip of economic crisis, can be explained by its value as a source of both agricultural and livestock production. Ninety percent of its area, 526 hectares, consisted of agricultural spaces, chiefly unirrigated crops: wheat, rye, barley, chickpeas, broad beans, etc., as well as vines, olives and fruit and vegetables, in the part located closest to the river Manzanares. Its plentiful harvests met the palace's own needs and provided enough for sowing and feeding the livestock – sheep, cattle and pigs – that was kept there and generated sizeable profits, and, of course, was sold in Madrid. The resulting income made it possible to maintain the site as it covered the payment of salaries, daily wages and other expenses and provided revenues for the royal treasury.

After it took shape as a royal site, plans to boost the production and reduce expenditure by adapting the staff of officials to its running were carried out during the reigns of both Charles IV and Joseph I. Leases were a prominent part of the management systems as they made it possible to curb expenditure and obtain cash quickly. That is why Charles IV regarded La Florida as the best estate for crop cultivation.[83] Joseph I wished to continue in this direction and implement programmes for augmenting income and reducing expenditure as at other royal sites; however, the vicissitudes of the war and the political instability it brought made it impossible to put into practice a policy of continuation, even though La Florida remained at the apex of the income-producing royal possessions during the French rule.

Abbreviations

AGP: Archivo General de Palacio
AG: Administración General

83 AGP, AG, caja 10,405, exp. 12.

AHN: Archivo Histórico Nacional
ARAS: Archivo Real Alcázar de Sevilla

Acknowledgements

This paper is part of the projects: "Madrid, Sociedad y Patrimonio: pasado y turismo cultural" (H2019/HUM-5989) of the del Programa de actividades de I+D entre grupos de investigación de la CAM en Ciencias Sociales y Humanidades 2019, cofounded by the FSE; "Corte y sitios reales: espacios de poder, representación y producción (1650–1750)," funded as part of the internal programme of the Universidad Rey Juan Carlos, and "Protection, production and environmental change: the roots of Modern Environmentalism in the Iberian Peninsula (XVI–XVIIIth centuries)" of Gerda Henkel Stiftüng (project reference AZ 60/V/19).

Bibliography

Antigüedad del Castillo-Olivares, María Dolores. *El patrimonio artístico de Madrid durante el Gobierno Intruso (1808–1813)*. Madrid: UNED, 1999.
Artola, Miguel. *La Hacienda del Antiguo Régimen*. Madrid: Alianza, 1982.
Cos-Gayón, Fernando. *Historia jurídica del Patrimonio Real*. Madrid: Imprenta de Enrique de la Riva, 1881.
Die Maculet, Rosario. "Lejos de la Corte. El "destierro alicantino de la princesa Pío en la segunda mitad del siglo XVIII." *Revista de Historia Moderna* 30 (2012): 67–86.
Fernández Talaya, María Teresa. *El Real Sitio de La Florida y La Moncloa. Evolución histórica y artística de un lugar madrileño*. Madrid: Fundación Caja Madrid, 1999.
Friera Álvarez, Marta. *La desamortización de la propiedad de la tierra en el tránsito del Antiguo Régimen al Liberalismo (La desamortización de Carlos IV)*. Gijón: Caja Rural de Asturias-Fundación Foro Jovellanos del Principado de Asturias, 2007.
García López, Aurelio. *La Isabela y Baños de Sacedón*. Guadalajara: Bornova, 2003.
García Monerris, Encarna and Carmen García Monerris. *Las cosas del rey. Historia política de una desavenencia (1808–1874)* (Madrid: Akal, 2015).
Gómez Iglesias, Agustín. "La Dehesa de Amaniel o de la Villa." *Anuario del Instituto de Estudios Madrileños* 2 (1967): 33–81. https://doi.org/10.14198/RHM2012.30.05.
Labrador Arroyo, Félix. "Desamortización o reforma. Los Sitios Reales en un contexto de cambios (1790–1814)." *Memoria y Civilización* 25 (2022): 223–41. DOI: 10.15581/001.25.045.
Labrador Arroyo, Félix. "Entre la costumbre y la transformación: el Real Sitio de San Fernando (1790–1814)." In *Arqueología, Historia y Medio Ambiente: visiones cruzadas,*

coordinated by Luis Alberto Polo Romero and Koldo Trapaga-Monchet, 121–48. Madrid: Dykinson, 2019.

Martínez Leiva, Gloria. "La destrucción del patrimonio arquitectónico durante la Guerra de Independencia: El Buen Retiro y el Edificio Villanueva." In *La Guerra de la Independencia, Jornadas de Arte e Iconografía*, edited by José Manuel Pita Andrade, 239–68. Madrid: Fundación Universitaria Española, 2009.

Mercader Riba, Juan. *José Bonaparte. Rey de España. 1808–1813. Estructura del estado español bonapartista*. Madrid, CSIC: 1983.

Index